Springer Tracts in Nature-Inspired Computing

The book series is aimed at providing an exchange platform for researchers to summarize the latest research and developments related to nature-inspired computing in the most general sense. It includes analysis of nature-inspired algorithms and techniques, inspiration from natural and biological systems, computational mechanisms and models that imitate them in various fields, and the applications to solve real-world problems in different disciplines. The book series addresses the most recent innovations and developments in nature-inspired computation, algorithms, models and methods, implementation, tools, architectures, frameworks, structures, applications associated with bio-inspired methodologies and other relevant areas.

The book series covers the topics and fields of Nature-Inspired Computing, Bio-inspired Methods, Swarm Intelligence, Computational Intelligence, Evolutionary Computation, Nature-Inspired Algorithms, Neural Computing, Data Mining, Artificial Intelligence, Machine Learning, Theoretical Foundations and Analysis, and Multi-Agent Systems. In addition, case studies, implementation of methods and algorithms as well as applications in a diverse range of areas such as Bioinformatics, Big Data, Computer Science, Signal and Image Processing, Computer Vision, Biomedical and Health Science, Business Planning, Vehicle Routing and others are also an important part of this book series.

The series publishes monographs, edited volumes and selected proceedings.

More information about this series at http://www.springer.com/series/16134

Serdar Carbas · Abdurrahim Toktas · Deniz Ustun
Editors

Nature-Inspired Metaheuristic Algorithms for Engineering Optimization Applications

Springer

Editors
Serdar Carbas
Department of Civil Engineering
Karamanoglu Mehmetbey University
Karaman, Turkey

Abdurrahim Toktas
Department of Electrical and Electronics Engineering
Karamanoglu Mehmetbey University
Karaman, Turkey

Deniz Ustun
Department of Computer Engineering
Tarsus University
Tarsus, Turkey

ISSN 2524-552X ISSN 2524-5538 (electronic)
Springer Tracts in Nature-Inspired Computing
ISBN 978-981-33-6772-2 ISBN 978-981-33-6773-9 (eBook)
https://doi.org/10.1007/978-981-33-6773-9

This Springer imprint is published by the registered company Springer Nature Singapore Pte Ltd.
The registered company address is: 152 Beach Road, #21-01/04 Gateway East, Singapore 189721, Singapore

Preface

One of the best alternative ways to solve optimization problems mostly faced by designers and practitioners in engineering applications is to use randomization-based probabilistic metaheuristic algorithms built on basic aspect of natural phenomenon. Many of these nature-inspired metaheuristic algorithms have been proven their performance and reliability as they are popularly utilized not only in engineering, but also in many various disciplines to achieve optimum solutions of challenging design problems. Therefore, researchers either develop novel techniques or modernize the algorithmic performance of existing ones in order to accomplish optimal designs of such complicated applications. This book struggles to present the recent advancements concerning principal nature-inspired metaheuristic algorithms for different engineering applications in main two parts as Part I: *Civil and Structural Engineering* and Part II: *Electrical and Electronics, Computer, and Communication Engineering*. The metaheuristic algorithms are selected to implement in engineering applications, which include conventional algorithms, such as harmony search (HS) algorithm, teaching and learning-based optimization (TLBO) algorithm, particle swarm optimization (PSO) algorithm, artificial bee colony (ABC) algorithm, firefly algorithm that have been used for many years, as well as the so-called newly evolved algorithms, such as electrostatic discharge algorithm (ESDA), butterfly optimization algorithm (BOA), interactive search algorithm (ISA), drosophila food-search algorithm (DSO), grasshopper optimization algorithm (GOA), backtracking search algorithm (BSA), charged system search (CSS) algorithm. The effectiveness of mentioned algorithms is demonstrated by implementing both single and multi-objective engineering design applications by defining the ways of constraint handlings, if needed.

The organization of the book is outlined in a logical order so that the readers can easily follow the chapters. The chapters are not categorized in a historical order, but they are ordered according to involving algorithm(s) that are grouped in terms of the implemented engineering design applications to help readers acquire better conception.

This book comes in to prominence for introducing the fundamental principles, keystone points, and basic steps of the nature-inspired metaheuristic algorithms. In some chapters, elaborate overviews are even provided for the algorithms by supplying demo codes in appendices. Moreover, the robustness of the algorithms is ensured by convergence rates, statistical accuracy tests, computational effort analysis, and so forth.

It is highly desirable in this book that readers gain deep insight into the essence of various nature-inspired metaheuristic algorithms and can thus show the endeavours to reach the optimum solutions of the complicated engineering design applications. The complexness of the tackled problems arises from the mathematical proof of the convergence, the theoretical framework of parameter tuning and control, statistical measures of performance comparison, and characteristics of real-size applications.

Finally, this book aims to acquire broad apprehension and deep perception to achieve design optimization of the engineering applications via nature-inspired metaheuristic algorithms.

Karaman, Turkey — Serdar Carbas
Karaman, Turkey — Abdurrahim Toktas
Tarsus, Turkey — Deniz Ustun

Contents

Editors and Contributors

About the Editors

Serdar Carbas received his B.Sc. in the Department of Civil Engineering from Ataturk University, Erzurum, Turkey, and his M.Sc. and Ph.D. in the Department of Engineering Sciences from Middle East Technical University (METU), Ankara, Turkey. He was Visiting Scholar at the University of California, Los Angeles (UCLA), CA, USA (August 2011–September 2012). His current research fields cover the use of metaheuristic optimization techniques that are found quite effective in obtaining the solution of combinatorial optimization problems which are based on natural phenomena in the field of optimum design of structures. He has authored several book chapters and published more than 40 peer-reviewed research papers. He is Associate Professor at the Department of Civil Engineering in Karamanoglu Mehmetbey University, Karaman, Turkey. Also, he is Adjunct Associate Professor at the Civil Engineering Department of KTO Karatay University, Konya, Turkey.

Abdurrahim Toktas is Associate Professor at the Department of Electrical and Electronics Engineering, Karamanoglu Mehmetbey University, Karaman, Turkey. He received B.Sc. degree in Electrical and Electronics Engineering at Gaziantep University, Gaziantep, Turkey, in July 2002. He worked as Telecom Expert from November 2003 to December 2009 for Turk Telecom Company which is the national PSTN and wideband Internet operator. He received M.Sc. and Ph.D. degrees in Electrical and Electronics Engineering at Mersin University, Mersin, Turkey, in January 2010 and July 2014, respectively. He worked as Network Expert in the Department of Information Technologies at Mersin University from December 2009 to January 2015. He is an editorial board member of the Journal of Recent Advances in Electrical & Electronic Engineering. He is the author of more than ninety research items involving articles, conference proceedings, and projects. His current research interests include electromagnetic modelling, computational electromagnetics, microstrip/printed antenna designing, radar absorber material modelling, design of MIMO antennas, design of UWB antennas, optimization algorithms, machine learnings, and surrogate model.

Deniz Ustun received his B.Sc. degree from the Department of Computer Science Engineering, Istanbul University, Istanbul, Turkey, in 2001. Besides, he received his M.Sc. and Ph.D. degrees from the Department of Electrical and Electronics Engineering, Mersin University, Mersin, Turkey, in 2009 and 2017, respectively. From 2003 to 2017, he was a senior lecturer in the Department of Software Engineering, Mersin University, Mersin, Turkey. Formerly, he was Assistant Professor in the Department of Computer Engineering, Karamanoglu Mehmetbey University, Karaman, Turkey, between 2017 and 2020. He has been working for the Department of Computer Engineering, Tarsus University, Tarsus, Turkey, since March 2020, as Assistant Professor. His current research interests include heuristic and artificial intelligence-based optimization algorithms, surrogate models, machine learning, microstrip antennas, and so forth.

Contributors

Bahriye Akay Department of Computer Engineering, Erciyes University, Melikgazi, Kayseri, Turkey

Musa Artar Department of Civil Engineering, Bayburt University, Bayburt, Turkey

Ibrahim Aydogdu Department of Civil Engineering, Akdeniz University, Antalya, Turkey

Mahdi Azizi Department of Civil Engineering, University of Tabriz, Tabriz, Iran

Gebrail Bekdaş Department of Civil Engineering, Istanbul University-Cerrahpaşa, Avcılar, Istanbul, Turkey

Serdar Carbas Department of Civil Engineering, Faculty of Engineering, Karamanoglu Mehmetbey University, Karaman, Turkey

Ayse T. Daloglu Department of Civil Engineering, Karadeniz Technical University, Trabzon, Turkey

Bilge Kagan Dedeturk Graduate School of Natural and Applied Sciences, Erciyes University, Melikgazi, Kayseri, Turkey

Sadik Ozgur Degertekin Department of Civil Engineering, Dicle University, Diyarbakir, Turkey

Erkan Dogan Department of Civil Engineering, Celal Bayar University, Manisa, Turkey

Ferhat Erdal Department of Civil Engineering, Akdeniz University, Antalya, Turkey

Murat Emre Erkoc Department of Electrical and Electronics Engineering, Erciyes University, Kayseri, Melikgazi, Turkey

Shigeyuki Funabiki Department of Information and Communication System, Graduate School of Natural Science and Technology, Okayama University, Okayama, Japan

Zong Woo Geem College of IT Convergence, Gachon University, Seongnam, Korea

Huseyin Hakli Department of Computer Engineering, Necmettin Erbakan University, Konya, Turkey

Mehmet Sedat Hayalioglu Department of Civil Engineering, Dicle University, Diyarbakir, Turkey

Dervis Karaboga Department of Computer Engineering, Erciyes University, Melikgazi, Kayseri, Turkey

Nurhan Karaboga Department of Electrical and Electronics Engineering, Erciyes University, Kayseri, Melikgazi, Turkey

Aylin Ece Kayabekir Department of Civil Engineering, Istanbul University-Cerrahpaşa, Avcılar, Istanbul, Turkey

Luciano Lamberti Dipartimento di Meccanica, Matematica e Management, Politecnico di Bari, Bari, Italy

Ali Mortazavi Graduate School of Natural and Applied Sciences, Ege University, Izmir, Turkey

Sinan Melih Nigdeli Department of Civil Engineering, Istanbul University-Cerrahpaşa, Avcılar, Istanbul, Turkey

Tevfik Oguz Ormecioglu Department of Civil Engineering, Akdeniz University, Antalya, Turkey

Ramazan Ozcclik Dcpartmcnt of Civil Engineering, Akdeniz University, Antalya, Turkey

Fatih Mehmet Özkal Department of Civil Engineering, Atatürk University, Erzurum, Turkey

Akiko Takahashi Department of Information and Communication System, Graduate School of Natural Science and Technology, Okayama University, Okayama, Japan

Hirotaka Takano Department of Electrical, Electronic and Computer Engineering, Gifu University, Gifu, Japan

Siamak Talatahari Department of Civil Engineering, University of Tabriz, Tabriz, Iran;
Engineering Faculty, Near East University, Nicosia, Turkey

Vedat Togan Civil Engineering Department, Karadeniz Technical University, Trabzon, Turkey

Abdurrahim Toktas Department of Electrical and Electronics Engineering, Faculty of Engineering, Karamanoglu Mehmetbey University, Karaman, Turkey

Osman Tunca Department of Civil Engineering, Karamanoglu Mehmetbey University, Karaman, Turkey

Serdar Ulusoy Department of Civil Engineering, Turkish-German University, Istanbul, Turkey

Esra Uray Department of Civil Engineering, KTO Karatay University, Konya, Turkey

Deniz Ustun Department of Computer Engineering, Faculty of Engineering, Tarsus University, Tarsus, Mersin, Turkey

Melda Yücel Department of Civil Engineering, Istanbul University-Cerrahpaşa, Avcılar, Istanbul, Turkey

Chapter 1
Introduction and Overview: Nature-Inspired Metaheuristic Algorithms for Engineering Optimization Applications

Serdar Carbas, Abdurrahim Toktas, and Deniz Ustun

1.1 Introduction

Engineering is a field of occupation that carries out the researches and developments studies required to make life more comfortable, safe, and easy. In this context, each engineering branch designs differently to find solutions to real-life problems associated with itself. While constituting these designs, the main objective(s) is/are to supply the most appropriate service for the specified purpose(s). Here, obtaining the most suitable design(s) can only be constructed possible by conducting an optimization process. However, due to the existence of very complex mathematical operations in this design process and the difficulty of calculating them manually, the concept of optimization has remained mostly at an academic level until the last two decades. Nonetheless, numerical analysis methods that have developed very rapidly over time, thanks to improvements on computer algorithms, have led to a focus on optimization for real engineering applications. At the same time, two main factors directly related to these developments have given rise to the topic of optimization to be applicable in the practical engineering and science by stripping from the narrow scope in which it is confined. These factors are; (i) high-performance calculations can be implemented at

S. Carbas (✉)
Department of Civil Engineering, Faculty of Engineering, Karamanoglu Mehmetbey University, Karaman, Turkey
e-mail: scarbas@kmu.edu.tr

A. Toktas
Department of Electrical and Electronics Engineering, Faculty of Engineering, Karamanoglu Mehmetbey University, Karaman, Turkey
e-mail: atoktas@kmu.edu.tr

D. Ustun
Department of Computer Engineering, Faculty of Engineering, Tarsus University, Tarsus, Mersin, Turkey
e-mail: denizustun@tarsus.edu.tr

S. Carbas et al. (eds.), *Nature-Inspired Metaheuristic Algorithms for Engineering Optimization Applications*, Springer Tracts in Nature-Inspired Computing,
https://doi.org/10.1007/978-981-33-6773-9_1

a low cost and (ii) rapid advances in the solutions of design optimizations that require hundreds of design variables and should satisfy certain constraints. In order for these achievements to be easy, reliable, economic and understandable for engineers, researchers, and scientists, a wide range of optimization methods based on natural phenomena have been emerged. In this way, with the aid of these nature-inspired simulated novel optimization methods, so-called nature-inspired metaheuristic algorithms, the complicated engineering design applications have become relatively more simple, trusted, cost-effective and apprehensible for practitioners, researchers, and scientists [1–4].

This book makes available a comprehensive and conspicuous overview of the nature-inspired metaheuristic algorithms and their employment in engineering design applications in a single massive volume. The contributing authors have made significant attempts to scrutinize carefully of literally all relevant optimization topics and contents presented in the chapters. The book has been planned as not only a practical scientific resource for researchers, but also is obviously concentrated on engineering applications of the nature-inspired metaheuristic optimization algorithms for practitioners.

The book comprises of seventeen technical chapters grouped under two main parts that are emphasized main engineering disciplines in which the algorithms are used in their content and their application areas as Part I: Civil and Structural Engineering and Part II: Electrical and Electronics, Computer, and Communication Engineering. Under these parts, each chapter has distinct orientation and involves its own original content and ended up with a reference list by which the readers are directed into the chapter-specific focussed studies.

1.2 Parts

1.2.1 Part I: Civil and Structural Engineering

The first and the dominant part of the book includes thirteen chapters that all of which are included by civil and structural engineering disciplines.

Chapter 2: Harmony Search Algorithm for Structural Engineering Problems
This chapter covers detailed review of applications of Harmony Search (HS) algorithm and its different variants on structural engineering design problems, which has proven its algorithmic effectiveness on many different design problems in various fields of engineering. The HS algorithm becomes one of the confidential mature-inspired optimization instruments for almost whole divisions of optimal engineering design applications. And also, the authors exhibit an explicit novel review of the implementation of HS algorithm on structural engineering problems such as optimal designs of steel and reinforced concrete structures. Therefore, some instructive key concepts and results are discussed about HS. This chapter begins with an Introduction

which is followed by Metaheuristics and Harmony Search and Survey on applications in structural engineering sections. Thereafter, in a separate section the design examples are represented to show the robustness of the HS and its proposed variants through deliberating over the obtained optimal structural designs. After representation of Conclusions and References, an Appendix including the detail MATLAB code of the proposed HS algorithm is given at the end of this chapter.

Chapter 3: Teaching Learning Based Optimum Design of Transmission Tower Structures
This chapter concerns with the application of Teaching Learning Based Optimization (TLBO) algorithm for achieving the minimum structural weight of steel transmission tower structures. The member grouping is practised on the elements in order to reduce the structural analysing time. The design constraints are directly implemented from AISC-ASD (American Institute of Steel Construction—Allowable Stress Design) specification as stress and displacement constraints. The TLBO algorithm is encoded on SAP2000, which is a trademark structural analysis software, integrated with MATLAB by which an Open Application Programming Interface (OAPI) is created to obtain optimum structural designs with minimum weights. The comparison of yielded optimal designs via TLBO for steel truss-type transmission towers with those reported previously in the literature ensure the efficiency of the proposed nature-inspired metaheuristic algorithm on real-sized structural engineering design applications.

Chapter 4: Modified Artificial Bee Colony Algorithm for Sizing Optimization of Truss Structures
This chapter of the book focuses on overcoming the deficiency in the convergence capability of the Artificial Bee Colony (ABC) algorithm that has been identified as a result of a detailed literature review given in Introduction section. It is especially stated in this chapter that the ABC algorithm is proved its reliability and efficiency not only in obtaining the optimum designs of structural problems, but also in variety of engineering applications. So, the authors intend to improve the convergence performance of standard ABC algorithm to increase its robustness through some modifications. This new variant of ABC is so-called Modified Artificial Bee Colony (MABC) algorithm. In order for showing the prominence of proposed novel technique, four different steel truss structures, which have previously been optimally designed by other nature-inspired metaheuristics, are also resolved via proposed MABC. Finally, a very good algorithmic performance of the MABC is clearly observed from acquired optimum designs with respect to convergence rate to reach minimum structural design weights. This better performance of the proposed modified variant of the ABC is supported by some statistical indices in design examples, as well.

Chapter 5: Electrostatic Discharge Algorithm for Optimum Design of Real-Size Truss Structures
Electrostatic discharge algorithm (ESDA) can be defined as brand new nature-inspired metaheuristic algorithm. There are only few amounts of study dealt with

ESDA applications on optimization problems. Thus, especially for structural optimization field, this chapter is at the forefront of implementing ESDA to accomplish the design optimization of real-sized steel truss structures. The design constraints treated as displacement ratio, slenderness ratio, and combination of axial and bending strength necessity which are carried through American Institute of Steel Construction-Load and Resistance Factor Design (AISC-LRFD) structural specification. The main objective of the proposed ESDA is to arrive at minimum structural weights with selected steel structural members from a ready profile list while gratifying all design constraints at the same time. At the end, the competence of the proposed ESDA on optimum design of real-sized structural problems is discussed in a logical manner.

Chapter 6: Solving of Distinct Engineering Optimization Problems Using Metaheuristic Algorithms
Relatively, a vast amount of nature-inspired metaheuristic algorithms are principally utilized to solve both constrained and unconstrained optimization problems in this chapter. The Butterfly Optimization Algorithm (BOA), Interactive Search Algorithm (ISA), Drosophila Food-Search Algorithm (DSO), Teaching- and Learning-Based Optimization Algorithm (TLBO), and Firefly Algorithm (FA) are selected optimizers. They have been applied to compare the results of many different optimization problems taken from different engineering fields, such as four mathematical functions as the general engineering problems, a pressure vessel cost minimization and a gear system optimal design from mechanical engineering, an optimal design of truss systems regarding structural engineering and a time–cost trade-off problem from construction management engineering. The attained optimal solutions/designs from each algorithm are compared illustratively among each other in terms of convergence rate, standard deviation, mean and best outcomes.

Chapter 7: The Design of Trapezoidal Corrugated Web Beams Using Firefly Method
This chapter discusses a state-of-the-art approach to optimally design the beam members of the steel structures, which are very significant structural elements since they are assumed to bear external loads and transmit the shear forces. In order to increase the load bearing capacity and to save the used material, the web of the steel beam is formed as corrugated with trapezoidal and sinusoidal geometries. So, stability loss and broad deformations are prevented. The Firefly Algorithm (FA) is the selected optimizer of this study to reach minimum weights of the corrugated shaped steel beams. To do this, the web height, the plate thickness, the distance between the peaks of trapezoidal and sinusoidal shaped webs, the flange width and the thickness are treated as design variables of the structural dimensioning optimization problem. The evolved FA as design solver has been progressed in the optimization of new trade structural steel beam members to satisfy the practical needs efficiently.

Chapter 8: Designing Fuzzy Controllers for Frame Structures Based on Ground Motion Prediction Using Grasshopper Optimization Algorithm; A Case Study of Tabriz, Iran

This chapter discusses a contemporary nature-inspired metaheuristic algorithm, so-called Grasshopper Optimization Algorithm (GOA), to handle optimum design of fuzzy logic controllers of a real-sized high-level nonlinearly behaved steel structure having 20 stories in which the ground motion should be predicted due to the necessitation of structural seismic analysis. The optimum results displayed that the building responses are decreased under damaging earthquake affects.

Chapter 9: Optimization and Artificial Neural Network Models for Reinforced Concrete Members

This chapter introduces the concept of Artificial Neural Networks (ANNs), to bring out the estimation of design variables in an optimization process of two different reinforced concrete beams, such as T-shaped and carbon fibre-reinforced polymer. In this study, three consecutive levels are applied to optimum design. At first, two different nature-inspired metaheuristic algorithms, namely the Teaching- and Learning-Based Optimization (TLBO) and Jaya Algorithm (JA), are applied to obtain optimum structural designs of beam models. Secondly, in order for estimation of design variables, the ANNs are utilized as a decision-making model. At the last, the generated test models as various structural options for the beams are evaluated with respect to correctness and exactness. It is mainly concluded from this study that the ANNs are reliable tool to estimate and identify the values of design variables and parameters for design optimization of structural engineering problems.

Chapter 10: Statistical Investigation of the Robustness for the Optimization Algorithms

Each nature-inspired metaheuristic algorithm has different and tiresome procedure to understand parameters directly impacting the performance of that algorithm to obtain the optimal solutions of the design problems while meeting the considered design constraints. The selection of the parameters of an algorithm is not an easy task since the type of solved problem, the features and amounts of the design variables and constraints, the characteristics of the objective function, etc. have direct effect on this selection. Moreover, this selection procedure is often conducted within a time-consuming trial-and-error logic which does not guarantee that the selected parameters combination set is the best one. Besides, a design parameter set of a metaheuristic algorithm combined for attaining optimal design in a specific optimization problem may not cause showing the same success of that algorithm in a different types of optimization problem. Hence, a statistic-based selection process for design parameter set of a nature-inspired metaheuristic algorithm to reach the optimal design of a geotechnical structure is successfully proposed in this chapter to get rid of lots of boring trials leading to consuming of time very much.

Chapter 11: Optimum Design of Beams with Varying Cross-Section by Using Application Interface

In this chapter, two different nature-inspired metaheuristic optimization algorithms, Harmony Search (HS) and Backtracking Search (BS), are compared and discussed on a structural steel truss design problem that is to be aimed at attaining minimum truss weight. This is implemented to test the validity and accuracy of the proposed methodology. After being sure of properly encoding the metaheuristics, two steel beams with varying cross sections as rectangular and circular are optimized with minimum section weight under particular design constraints, which is the main objective of this study. To do this, those algorithms' coding language platform is linked with one of the Finite Element Analysis (FEA) package software via Application Programming Interface (API). According to literature survey, this chapter is an initiator where the BS algorithm is firstly used in structural design optimization.

Chapter 12: Metaheuristic-Based Structural Control Methods and Comparison of Applications

This chapter gives an opportunity to introduce optimum tuning of system parameters in order to control the structures under seismic effects in terms of two different methods, namely active tuned mass damper (ATMD) and active tendon control (ATC). The utilized optimizers are Harmony Search (HS) and Teaching- and Learning-Based Optimization (TLBO) algorithms that are well-known nature-inspired metaheuristic algorithms. By representing design examples and obtained uniqueness solutions, the authors discuss the effectiveness of proposed strategies properly.

Chapter 13: Evolutionary Structural Optimization—A Trial Review

This chapter presents an in-depth review of a broad range of structural evolutionary optimization applications. Various approximations about objective functions, design variables, design constraints, design formulations and termination criteria are designated. This chapter not only includes detailed review of evolutionary optimization algorithm for structural systems, but also contains some design examples to demonstrate the effectiveness and algorithmic ability across the optimum design of structures. This chapter is constructed as beginning with a detail Introduction section. Thereafter, concept of topology optimization, algorithms' keynotes, fundamentals of principles, design objectives and design constraints, parameters used in optimization, optimality decision, progression of the algorithm, supremacy of the proposed algorithm, and the conclusions are examined and consulted in separate sub-sections of this chapter.

Chapter 14: An Extensive Review of Charged System Search Algorithm for Engineering Optimization Applications

This chapter extensively reviews a variety of applications of the state-of-the-art nature-inspired metaheuristic algorithm, Charged System Search (CSS), in various engineering optimization fields. Addition to classical CSS algorithm, a modified variant of which is also identified briefly in this study. One of the most attractive side of this study is to address both hybrid and multi-objective approaches for proposed CSS algorithms. Moreover, an apprehended programming code of CSS is firstly supplied with this chapter. Also, how the proposed CSS algorithms exploit detailed search either due to the mathematical structure of the algorithm's operations or due to the structure of the optimization problems handled are defined, as well.

1.2.2 Part II: Electrical and Electronics, Computer, and Communication Engineering

This practical and very important part of the book consists of four massive chapters. The sections in these chapters introduce specialized nature-inspired metaheuristic algorithms for the engineering application areas covered. These algorithmic statements are then followed by the discussions of suitable solution strategies and also by accomplished optimal results.

Chapter 15: Artificial Bee Colony Algorithm and Its Application to Content Filtering in Digital Communication
This chapter surveys methodological development issues that arise in experimental research capability of a very famous nature-inspired metaheuristic algorithm, so-called Artificial Bee Colony (ABC) algorithm, on interesting engineering design optimization application. The presented ABC is implemented to optimization of filtering unsolicited content in digital communication. Before obtaining the optimal designs of a digital communication problem, the natural phenomenon lies behind the ABC is defined in detail to illustrate the main analogy in this chapter.

Chapter 16: Multi-objective Design of Multilayer Microwave Dielectric Filters Using Artificial Bee Colony Algorithm
This chapter presents an attractive inspection of multilayer microwave dielectric filters (MMDFs) that is a thought-provoking subject electromagnetic design in electrical and electronics engineering and, furthermore, it comes under the subject of Pareto optimality-based multi-objective optimization problems. This chapter expresses the concept of multi-objective ABC algorithm based on Pareto optimality and the design optimization of tackled MMDFs problem via proposed analogy in detail.

Chapter 17: Multi-objective Sparse Signal Reconstruction in Compressed Sensing
This chapter of the book gives significant evaluations on multi-objective optimization in brief but an exclusive perspective for sparse reconstruction method based on NSGA-II. The proposed multi-objective method is designed and tested on compressed sensing-based electrocardiogram signal compression. The design example is handled to display the efficiency and validity of NSGA-II-based sparse reconstruction method which is classified as a biomedical engineering application.

This chapter also provides significant opportunities for optimization researchers on the basis of internet technology. These opportunities are classified into increased information transfer, alternative software distribution channels, and the creation of an online optimization community. Although past academic and professional organization efforts are now being supplemented by extensive commercial operations, there are still tremendous opportunities for further efforts, particularly in the creation of online, specialized communities. The chapter is well organized to reflect the essence of the proposed methodology. The essential rationality of multi-objective optimization methods is specified in the second section. The basic characteristics of compressed sensing and its multi-objective reconstruction are delineated in the third section. After then, in the fourth section, the NSGA-II applied sparse signal reconstruction is clearly interpreted. Eventually, the conclusions are depicted in the fifth section.

Chapter 18: Optimal Allocation of Flexible Alternative Current Transmission Systems: An Application of Particle Swarm Optimization
This final chapter of the book evinces the application of Particle Swarm Optimization (PSO) to optimize the allotment of flexible alternative current transmission system (FACTS) devices for voltage control in electric power grids. To achieve this, a static VAR compensator (SVC) is underlined as an ordinary FACTS device. The applicability of proposed optimum design conception is proved by virtue of some computer simulations.

1.3 Concluding Remarks

The editors of this book achieved a compelling task of describing the whole range and then inviting, coordinating, systemizing, and operating contributions by over 35 experts under two parts and totally 18 chapters. The immense issue areas of the nature-inspired metaheuristic optimization methods and its engineering applications are all managed in enough detail. Moreover, the chapters of the book are entirely well organized and represented by the authors in direction of a blind peer-review process. Therefore, the sub-sections of the chapters are arranged and established according to the review suggestions.

This book has a mighty algorithmic and practical aspects. This is obviously a main opt for researchers and practitioners who think to utilize the book as a reference guide and/or resource. In lots of chapters, the introduced nature-inspired metaheuristic algorithms are intensified by flowcharts and pseudo-codes which are of serviced to the readers accommodated, enforced, and customized with relative ease. Each asserted algorithm is further validated and ensured by detailed defined design examples; most of them are selected as real-life design optimization engineering applications.

Finally, it can be stated that this book can be suggested to a broad reader, and it shall be in the service of wide range of audiences with particular aims, concerns, and interests. This book may also be utilized as a widespread goal since it holds together broad-ranging topics and contents about the nature-inspired metaheuristic algorithms for engineering optimization problems. For many years, this book will proudly serve to lecturers, graduate students, academic researchers, optimum model designers, practitioners, engineers, and technical decision-makers.

References

1. Yang X-S, Dey N, Fong S (eds) (2020) Springer tracts in nature-inspired computing (STNIC). Springer, Cham
2. Yang X-S (ed) (2018) Nature-inspired algorithms and applied optimization. Springer, Cham
3. Yang X-S (2010) Engineering optimization: an introduction with metaheuristic applications. Wiley, Hoboken, NJ
4. Rao SS (2019) Engineering optimization: theory and practice, 5th edn. Wiley, Hoboken, NJ

Part I
Civil and Structural Engineering

Chapter 2
Harmony Search Algorithm for Structural Engineering Problems

Aylin Ece Kayabekir, Gebrail Bekdaş, Melda Yücel, Sinan Melih Nigdeli, and Zong Woo Geem

2.1 Introduction

In structural engineering, optimization is an essential subject to find economical solutions by adjusting section dimensions and the amount of material or to provide a better seismic behavior by using optimally tuned structural control applications. In addition to these, it is possible to build lighter space structure or ecological structures by considering CO_2 emission as an objective of the process.

In all branches of life and engineering, we use human brain to generate a product optimally. For that reason, the age of optimization is the same as the beginning of the first life. In the optimization process made by the human mind, people gain experience after several trials that are failing or partially succeeded. By the time, people express their knowledge with each other, and now, it is possible to create technological equipment. In new era, it is the time to generate advance solution by skipping long learning and experience processes to avoid unnecessary material and workmanship resources due to providing high demands of the populated world. For

A. E. Kayabekir · G. Bekdaş · M. Yücel · S. M. Nigdeli
Department of Civil Engineering, Istanbul University-Cerrahpaşa, 34320 Avcılar, Istanbul, Turkey
e-mail: ecekayabekir@gmail.com

G. Bekdaş
e-mail: bekdas@istanbul.edu.tr

M. Yücel
e-mail: melda.yucel@yahoo.com.tr

S. M. Nigdeli
e-mail: melihnig@istanbul.edu.tr

Z. W. Geem (✉)
College of IT Convergence, Gachon University, Seongnam 13120, Korea
e-mail: zwgeem@gmail.com

S. Carbas et al. (eds.), *Nature-Inspired Metaheuristic Algorithms for Engineering Optimization Applications*, Springer Tracts in Nature-Inspired Computing,
https://doi.org/10.1007/978-981-33-6773-9_2

that reason, soft computing techniques must be applied, and iterative computations can be provided via algorithm processes.

The first published material about structural optimization is the work of Schmit [1] presented in 1960. Schmit defined systematic synthesis in additional to usage of tools in analysis of structures. Shortly, optimization is a decision-making problem that is systematically done. There are several options with different stresses and displacements. The goal is to choose a solution by considering an objective or multiple ones from the suitable options providing limited values of stresses, displacements, or another needed factors.

In a structural engineering problem for a steel frame structure, the number of combinations of steel profile options is 9.536743×10^{23} as reported by Saka et al. [2]. The decision of the best solution cannot be virtualized via human mind or mathematical optimization techniques. Although several mathematical optimization techniques [3–9] have been proposed, the existing of design constraints that are generally related to stresses and displacements makes the optimization problem nonlinear and complex. In that case, numerical iterations are needed, and metaheuristic algorithms are used to converge the best solution quickly for big structural design problems or problems using complex vibration behavior in structural control.

Metaheuristics are heuristic techniques that are inspired from a phenomenon. In an inspired phenomenon, a goal is provided by several features, and the final goal is provided as an optimization problem. Harmony search (HS) [10] as a music-inspired metaheuristic algorithm imitates the performances of a musician trying to find the pleasing harmony for listeners. The best note or optimum notes are found via the features such as randomly generating a note, modifying the known notes a little bit, and putting forward of liked notes.

In this chapter, a state-of-the-art review about HS employed structural optimization techniques is presented. An optimization problem about carbon fiber reinforced concrete polymer (CFRP) retrofit is presented with HS results. Since the problem is a discrete optimization, classical HS is dominantly effective in finding the optimum result. In order to show the advantages of modification of an algorithm, different variants of HS are presented with a structural engineering benchmark problem, and the most recent applications of HS are briefly described.

2.2 Metaheuristics and Harmony Search

The metaheuristic word is generated via the Greek verb "heuriskein" that means to find or discover, and the word "meta" which means upper, advanced, etc.

The fact that most of the real-world applications have many more complex factors and parameters as well as the constraints that will affect the behavior of the system requires a different approach from basic scientific thinking in terms of optimum resource use in any field [11]. On the other hand, the complexity of the problem of interest makes it impossible to look for any possible solution or combination of solutions, and therefore, finding good and appropriate solutions to the problem within an acceptable time frame becomes the main goal. In this respect, the uncertainty and irregular structure in the behavior of metaheuristic approaches, which are intuitive and more developed and differentiated, can be a solution for various applications and problems.

Many classical heuristic optimization methods rely on local search procedures that are iterative around the existing solution to get a better solution. This search usually ends when the first local optimum is achieved. However, the randomization and reboot approaches used for these solutions to have more optimum values are often ineffective. Metaheuristic methods, on the other hand, generally combine some heuristic approaches to provide better solutions than those found with local search [12].

The main components of any metaheuristic algorithm are exploitation and exploration. Exploration means searching the whole area of search and producing various solutions, and exploitation means searching for the best solutions available in any region, using the knowledge of the chosen solution, focusing on the search (local search) in that region and choosing the best candidate solutions. While randomization is beneficial in exploration solutions, it prevents being caught in a local optimum. The exploration stage enables the algorithm to search the solution area more efficiently [13]. In the other word, the search area is explored to find a best solution; the surrounding of this solution is searched; meanwhile, the exploration continues to find better solutions.

On the other hand, these methods are known as algorithms developed by taking inspiration from various natural features and special abilities that some living things possess and use in their lives, some physical and chemical events that substances undergo, various evolutionary/genetic processes, natural phenomena, and some natural systems. For example, human memory in taboo search, foraging behavior of ants in ant colony algorithm, true genetic structure in genetic algorithm, etc., are simulated. These methods also have various distinctions like in nature.

For example, methods based on swarm intelligence, which include methods such as particle swarm optimization, cuckoo search (CS), and firefly algorithm, express the intelligence used by living things to act in a coordinated manner, especially in finding food. These metaheuristic algorithms are often preferred because of their simple, easy application, and ability to solve various, often highly nonlinear problems [14, 15]. Another example is the methods using memory. With methods such as tabu search, ant colony optimization, genetic algorithm, and harmony search, it is possible to save the solutions using memory, so that the appropriate solutions obtained in the search process can be used in the next steps [16].

Several well-known and used in structural engineering applications are listed in Table 2.1 with the inspiration used in generation of the unique features of these algorithms.

Table 2.1 Well-known metaheuristic algorithms and the employed inspirations

Algorithm	Developer	Year	Inspiration
Genetic algorithm (GA) [17]	Holland	1975	The features of biological evolution that are reproduction, mutation, recombination, and selection
Simulated annealing (SA) [18]	Kirkpatrick, Gelatt, and Vecchi	1983	The process of annealing in material production
Tabu search (TS) [19]	Glover	1986	The human memory
Particle swarm optimization (PSO) [20]	Kennedy and Eberhart	1995	The movement of swarms including birds, fishes, etc.
Ant colony optimization (ACO) [21]	Dorigo, Maniezzo, and Colorni	1996	The behavior of ants seeking a path
Differential evolution (DE) [22]	Storn and Price	1997	Mutation, crossover, and selection
Harmony search (HS) [10]	Geem, Kim and Loganathan	2001	Efforts in musical performances to find the best harmony to gain the appreciation of the audience
Big bang-Big Crunch (BB-BC) [23]	Erol and Eksin	2006	Big bang-big crunch theory that formed the universe
Artificial bee colony algorithm (ABC) [24]	Karaboga and Basturk	2007	Intelligent foraging behavior of honeybees
Cuckoo search (CS) [25]	Yang and Deb	2010	Brood parasitic behavior of some cuckoo species
Firefly algorithm (FA) [26]	Xin-She Yang	2010	Flashing light property of fireflies
Bat algorithm (BA) [27]	Xin-She Yang	2010	Echolocation behavior of bats
Teaching-learning based optimization (TLBO) [28]	Rao, Savsani, and Vakharia	2011	Simulation of teaching and learning processes of education
Flower pollination algorithm (FPA) [29]	Xin-She Yang	2012	Pollination types and process of flowering plants
Ray optimization (RO) [30]	Kaveh and Khayatazad	2012	Refraction property of light rays
Krill Herd algorithm (KHA) [31]	Gandomi and Alavi	2012	Herding behavior of krill
Gray wolf optimizer (GWO) [32]	Mirjalili, Mirjalili, and Lewis	2014	Hunting methods of gray wolf according to hierarchical leadership
Jaya algorithm (JA) [33]	Rao	2016	Victory (Jaya means "Victory" in Sanskrit language)

(continued)

Table 2.1 (continued)

Algorithm	Developer	Year	Inspiration
Billiards-inspired optimization algorithm [34]	Kaveh, Khanzade, and Rastegar Moghaddam	2020	The billiards game

2.2.1 *Mathematical Representation of Engineering Optimization Problems*

The aim of optimization is to minimize or maximize an objective function. Also, several numbers of objectives can be considered. For ith objective function, the mathematical representation is given for a problem including M number of objectives as Eq. (2.1).

$$\text{Minimize } f_i(x), \quad x \in R^n, \quad (i = 1, 2, \ldots, M) \tag{2.1}$$

In structural design and optimization, objective function is generally taken as the optimum cost for design optimization of structures. A response of structure is the general objective in structure control. Both cost and response are minimized. In several cases, objective functions are maximized, but general way is the minimization of objective functions in structural engineering.

In multi-objective optimization, the second objective may be related to ecological or comfort conditions. The consideration of minimization of CO_2 emission is one of the popular second objective in structural optimization. The stress and displacement requirements can be also considered as multiple objectives. These quantities can be also considered as design constraints. The boundary allowed limits of design variables can be also considered as design constraints.

As summary, the optimization is subjected to equality constraints ($h_j(x)$) defined as Eq. (2.2) and inequality constraints ($g_k(x)$) defined as Eq. (2.3). J and K are the number of equality and inequality constraints, respectively.

$$h_j(x), \quad (j = 1, 2, \ldots, J) \tag{2.2}$$

$$g_k(x) \le 0, \quad (k = 1, 2, \ldots, K) \tag{2.3}$$

During the optimization process, a set of design variables is randomized as given in Eq. (2.4) for n number of variables.

$$x = (x_1, x_2, \ldots, x_n)^T \tag{2.4}$$

The violation of design constraints is generally considered as a penalty function that negatively affects objective functions. In optimum structural design, member dimension, group members for section profiles, properties of reinforcements in reinforced concrete (RC) structures, and material classes may be variables while some of them are considered as design constants or calculated values via classical methods like reinforcement area of RC members.

In structural control, the control system properties that are tuned are the design variables to minimize the response of structures.

2.2.2 Harmony Search (HS)

The aim of a musician to find the best harmony. Musicians make several notes from various instruments to match the harmony. The existing good harmony is memorized, and it is decided according to admiration of audience. As summary, the operations used in the development of HS is shown in Fig. 2.1.

In the HS, four different parameters are used. These parameters are listed as follows:

- Harmony memory considering rate (HMCR): It is the probability of using one of the existing values in memory to generate a new harmony.
- Pitch adjusting rate (PAR): It is the probability of the using slight variation if existing HM is used.
- Bandwidth (b_w): It is the parameter to adjust the maximum amount of change with PAR.
- Harmony memory size (HMS): It is the size of memory (the number of stored harmonies).

Briefly, an initial harmony memory (HM) is generated as a matrix called harmony memory matrix (HMM). The randomly generated design variables are stored in HMM. Then, a new harmony vector for a set of design variables is randomized. In randomization, global or local generation is used. The global generation is the usage of all solution ranges as formulated in Eq. (2.5).

$$x_i^{t+1} = x_{\min} + \text{rand}(1)(x_{\max} - x_{\min}) \quad \text{if HMCR} > r_1 \tag{2.5}$$

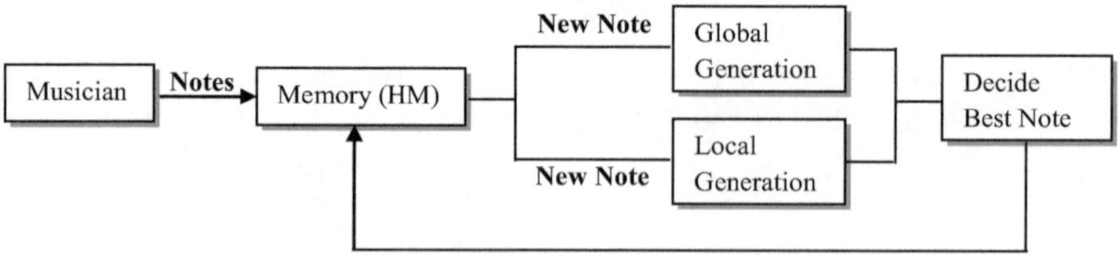

Fig. 2.1 Operation of HS

x_i^{t+1} represents the ith set of design variables in $(t + 1)$th iteration in HMM. r_1, r_2, and r_3 are randomly generated numbers between 0 and 1. The minimum and maximum of design variables are shown as $x_{\min}$ and $x_{\max}$, respectively. The random number between 0 and 1 used in the generation of new variables is shown as rand (1). According to the value of HMCR and PAR, local generation is done via Eqs. (2.6)–(2.8).

$$x_i^{t+1} = x_j^t \quad \text{if HMCR} \leq r_1 \text{ and PAR} > r_2 \tag{2.6}$$

$$x_i^{t+1} = x_{j+\mathrm{ni}}^t \quad \text{if HMCR} \leq r_1 \text{ and PAR} \leq r_2 \text{ (for discrete variables)} \tag{2.7}$$

$$x_i^{t+1} = x_j^t + b_w(r_3) \quad \text{if HMCR} \leq r_1 \text{ and PAR} \geq r_2 \text{ (for continuous variables)} \tag{2.8}$$

For discrete variables, a neighboring index (ni) is used instead of b_w. Via using ni, a neighboring value is chosen, while the surrounding of a randomly chosen existing vector (x_j^t) is searched for continuous variables.

In each iteration, a new harmony is generated. Iterations continue until stopping criterion is provided. This criterion may be number of maximum iterations or a desired value for objective function or a value that considers the last two values of objective function. In each iteration, better solution than existing ones are updated, and it is decided according to the values of objective function. The flowchart of the classical form of HS is presented as Fig. 2.2.

As known, HMCR and PAR are effective in the performance of HS since these parameters control the used and stored variables in HM. Despite using fixed values for HS parameters, adaptive HS approached have been proposed to determine HMCR and PAR [35–40]. Since these methods only provide the basic values, Geem and Sim generated a parameter-setting-free HS proposing the use of the number of times that the values stored in HM [41]. Also, Jeong et al. proposed an advance parameter-setting-free HS to avoid the use of one of parameters after some progress of the HS [42].

As a recent improvement of HS, Tuo et al. increased the performance of HS by using including take-k and take-all strategies, and the new variant has improved success rate and average convergence rate [43].

Another improvement in metaheuristic algorithms is the usage two or more algorithms by combining the effective features together. In this manner, HS has been combined with several algorithms including BA [44], PSO [45], ABC [46], FPA [47], SA [48], ACO [49], CS [50, 51], and FA [52].

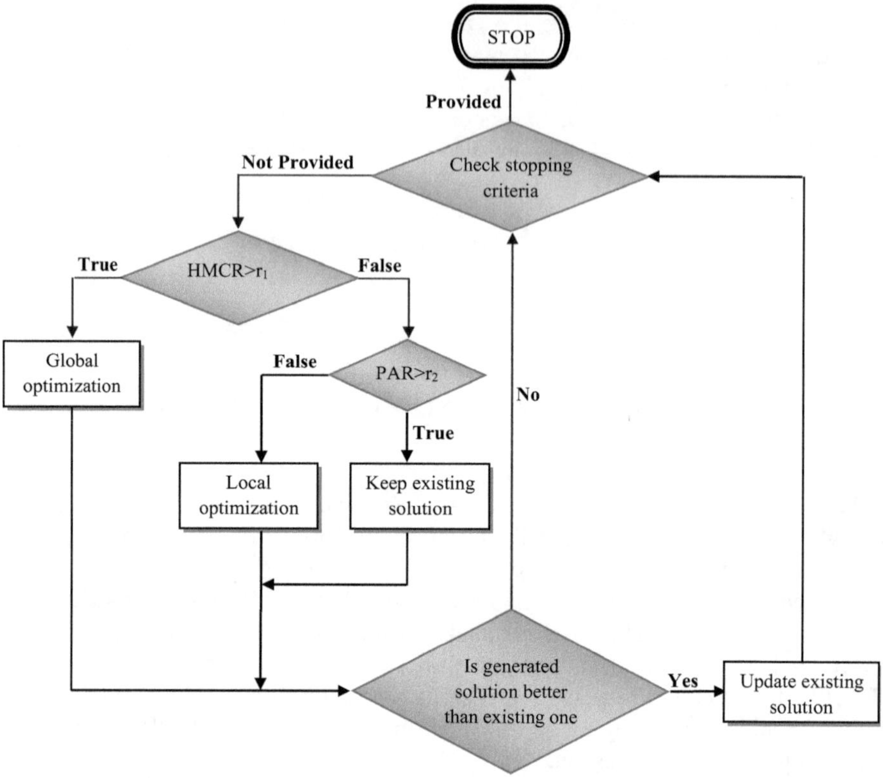

Fig. 2.2 Flowchart of HS algorithm

2.3 Survey on Applications in Structural Engineering

In this section, recent approaches on structural engineering are mentioned. The studies are grouped as steel structures, reinforced concrete structures, structural control, and others. Previously, several literature surveys were published for HS-based structural optimization [2, 53–56].

2.3.1 Steel Structures

In optimization, both truss and frame structures have been investigated via employing HS.

Truss structures that are commonly used as bridges, roof, and towers consist of structural bar members together connected triangle to assembly behave as a single structure. The bar members are straight and connected at joints with each other.

In sizing optimization of truss structures, the aim is to maintain weight minimization to also provide cost optimization by selecting optimum cross-sectional area of bars that are generally grouped. Lee and Geem [57] proposed HS as a structural optimization method and optimized planar and space truss structures as a sizing optimization problem considering continuous design variables. Then, Lee et al. considered discrete design variables for HS based optimum truss sizing problem that provided construction-friendly fixed optimum values [58]. Kaveh and Talatahari generated a hybrid algorithm combining PSO, ACO, and HS for sizing optimization of trusses. Another sizing optimization of trusses was conducted by Degertekin, and two improved HS called efficient HS (EHS) and self-adaptive HS (SAHS) was proposed. EHS dynamically updates PAR and b_w, and SAHS only updates PAR while the formulation of HS was modified without b_w [59]. In order to improve exploitation capabilities of HS, it is hybridized with mine blast algorithm for optimum sizing of trusses including continuous [60] and discrete [61] variables. For discrete sizing optimization of trusses, the randomization function was replaced with global-best PSO search and neighborhood search by Chang et al. [62]. Cao et al. [63] proposed four different penalty-free constraint-handling techniques that are used in sizing optimization of truss structures. The names of these techniques are death penalty, Deb rule, filter method, and mapping strategy. The importance of these method is the increase of search capacity, search stability, and computational efficiency. Differently from the sizing optimization of trusses, subspace HS with improved Deb rule was proposed to conduct size and layout optimization of trusses [63].

In optimization of steel frames, the profile type of beams and columns is optimized subjected to design constraints that are reported in design regulations. The constraints are related to serviceability and strength requirements, and the general goal of the optimization is to minimize the material cost. Saka employed HS for steel sway frames by considering British Code [64]. A general steel structures optimization using HS was conducted by Saka by considering moment resisting steel frame, grillage systems, geodesic domes, and cellular beam via British and American design codes [65]. Hasançebi et al. [38] employed adaptive HS for optimization of steel frames designed according to stress, stability, displacement, and geometric constraints given in ASD-AISC regulation [66]. Artar and Daloğlu [67] investigated steel frames with semi-rigid connecting, and two metaheuristic such as GA and HS were employed in optimization by using MATLAB [68] and SAP2000 [69] as open application programming interface (OAPI) and factors in ASD-AISC [66]. Arafa et al. optimized nonlinear steel frames via HS by using Frye–Morris polynomial model for semi-rigid connections [70]. Carbas and Saka optimized geometrically nonlinear latticed steel domes via HS [71] and improved HS [72] that has an adaptive strategy on selection of HMCR and PAR in each iteration. Carbas and Aydoğdu employed classical and improved versions of HS on optimization of cold-formed steel structures by choosing the cold-formed thin-walled cross sections [73]. Erdal et al. employed HS and particle swarm optimizers in optimum design of cellular beams that are generally used as primary or secondary floor beams to maintain service integration and support long spans [74].

2.3.2 Reinforced Concrete (RC) Structures

Reinforced concrete (RC) is a composite structure formed via concrete and steel. Steel reinforcement bars (rebar) are generally used to provide strength in the parts where the type of stress is tension. As known, concrete is a brittle material with sudden fracture behavior and low tensile strength.

RC structures also need a ductile behavior to carry loads without sudden loss of strength. This case is especially critical under earthquake loads. For that reason, rebars are also used in compression zone when the required reinforcement for the design load is more than the allowed value that is a little less than the amount of rebar providing a balanced situation where the concrete fracture and steel yielding occur in the same time. Under shear forces and when ductile behavior is provided for the column that are generally affected by axial compressive forces, rebars are positioned as stirrups in RC structures.

Differently from the providing a ductile response, serviceability, adherence, and constructivity requirements are formulated in design regulations. The consideration of all factors in an optimization problem for RC members or structures is a nonlinear problem, and metaheuristic methods are generally used in this area. A detailed state-of-the-art review and explanations of metaheuristic-based methodologies for RC beams, columns, footings, frames, retaining walls, carbon fiber reinforced polymer implemented beam, and post-tensioned axially symmetric cylindrical walls are collected in a book written by Kayabekir et al. [75].

Beams are the most investigated members of RC structures for optimization. Akın and Saka [76] optimized RC continuous beams via HS according to design constraints related to flexural and shear strength, serviceability, allowed rebar percentage, and spacing range defined in ACI318: Building Code Requirements for Structural Concrete [77]. Jaberipour and Khorram [78] proposed an improved HS for mixed-discrete engineering problems including a RC beam optimization problem using three design variables that are integer, discrete, and continuous. A detailed optimization of T-shaped beam having rebar in both tensile and compressive zones was done by Bekdaş and Nigdeli [79]. A modified parameter-setting-free HS was proposed by Shaqfa and Orban for multi-objective optimization of RC beams considering weight and cost [80].

HS was employed in optimum design of RC column including seven design variables related to dimension and rebar properties, and the results proved that HS outperforms SA [81]. The effect of slenderness is important for RC columns, and Bekdaş and Nigdeli considered the slenderness effect via ACI318 [77] in the methodology employing HS [82]. Nigdeli et al. proposed a HS methodology for RC columns that are subjected to flexural moments in two directions [83]. Medeiros and Kripta proposed a modified HS and re-initialization of HM for solving RC column optimization problem [84]. Nigdeli and Bekdaş investigated the optimum design of RC shear walls that have design variables such as thickness, web reinforcements in column heading, and wall and stirrups [85].

As the combination of beams and columns, RC plane frames were optimized via HS by selecting optimum values from the design variables pools for grouped beams and columns [86]. Bekdaş and Nigdeli proposed a modified HS for RC frames subjected to static and dynamic earthquake forces via time history analysis [87]. Boscardin et al. employed a variant of HS to present a RC frame optimization methodology that has automated grouping feature for columns [88].

Axially symmetric cylindrical walls that are used as liquid tanks were also considered in metaheuristic-based optimization using HS. The compressive strength of concrete and thickness of the tank were optimized by Bekdaş [89]. Also, the optimization of post-tensioning cylindrical walls was done by including intensity and location of post-tensioning forces additional to classical design variables in the optimum RC design [90]. For post-tensioning axially symmetric cylindrical walls, the maximum longitudinal moment on the wall also decreased by optimizing post-tensioning forces via four metaheuristics including HS [91]. HS was also hybridized with FPA and TLBO as two approaches to optimize post-tensioning axially symmetric cylindrical RC walls [92].

In optimum design of RC retaining walls that include both structural and geotechnical limit states formulated as design constraints, Moline-Moreno et al. optimized buttressed earth-retaining walls by using a hybrid algorithm combining HS and threshold accepting [93]. HS also used in optimization of geosynthetic-reinforced earth walls as an alternative of RC [94]. Bekdaş and Nigdeli employed HS-based methodology for RC spread footings that are also subjected to structural and geotechnical state limits [95]. Bekdaş et al. optimized cantilever solider pile retaining walls that are embedded in frictional soil, and HS was employed in this study [96]. As recent studies, multi-objective optimization was done via HS for cantilever solider piles [97] and retaining walls [98].

Arama et al. [97] conducted a parametric study for CO_2 and cost optimization of cantilever soldier pile walls. In the optimization process, HS was employed to investigate the optimum design for cost and CO_2 emissions objectives. In this process, diameter of soldier pile, diameter, and number of reinforcing bars were defined as design variables. Besides, the shear and flexural strength capacities of section, minimum and maximum cross section of the reinforcing bars were considered as design constants. The optimization was performed for four different objectives. The first objective function was defined to obtain optimum design with minimum cost. The second objective function aimed most environmentally friendly solution and was defined as minimum CO_2 emission. The third function was a multi-objective function including cost and CO_2 minimization. For the similar objective, the last function was determined by using a well-accepted study done by Aydoğdu and Akin [99] in 2015. In the design of cantilever soldier pile wall, beams on elastic soil assumption were considered, and the piles were modeled as embedded in a constant type of pure frictional soil formation. For the calculation of lateral earth trust due to the soil mass, Rankine's earth pressure theory was utilized. The study contains over the two thousand parametric analyses done for various values of design parameters such as excavation depths, unit costs, and CO_2 emissions of the materials by using HS.

Similar investigation was also done by Kayabekir et al. [98] for a different structural type of retaining systems called cantilever retaining wall. In this study, optimum relationship between the cost and the CO_2 emission was investigated by using HS algorithm. In the optimization process, optimum values of eight design variables related to cross-sectional dimensions and reinforcement details were investigated for four different objective functions. The equations of objective functions were the same as the study presented in the previous paragraph. In the wall design, totally sixteen design constraints including stability and reinforced concrete design safety conditions were checked. During the design process, wall was considered as embedded in a pure cohesionless soil formation, lateral pressures caused by soil was calculated according to Rankine earth pressure theory, and a constant value was taken for shear strength angle of the soil formation. Several parametric analyzes were also presented for different excavation depths, surcharge loads, unit weights of the backfill soil, costs, and CO_2 emissions of the concrete and the reinforcement bars.

Nowadays, when the awareness and sensitivity of environmental issues are increasing, it is seen that scientists have started to use emission values as well as cost in optimum design processes and purpose functions accordingly. In both studies conducted for this motivation, the studies are noteworthy in terms of using only cost, emission, and both values for many cases. Although it is seen from the analysis results that taking into account the emission values during analysis has an effect on the cost, it is clearly understood that the environmental benefit to be obtained by considering the emission values is quite remarkable.

As a recent and interesting study, Toklu et al. [100] employed HS for optimization of nozzle movement for additive manufacturing of concrete structures and elements via three-dimensional printing techniques.

Garcia-Segura et al. hybridized HS with threshold optimization to find a sustainable optimum design of post-tensioning concrete box-girder pedestrian bridges [101].

2.3.3 Structural Control

During string ground motions, winds, and traffic, structures may be subjected to unsteady vibrations. Due to that, control methods are used in the structures. In general, structural control is divided into two types; namely passive and active. Additional to these two main types, hybrid and semi-active methods also exist.

In passive control of structures, the mechanical components of the control systems are used to reduce structural responses by dispatching energy. In active control, an external energy source is generally needed to generate time varying control forces.

All structural control systems are needed to be perfectly tuned for efficiency. Since the mathematical formulations of these systems are complex, metaheuristics are the best option to be used in optimization.

The most optimized passive control system is tuned mass dampers (TMDs) that are generally positioned on high and tower-type structures. The parameters such as mass, stiffness, and damping must be tuned by considering the feasibility factors according to the controlled main structure. Until now, HS-based methodologies have been proposed for TMDs.

In search of optimum mass, stiffness, and damping coefficient of TMDs, HS-based methodology was developed by Bekdas and Nigdeli for the seismic structures [102] that are optimized via harmonic excitations. Mass ratio factor for optimum TMDs via HS and closed form TMD formulas were done by Bekdaş and Nigdeli [103]. Nigdeli and Bekdaş employed HS in optimum tuning of TMDs that prevent brittle fracture of RC structures [104]. Nigdeli and Bekdaş also investigated different objectives for TMD optimization problem employing HS [105]. Bekdas and Nigdeli proposed the HS-based optimized TMDs to prevent the pounding of adjacent structures [106]. Nigdeli and Bekdas proposed an optimization methodology that uses impulsive motions of near-fault excitations employing HS [107]. Nigdeli and Bekdas proposed two different control types for adjacent structures and employed HS on optimization of mass dampers that are connected to both structures [108]. Differently from the time domain analysis, a HS-based approach using frequency domain analysis was proposed for TMDs [109]. Soil-structure interaction (SSI) on optimum TMD parameters was investigated by Bekdaş and Nigdeli by using HS and BA [110]. A hybrid HS that is combined with FPA was proposed by Nigdeli and Bekdaş for tuning of passive mass dampers [111]. Zhang and Zhang proposed an improved HS for TMD system of high-rise intake towers [112]. An adaptive dynamic HS that is based on the dynamical parameters and using previous results of the harmony memory with a simple formulation was for optimum design of TMDs [113]. Bekdaş and al. [114] employed several metaheuristics such as HS, FPA, TLBO, and JA for frequency domain optimization of TMDs for structures considering SSI.

Differently from TMDs, HS was employed in optimum locating of structural dampers [115] and seismic isolation systems [116].

As a recent study, an improved HS was proposed for tuning of active tuned mass dampers (ATMDs) that controlled by optimized proportional–integral–derivative (PID) type controllers additional to optimized mass damper parameters [117].

2.3.4 Others

Structures that are generated via composites are another area of structural engineering that needs optimization. With several modifications of HS that allows diversity in keeping different memory size during optimization, HS was employed for optimization of laminated composite structures [118]. Kaveh and Shakouri Mahmud Abadi [119] optimized a composite floor system that is designed by the LRFD-AISC method, using a unit consisting of a reinforced concrete slab and steel beams and employing an improved HS. Cakıroglu et al. [120] optimized shear and lateral–torsional buckling of steel plate girders via HS. Also, Cakıroglu et al. [121] proposed HS on optimization of dispersed laminated composite plates.

Total potential optimization using metaheuristic algorithms (TPO/MA) has been proposed as an iterative analysis method by Toklu [122]. As known, all mechanical systems are in stable equilibrium in the state of minimum total potential energy. Through this theory, the minimization of the total potential energy of a system as an optimization problem can give the exact solution of complex structural analysis problems considering the deflections as design variables. TPO/MA has been applied successfully for the analysis of several structural systems, and HS is one of the most employed methods in TPO/MA. The first application area of TPO/MA is the truss structures, and it was employed [123] including the consideration of the finding of non-unique solutions [124] and material nonlinearity [125]. Also, TPO/MA was also used for cable [126] and tensegric structures [127]. Kayabekir et al. [128] proposed a novel HS for the analysis of plane stress systems via total potential optimization.

As the other applications in structural engineering, HS was also used in the development of computer-support tool to optimize bridges automatically [129], a construction site layout problem [130], and selecting and scaling real ground motion records [131].

2.4 The Optimization Problems

After the literature review, one discrete and one continuous optimization problem are presented in the study. The advantages of the variants of HS that are presented in Sect. 2.4.2.1 are shown with the continuous optimization problem.

2.4.1 Optimization of Design Variables for CFRP Used for Increasing the Shear Force Capacity of RC Beams

The optimization is related to the minimization of the required CFRP area per meter; A. The design variables are taken as width; w_f, spacing; s_f, and angle; β of CFRP. In this problem, it is aimed to increase the capacity of RC beam with an additional shear force for a RC cross section with breadth; b_w, effective depth; d, and flange height; h_f. Figure 2.3 shows a T-shaped cross section where df is the depth of the RC member covered by CFRP formulized as Eq. (2.9).

$$d_f = d - h_f \tag{2.9}$$

The optimization objective function of the optimization problem is given as Eq. (2.10). The three constraints ($g_1(x) - g_2(x)$) that are generated via ACI 318: Building code requirements for structural concrete [77] are given as Eqs. (2.11)–(2.13).

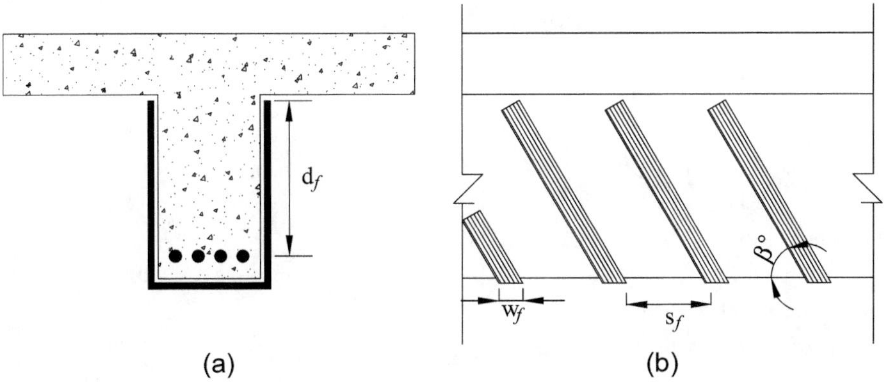

Fig. 2.3 T-shaped section with CFRP application

$$A = \frac{w_f\left(\frac{2d_f}{\sin\beta} + b\right)}{s_f + w_f} \times 1000 \tag{2.10}$$

$$g_1(x) : s_f \leq d/4 \tag{2.11}$$

$$g_2(x) : V_{\text{additional}} < 0.7R\frac{(2t_f w_f f_{fe})(\sin\beta + \cos\beta)d_f}{s_f + w_f} \tag{2.12}$$

$$g_3(x) : \frac{(2t_f w_f f_{fe})(\sin\beta + \cos\beta)}{s_f + w_f} \leq \frac{2\sqrt{f_c'}b_w d}{3} - V_s \tag{2.13}$$

The design variables are discrete practical design. β was assigned with values that are multiples of 5° while w_f and s_f are multiples of 10 mm. For violating design constraints, the objective function is taken as 10^{16} mm^2 per m. The numerical values of design constraints and the ranges of design variables are given in Table 2.2.

The optimization process was repeated for 50 cycles and done for 10,000 maximum iterations. The final objective function values of all cycles were found as the same. For that reason, the classical form of HS is effective for this discrete optimization problem. The optimum results that are found are presented as Table 2.3.

2.4.2 *Optimization of Design Variables for I-Beam Vertical Deflection Minimization*

The optimization problem that was presented in Gold and Krishnamurty [132] was solved for minimization of vertical deflection of beams with I-section. This problem is a continuous optimization problem, and the presented variants of HS were tested

Table 2.2 Design constants and ranges of CFRP problem

Definition	Symbol	Unit	Value
Breadth	b_w	mm	200
Height	h	mm	500
Effective depth	d	mm	450
Thickness of CFRP	t_f	mm	0.165
Reduction factor	R	–	0.5
Thickness of slab	h_f	mm	100
Comp. strength of concrete	f_c'	MPa	20
Effective tensile strength of CFRP	f_{fe}	MPa	3790
Width of CFRP	w_f	mm	10–100
Specific of CFRP	s_f	mm	0-d/4
Angle of CFRP	β	°	0–90
Additional shear force	$V_{\text{additional}}$	kN	100
Shear force capacity of rebar	V_s	kN	50

Table 2.3 Optimum results of CFRP problem

	Value
w_f (mm)	50
s_f (mm)	50
β (°)	65
A (mm^2)	486182.27

for this example. For that reason, the improvements of the algorithms can be observed via the minimization of the objective function.

2.4.2.1 Variants of HS Used for the Problem

In this chapter, four different HS methods were used. The first one is the classical form of HS.

As the second one, the parameters PAR and HMCR are modified in adaptive HS (AHS) as given in Eqs. (2.14) and (2.15). MI and IN represent the maximum iteration number and current iteration number, respectively. The initial values of algorithm parameters at the start of the optimization process are given as PAR_{in} and HMCR_{in}.

$$\text{PAR} = \text{PAR}_{\text{in}}\left(1 - \frac{\text{IN}}{\text{MI}}\right) \tag{2.14}$$

$$\text{HMCR} = \text{HMCR}_{\text{in}}\left(1 - \frac{\text{IN}}{\text{MI}}\right) \tag{2.15}$$

The third one is a modified version of HS (MHS). In the MHS, the selection of an existing solution is done in two ways. The first way is to choose the best-existing solution with a probability of using the best solution called the best solution considering rate (BSCR) compared with a random number (r_3). Like HMCR, it is also tested with a random number for the decision of using. The first way is formulized as Eq. (2.16) if the best existing solution is shown as x_{best}. As the second way, a random solution is chosen as Eq. (2.8) for the continuous case of classical HS. As the last algorithm, MHS was used with Eqs. (2.17) and (2.18) as modified adaptive HS (MAHS).

$$x_i^{t+1} = x_{\text{best}} + b_w(r_3) \quad \text{if HMCR} \le r_1,\ \text{PAR} \ge r_2, \text{BSCR} \le r_3 \tag{2.16}$$

2.4.2.2 Optimization of Design Variables for I-Beam Vertical Deflection Minimization

As shown in Fig. 2.4, section properties of beam are design variables. Two loads that are vertical (P) and horizontal (Q) loads, respectively, act to beam's middle of span and web which are the design constants with the length of the beam (L).

The vertical deflection of an I-beam is calculated according to Eq. (2.17). In the numerical example, design load (P), length of the beam (L), modulus of elasticity (E), and vertical load (Q) are taken as 600 kN, 200 cm, 20,000 kN/cm^2, and 50 kN, respectively.

$$f(x) = \frac{PL^3}{48EI} \tag{2.17}$$

The objective function of the problem is written as Eq. (2.18) if the values of design constants and representation of the moment of inertia (I) of the I-beam with geometrical dimensions are defined.

$$\text{Minimize } f\left(b, h, t_w, t_f\right) = \frac{5000}{\frac{t_w\left(h-2t_f\right)^3}{12} + \frac{bt_f^3}{6} + 2bt_f\left(\frac{h-t_f}{2}\right)^2} \tag{2.18}$$

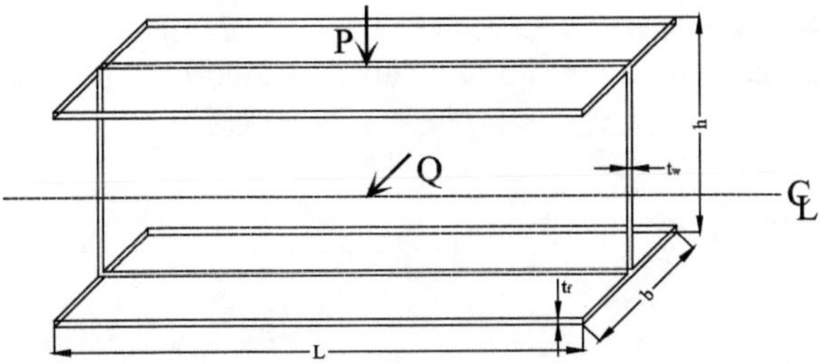

Fig. 2.4 I-beam section properties and design loads

The ranges of the design variables are given in cm as follows:

$$10 \leq h \leq 80, \tag{2.19}$$

$$10 \leq b \leq 50, \tag{2.20}$$

$$0.9 \leq t_w \leq 5 \tag{2.21}$$

$$0.9 \leq t_f \leq 5. \tag{2.22}$$

The problem has two design constraints. For the first one, the cross section of I-beam must be less than 300 cm^2 as given in Eq. (2.23). The second design constraint related to the allowable bending stress of the beam is formulated as Eq. (2.24).

$$g_1 = 2bt_w + t_w(h - 2t_f) \leq 300 \tag{2.23}$$

$$g_2 = \frac{18{,}000h}{t_w\left(h - 2t_f\right)^3 + 2bt_w\left(4t_f^2 + 3h\left(h - 2t_f\right)\right)} + \frac{15{,}000b}{t_w^3\left(h - 2t_f\right) + 2t_w b^3} \leq 6 \tag{2.24}$$

The optimization process was repeated for 50 cycles and done for 10,000 maximum iterations. It is possible to see that the adaptive methods outperform the others. Also, the iteration that is the number in which the objective function lastly updated are also given. It is seen that the results are not updated after 3624 and 3401 iterations for non-adaptive methods. In that case, active changing of algorithm parameters provides an improvement of the results until the end of the optimization without trapping a local optimum. The optimum results obtained for all methods are presented in Table 2.4.

2.4.2.3 The Optimization Code

In order to see the detailed loops done in the optimization process, the computer code written in MATLAB [68] for MAHS is also presented in Appendix. In the code, it is possible to see comment lines for the description of the operations. Also, it is possible to see the chosen values of algorithm parameters in the code.

Table 2.4 Optimum results of I-beam vertical deflection minimization

	HS	AHS	MHS	MAHS
h (cm)	80.00	80.00	80.00	80.00
b (cm)	50.00	50.00	50.00	50.00
t_w (cm)	0.9	0.9	0.9	0.9
t_f (cm)	2.3217912	2.3217922	2.3217904	2.3217922
f (min)	0.01307741238	0.01307741191	0.01307741279	0.01307741190
f (mean)	0.01307743327	0.01307741309	0.01307743514	0.01307741288
Standard deviation	2.487907452167E−07	1.072221345790E−08	2.381808009822E−07	9.243812087906E−09
Iteration	3624	9994	3401	9945

2.5 Conclusions

In this chapter, several structural engineering applications that are optimized via HS are mentioned. From this review, it is possible to use the variety of the problems. Also, the same optimization problem can be solved according to different demands (described as objectives in optimization) and different scenarios. This situation proves that the subject and research area will be always a living interest for engineers and academics.

Two structural engineering applications were optimized via HS. The first example is the optimum design CFRP strips that are implemented to increase the shear force capacity of RC beams. This problem is a discrete optimization problem. According to results of the classical form of HS, the algorithm is effective to find the same optimum result for every cycle of repeated optimization processes. The second example is presented for a continuous optimization problem that is optimized to minimize the vertical deflection of an I-beam. Classical and modified HS presented in the paper were used. Also, adaptive versions of these algorithms were also employed to the problem. Since the optimization problem is a continuous one, all methods gave different best optimum results. In that case, it is possible to see the efficiency of the modifications on this problem. When the optimum results are investigated, the adaptive methods that are using the active updating of optimum results are more effective than others in minimization of the objective function. Although the maximum number of iterations is 10,000, the optimum results of non-adaptive ones are not updated after 3624 and 3401 iterations for the classical and modified HS, respectively. It is understood that adaptive methods provided continuous updating of current best results of iterations until the end of the optimization. When the standard deviation results of the methods were examined, adaptive methods are robust optimization tools for structural engineering.

On the side of the HS, this algorithm is also a fruitful tool for structural optimization by modifications and hybridization with the features of other metaheuristics. In future, it is not a surprise to see novel HS-based structural optimization methods with an increase in numbers.

Appendix

```
% ==========================================================
% APPLICATION: Minimization of Vertical Displacement of
Beam with I Section by Using Harmony Search Algorithm (HS)
% ==========================================================
clear all

% 1. DESIGN CONSTANTS WITH VALUES FOR VERTICAL
DISPLACEMENT PROBLEM
% E(kN/cm^2): elasticity modulus of material
% L(cm): beam length (span)
% P(kN): vertical design load
% Q(kN): horizontal design load

P=600;  Q=50;  L=200;  E=20000;

% 2. DESIGN VARIABLES OF MINIMIZATION PROBLEM
% h: beam height
% b: flange width of beam
% tw: web thickness
% tf: flange thickness

% 3. BOUND VALUES FOR DESIGN VARIABLES
% minh: lower bound of beam height
% maxh: upper bound of beam height

% minb: lower bound of width of beam flange
% maxb: upper bound of width of beam flange

% mintw: lower bound of web thickness of beam
% maxtw: upper bound of web thickness of beam

% mintf: lower bound of flange thickness
% maxtf: upper bound of flange thickness

minh=10;
maxh=80;

minb=10;
maxb=50;
```

```
mintw=0.9;
maxtw=5;

mintf=0.9;
maxtf=5;

% 4. HARMONY SEARCH (HS) PARAMETERS
% stopping: stopping criteria of algorithm (maximum
iteration number)
% HMS: capacity of harmony memory (population number)
% HMCR: harmony memory consideration rate
% PAR: pitch adjusting rate

stopping=10000;
HMS=30;
HMCR=0.5;
PAR=0.2;

% NOTE: Firstly, the changing of HS parameters according
to iterations were performed via below code lines !!!
for xx=1:stopping
  HMCR_modified(xx)=HMCR*(1-(xx/stopping));
  PAR_modified(xx)=PAR*(1-(xx/stopping));
end

% NOTE: Multiple cycles were carried out as 50 in order to
observe and evaluate the success of optimization process
!!!

% Cycle: total application number used for each
minimization process
% Result: matrix utilized for the storage of optimum
results with some values provided in each cycle

Cycle=50;
Result=zeros(9,Cycle);

for process=1:Cycle
% OPT: Initial solution matrix
% OPT_YM: new (generated in each iteration process)
solution matrix

% NOTE: These matrices, which are created to size of
population, were also generated in the way of including
all design variables, objective function and design
```

```
constraints!!
OPT=zeros(7,HMS);
OPT_YM=zeros(7,HMS);

% 5. GENERATION OF INITIAL SOLUTION MATRIX
for i=1:HMS
% 5.1. Determination of values belonging design variables
within the range of defined bound limits as random

% NOTE: rand() is a MATLAB function that it is used to
generate random number.
h=minh+(maxh-minh)*rand();
b=minb+(maxb-minb)*rand();
tw=mintw+(maxtw-mintw)*rand();
tf=mintf+(maxtf-mintf)*rand();

% 5.2. Objective function of minimization problem
Fx=5000/(((1/12)*tw*(h-
2*tf)^3)+(((1/6)*b*tf^3)+(2*b*tf*((h-tf)/2)^2)));

% 5.3. Design constraints
% g1: Constraint for cross-section of beam (The cross-
section should be less than or equal to 300 cm^2)
% g2: Constraint for moment stress (The moment stress
should be less than or equal to 6 kN/cm^2)
g1=2*b*tf+tw*(h-2*tf);
g2=(18000*h)/((tw*(h-2*tf)^3)+(2*b*tw*((4*tf^2)+(3*h*(h-
2*tf)))))+(15000*b)/(((h-2*tf)*tw^3)+(2*tw*b^3));

% 5.4. OPT: Initial solution matrix
OPT(1,i)=h;
OPT(2,i)=b;
OPT(3,i)=tw;
OPT(4,i)=tf;
OPT(5,i)=Fx;
OPT(6,i)=g1;
OPT(7,i)=g2;

% 5.5. Penalization of the solutions with a big number
(for instance 10^6) that they cannot provide the
constraint functions of problem
if OPT(6,i)>300
OPT(5,i)=10^6;
end
if OPT(7,i)>6
```

```
OPT(5,i)=10^6;
end
end

% 6. ITERATION PROCESS
for iter=1:stopping

% 6.1. GENERATION OF NEW SOLUTION MATRIX ACCORDING TO HS
RULES

% Controlling of HMCR to determine of solution type

% number: random number generated in the range of [0,1]
number=rand();

if number<HMCR_modified(iter)

% 6.1.a. GENERATION A NEW HARMONY VECTOR (RANDOMINIZATION)
for ii=1:HMS
% 6.1.a.1. Determination of values belonging design
variables within the range of defined bound limits as
random (this is similar to initial process of
determination of design values.)
h=minh+(maxh-minh)*rand();
b=minb+(maxb-minb)*rand();
tw=mintw+(maxtw-mintw)*rand();
tf=mintf+(maxtf-mintf)*rand();

% 6.1.a.2. Controlling of lower and upper bounds of new
generated design variables
if h>maxh
    h=maxh;
end
if h<minh
    h=minh;
end

if b>maxb
    b=maxb;
end
if b<minb
    b=minb;
end

if tw>maxtw
```

```
    tw=maxtw;
end
if tw<mintw
    tw=mintw;
end

if tf>maxtf
    tf=maxtf;
end
if tf<mintf
    tf=mintf;
end

% 6.1.a.3. Objective function of minimization problem
Fx=5000/(((1/12)*tw*(h-2*tf)^3)+(((1/6)*b*tf^3)+
(2*b*tf*((h-tf)/2)^2)));

% 6.1.a.4. Design constraints
% g1: Constraint for cross-section of beam (The cross-
section should be less than or equal to 300 cm^2)
% g2: Constraint for moment stress (The moment stress
should be less than or equal to 6 kN/cm^2)
g1=2*b*tf+tw*(h-2*tf);
g2=(18000*h)/((tw*(h-2*tf)^3)+(2*b*tw*((4*tf^2)+(3*h*(h-
2*tf)))))+(15000*b)/(((h-2*tf)*tw^3)+(2*tw*b^3));

% 6.1.a.5. OPT_YM: New solution matrix
OPT_YM(1,ii)=h;
OPT_YM(2,ii)=b;
OPT_YM(3,ii)=tw;
OPT_YM(4,ii)=tf;
OPT_YM(5,ii)=Fx;
OPT_YM(6,ii)=g1;
OPT_YM(7,ii)=g2;

% 6.1.a.6. Penalization of the solutions with a big number
(for instance 10^6) that they can not provide the
constraint functions of problem
if OPT_YM(6,ii)>300
    OPT_YM(5,ii)=10^6;
end
if OPT_YM(7,ii)>6
    OPT_YM(5,ii)=10^6;
end
```

```
% 6.1.a.7. Updating of matrix with values of variables, which pertain to better solution, by comparing of new developed vectors with the current ones in terms of objective function minimization
if OPT(5,ii)>OPT_YM(5,ii)
    OPT(:,ii)=OPT_YM(:,ii);
end
end

else
% 6.1.b. PLAY NOTES FROM MEMORY
for iii=1:HMS

% NOTE: In modified version of HS algorithm, a parameter called as BSCR (best solution consideration rate) is utilized to generate of new values of design variables via below code lines.

% BSCR: best solution consideration rate
BSCR=0.3;

% 6.1.b.1. Determination of a required solution as the best solution from initial solution matrix

% bc: objective function value of the best solution
% rc: rank of the best solution within the solution matrix

% NOTE: The best solution takes part in 5th line of matrix as objective function!!!
[bc,rc]=min(OPT(5,:));

if rand()>BSCR
h=OPT(1,rc)+PAR_modified(iter)*(rand()-0.5)*(maxh-minh);
b=OPT(2,rc)+PAR_modified(iter)*(rand()-0.5)*(maxb-minb);
tw=OPT(3,rc)+PAR_modified(iter)*(rand()-0.5)*(maxtw-
mintw);
tf=OPT(4,rc)+PAR_modified(iter)*(rand()-0.5)*(maxtf-
mintf);

else
% xk: a solution, which is random selected from harmony memory
xk=ceil(rand()*HMS);

% Determination of values belonging design variables
```

```
according to the option as "play notes from memory"
h=OPT(1,xk)+PAR_modified(iter)*(rand()-0.5)*(maxh-minh);
b=OPT(2,xk)+PAR_modified(iter)*(rand()-0.5)*(maxb-minb);
tw=OPT(3,xk)+PAR_modified(iter)*(rand()-0.5)*(maxtw-
mintw);
tf=OPT(4,xk)+PAR_modified(iter)*(rand()-0.5)*(maxtf-
mintf);
end

% 6.1.b.2. Controlling of lower and upper bounds of new
generated design variables
if h>maxh
    h=maxh;
end
if h<minh
    h=minh;
end

if b>maxb
    b=maxb;
end
if b<minb
    b=minb;
end

if tw>maxtw
    tw=maxtw;
end
if tw<mintw
    tw=mintw;
end

if tf>maxtf
    tf=maxtf;
end
if tf<mintf
    tf=mintf;
end

% 6.1.b.3. Objective function of minimization problem
Fx=5000/(((1/12)*tw*(h-2*tf)^3)+(((1/6)*b*tf^3)+
(2*b*tf*((h-tf)/2)^2)));

% 6.1.b.4. Design constraints
% g1: Constraint for cross-section of beam (The cross-
```

```
section should be less than or equal to 300 cm^2)
% g2: Constraint for moment stress (The moment stress
should be less than or equal to 6 kN/cm^2)
g1=2*b*tf+tw*(h-2*tf);
g2=(18000*h)/((tw*(h-2*tf)^3)+(2*b*tw*((4*tf^2)+(3*h*(h-
2*tf)))))+(15000*b)/(((h-2*tf)*tw^3)+ (2*tw*b^3));

% 6.1.a.5. OPT_YM: New solution matrix
OPT_YM(1,iii)=h;
OPT_YM(2,iii)=b;
OPT_YM(3,iii)=tw;
OPT_YM(4,iii)=tf;
OPT_YM(5,iii)=Fx;
OPT_YM(6,iii)=g1;
OPT_YM(7,iii)=g2;

% 6.1.a.6. Penalization of the solutions with a big number
(for instance 10^6) that they can not provide the
constraint functions of problem
if OPT_YM(6,iii)>300
    OPT_YM(5,iii)=10^6;
end
if OPT_YM(7,iii)>6
    OPT_YM(5,iii)=10^6;
end

% 6.1.b.7. Updating of matrix with values of variables,
which pertain to better solution, by comparing of new
developed vectors with the current ones in terms of
objective function minimization
if OPT(5,iii)>OPT_YM(5,iii)
    OPT(:,iii)=OPT_YM(:,iii);
end
end  % end of all population
end    % end of if-else process

% OPT_process: The empty and transitory vector, which is
generated for storage of optimum values of the best
solution obtained in each iteration step
OPT_process=zeros(8,1);

% 7. For each iteration stage, determination of the best
solution, which provides the minimum value for vertical
deflection
```

```
% oz: vertical deflection value of the best solution
% fmin: rank of the best solution

% NOTE: The best solution takes part in 5th line of matrix
as objective function!!!
[oz,fmin]=min(OPT(5,:));

OPT_process(1,1)=OPT(1,fmin);   % h
OPT_process(2,1)=OPT(2,fmin);   % b
OPT_process(3,1)=OPT(3,fmin);   % tw
OPT_process(4,1)=OPT(4,fmin);   % tf
OPT_process(5,1)=OPT(5,fmin);   % min (Fx) (objective
function)
OPT_process(6,1)=OPT(6,fmin);   % g1
OPT_process(7,1)=OPT(7,fmin);   % g2
OPT_process(8,1)=iter;         % the current iteration step

% OPT_CEV: Permanent matrix, which is generated for
storage of solution values provided in each iteration step
OPT_CEV(:,iter)=OPT_process(:,:);

end      % end of all iteration steps

% 8. Determination of the best solution provided in the
current cycle
% NOTE: This operation was realized to find the best
iteration step in terms of providing of minimum deflection
in each cycle

% KK: Minimum vertical deflection value provided via
selected best solution from all iterations
% MM: Rank of the best solution

% NOTE: The best solution takes part in 5th line of matrix
as objective function!!!
[KK,MM]=min(OPT_CEV(5,:));

Result(1,process)=OPT_CEV(1,MM);   % h
Result(2,process)=OPT_CEV(2,MM);   % b
Result(3,process)=OPT_CEV(3,MM);   % tw
Result(4,process)=OPT_CEV(4,MM);   % tf
Result(5,process)=OPT_CEV(5,MM);   % min (Fx) (objective
function)
Result(6,process)=OPT_CEV(6,MM);   % g1
Result(7,process)=OPT_CEV(7,MM);   % g2
```

```
Result(8,process)=MM;                % the best iteration,
which provides the minimum deflection within the current
cycle
Result(9,process)=process;           % current cycle

end    % end of 50 cycles

% CALCULATION: Matrix generated for representation of all
best solutions from 50 cycles
CALCULATION=Result';
```

References

1. Schmit LA (1960) Structural design by systematic synthesis. In: Proceedings of the second national conference on electronic computation, ASCE, Sept 1960
2. Saka MP, Hasançebi O, Geem ZW (2016) Metaheuristics in structural optimization and discussions on harmony search algorithm. Swarm Evol Comput 28:88–97
3. Lamberti L, Pappalettere C (2011) Metaheuristic design optimization of skeletal structures: a review. Comput Technol Rev 4(1):1–32
4. Arora JS, Wang Q (2005) Review of formulations for structural and mechanical system optimization. Struct Multidiscipl Optim 30(4):251–272
5. Saka MP (2003) Optimum design of skeletal structures: a review. Prog Civil Struct Eng Comput 10:237–284
6. Saka MP (2007) Optimum design of steel frames using stochastic search techniques based on natural phenomena: a review. Civil engineering computations: tools and techniques. Saxe-Coburg Publications, Stirlingshire, UK, pp 105–147
7. Saka MP (2007) Optimization in structural engineering. In: Geem ZW (ed) Optimization in civil & environmental engineering. Old City Publishing, Philadelphia, pp 59–123
8. Saka MP, Geem ZW (2013) Mathematical and metaheuristic applications in design optimization of steel frame structures: an extensive review. Math Probl Eng. Article ID 271031
9. Venter G (2010) Review of optimization techniques. In: Blockley R, Shyy W (eds) Encyclopedia of aerospace engineering. Wiley, Hoboken, NJ
10. Geem ZW, Kim JH, Loganathan GV (2001) A new heuristic optimization algorithm: harmony search. Simulation 76:60–68
11. Koziel S, Yang XS (2011) Computational optimization, methods and algorithms, vol 356. Springer, Heidelberg, Berlin. ISBN: 978-3-642-20858-4
12. Onwubolu GC, Babu BV (2004) New optimization techniques in engineering, vol 141. Springer, Heidelberg, Berlin. ISBN: 978-3-540-39930-8
13. Gandomi AH, Yang XS, Talatahari S, Alavi AH (2013) Metaheuristic algorithms in modeling and optimization. Metaheuristic applications in structures and infrastructures. Elsevier, pp 1–24. ISBN: 9780123983640
14. Yang XS, Koziel S, Leifsson L (2013) Computational optimization, modelling and simulation: recent trends and challenges. In: International conference on computational science, ICCS 2013, 5–7 Haziran 2013 Barselona-İspanya. Procedia Computer Science, pp 855–860
15. Erdoğmuş P (2016) Doğadan esinlenen optimizasyon algoritmaları ve optimizasyon algoritmalarının optimizasyonu. Düzce Üniversitesi Bilim ve Teknoloji Dergisi 4(1):293–304
16. Onan A (2013) Metasezgisel yöntemler ve uygulama alanları. Çukurova Üniversitesi İktisadi ve İdari Bilimler Fakültesi Dergisi 17(2):113–128

17. Holland JH (1975) Adaptation in natural and artificial systems. University of Michigan Press, Ann Arbor, MI
18. Kirkpatrick S, Gelatt CD, Vecchi MP (1983) Optimization by simulated annealing. Science 220(4598):671–680
19. Glover F (1986) Future paths for integer programming and links to artificial intelligence. Comput Oper Res 13(5):533–549
20. Kennedy J, Eberhart RC (1995) Particle swarm optimization. In: Proceedings of IEEE international conference on neural networks, no IV, 27 Nov–1 Dec, Perth Australia, pp 1942–1948
21. Dorigo M, Maniezzo V, Colorni A (1996) The ant system: optimization by a colony of cooperating agents. IEEE Trans Syst Man Cybern B 26:29–41
22. Storn R, Price K (1997) Differential evolution–a simple and efficient heuristic for global optimization over continuous spaces. J Global Optim 11(4):341–359
23. Erol OK, Eksin I (2006) A new optimization method: big bang–big crunch. Adv Eng Softw 37(2):106–111
24. Karaboga D, Basturk B (2007) A powerful and efficient algorithm for numerical function optimization: artificial bee colony (ABC) algorithm. J Global Optim 39(3):459–471
25. Yang XS, Deb S (2010) Engineering optimisation by cuckoo search. Int J Math Model Numer Optim 1(4):330–343
26. Yang XS (2010) Firefly algorithm, stochastic test functions and design optimisation. Int J Bio-Inspir Comput 2(2):78–84
27. Yang XS (2010) A new metaheuristic bat-inspired algorithm. In: Nature inspired cooperative strategies for optimization (NICSO 2010). Springer, Berlin, Heidelberg, pp 65–74
28. Rao RV, Savsani VJ, Vakharia DP (2011) Teaching–learning-based optimization: a novel method for constrained mechanical design optimization problems. Comput Aided Des 43(3):303–315
29. Yang XS (2012) Flower pollination algorithm for global optimization. In: International conference on unconventional computing and natural computation. Springer, Berlin, Heidelberg, pp 240–249
30. Kaveh A, Khayatazad M (2012) A new meta-heuristic method: ray optimization. Comput Struct 112:283–294
31. Gandomi AH, Alavi AH (2012) Krill herd: a new bio-inspired optimization algorithm. Commun Nonlinear Sci Numer Simul 17(12):4831–4845
32. Mirjalili S, Mirjalili SM, Lewis A (2014) Grey wolf optimizer. Adv Eng Softw 69:46–61
33. Rao R (2016) Jaya: a simple and new optimization algorithm for solving constrained and unconstrained optimization problems. Int J Ind Eng Comput 7(1):19–34
34. Kaveh A, Khanzadi M, Moghaddam MR (2020) Billiards-inspired optimization algorithm; a new meta-heuristic method. In: Structures, vol 27. Elsevier, pp 1722–1739
35. Mahdavi M, Fesanghary M, Damangir E (2007) An improved harmony search algorithm for solving optimization problems. Appl Math Comput 188(2):1567–1579
36. Kumar V, Chhabra JK, Kumar D (2016) Automatic data clustering using parameter adaptive harmony search algorithm and its application to image segmentation. J Intell Syst 25(4):595–610
37. dos Santos Coelho L, de Andrade Bernert DL (2009) An improved harmony search algorithm for synchronization of discrete-time chaotic systems. Chaos Solitons Fractals 41(5):2526–2532
38. Hasançebi O, Erdal F, Saka MP (2010) Adaptive harmony search method for structural optimization. J Struct Eng 136(4):419–431
39. Papa JP, Scheirer W, Cox DD (2016) Fine-tuning deep belief networks using harmony search. Appl Soft Comput 46:875–885
40. Choi YH, Lee HM, Yoo DG, Kim JH (2017) Self-adaptive multi-objective harmony search for optimal design of water distribution networks. Eng Optim 49(11):1957–1977
41. Geem ZW, Sim KB (2010) Parameter-setting-free harmony search algorithm. Appl Math Comput 217(8):3881–3889

42. Jeong YW, Park SM, Geem ZW, Sim KB (2020) Advanced parameter-setting-free harmony search algorithm. Appl Sci 10(7):2586
43. Tuo S, Zhang J, Yuan X, Zhang Y, Liu Z (2016) FHSA-SED: two-locus model detection for genome-wide association study with harmony search algorithm. PLoS ONE 11(3):e0150669
44. Wang G, Guo L (2013) A novel hybrid bat algorithm with harmony search for global numerical optimization. J Appl Math 2013
45. Pandi VR, Panigrahi BK (2011) Dynamic economic load dispatch using hybrid swarm intelligence based harmony search algorithm. Expert Syst Appl 38(7):8509–8514
46. Wu B, Qian C, Ni W, Fan S (2012) Hybrid harmony search and artificial bee colony algorithm for global optimization problems. Comput Math Appl 64(8):2621–2634
47. Abdel-Raouf O, El-Henawy I, Abdel-Baset M (2014) A novel hybrid flower pollination algorithm with chaotic harmony search for solving Sudoku puzzles. Int J Mod Educ Comput Sci 6(3):38
48. Assad A, Deep K (2018) A hybrid harmony search and simulated annealing algorithm for continuous optimization. Inf Sci 450:246–266
49. Kaveh A, Talatahari S (2009) Hybrid algorithm of harmony search, particle swarm and ant colony for structural design optimization. In: Harmony search algorithms for structural design optimization. Springer, Berlin, Heidelberg, pp 159–198
50. Gheisarnejad M (2018) An effective hybrid harmony search and cuckoo optimization algorithm based fuzzy PID controller for load frequency control. Appl Soft Comput 65:121–138
51. Feng Y, Wang GG, Gao XZ (2016) A novel hybrid cuckoo search algorithm with global harmony search for 0-1 knapsack problems. Int J Comput Intell Syst 9(6):1174–1190
52. Guo L, Wang GG, Wang H, Wang D (2013) An effective hybrid firefly algorithm with harmony search for global numerical optimization. Sci World J 2013
53. Manjarres D, Landa-Torres I, Gil-Lopez S, Del Ser J, Bilbao MN, Salcedo-Sanz S, Geem ZW (2013) A survey on applications of the harmony search algorithm. Eng Appl Artif Intell 26(8):1818–1831
54. Yang XS, Bekdaş G, Nigdeli SM (2016) Review and applications of metaheuristic algorithms in civil engineering. In: Metaheuristics and optimization in civil engineering. Springer, Cham, pp 1–24
55. Nasir M, Sadollah A, Yoon JH, Geem ZW (2020) Comparative study of harmony search algorithm and its applications in China, Japan and Korea. Appl Sci 10(11):3970
56. Ala'a A, Alsewari AA, Alamri HS, Zamli KZ (2019) Comprehensive review of the development of the harmony search algorithm and its applications. IEEE Access 7:14233–14245
57. Lee KS, Geem ZW (2004) A new structural optimization method based on the harmony search algorithm. Comput Struct 82(9–10):781–798
58. Lee KS, Geem ZW, Lee SH, Bae KW (2005) The harmony search heuristic algorithm for discrete structural optimization. Eng Optim 37(7):663–684
59. Degertekin SO (2012) Improved harmony search algorithms for sizing optimization of truss structures. Comput Struct 92:229–241
60. Sadollah A, Sayyaadi H, Lee HM, Kim JH (2018) Mine blast harmony search: a new hybrid optimization method for improving exploration and exploitation capabilities. Appl Soft Comput 68:548–564
61. Sadollah A, Bahreininejad A, Eskandar H, Hamdi M (2012) Mine blast algorithm for optimization of truss structures with discrete variables. Comput Struct 102:49–63
62. Cheng MY, Prayogo D, Wu YW, Lukito MM (2016) A Hybrid Harmony Search algorithm for discrete sizing optimization of truss structure. Autom Constr 69:21–33
63. Cao H, Qian X, Zhou YL, Yang H (2018) Applicability of subspace harmony search hybrid with improved deb rule in optimizing trusses. J Comput Civil Eng 32(4):04018021
64. Saka MP (2009) Optimum design of steel skeleton structures. In: Music-inspired harmony search algorithm. Springer, Berlin, Heidelberg, pp 87–112
65. Saka MP (2009) Optimum design of steel sway frames to BS5950 using harmony search algorithm. J Constr Steel Res 65(1):36–43

66. AISC A (1989) Manual of steel construction–allowable stress design. American Institute of Steel Construction (AISC), Chicago
67. Artar M, Daloğlu AT (2018) Optimum weight design of steel space frames with semi-rigid connections using harmony search and genetic algorithms. Neural Comput Appl 29(11):1089–1100
68. MATLAB (2009) The language of technical computing. The Mathworks, Natick
69. SAP2000 (2008) Integrated finite elements analysis and design of structures. Computers and Structures Inc, Berkeley
70. Arafa M, Khalifa A, Alqedra M (2016) Design optimization of semi-rigidly connected steel frames using harmony search algorithm. J Eng Res Technol 2(2)
71. Carbas S, Saka MP (2009) Optimum design of single layer network domes using harmony search method. Asian J Civil Eng 10(1):97–112
72. Çarbaş S, Saka MP (2012) Optimum topology design of various geometrically nonlinear latticed domes using improved harmony search method. Struct Multidiscipl Optim 45(3):377–399
73. Carbas S, Aydogdu I (2017) Utilization of harmony search algorithm in optimal structural design of cold-formed steel structures. In: Del Ser J (ed) Harmony search algorithm. ICHSA 2017. Advances in intelligent systems and computing, vol 514. Springer, Singapore
74. Erdal F, Doğan E, Saka MP (2011) Optimum design of cellular beams using harmony search and particle swarm optimizers. J Constr Steel Res 67(2):237–247
75. Kayabekir AE, Bekdaş G, Nigdeli SM (eds) (2020) Metaheuristic approaches for optimum design of reinforced concrete structures: emerging research and opportunities: emerging research and opportunities. IGI Global
76. Akin A, Saka MP (2010) Optimum detailed design of reinforced concrete continuous beams using the harmony search algorithm. In: The tenth international conference on computational structures technology, Paper 131, Stirlingshire, UK
77. ACI-318 (2005) Building code requirements for structural concrete and commentary, metric version. American Concrete Institute
78. Jaberipour M, Khorram E (2011) A new harmony search algorithm for solving mixed–discrete engineering optimization problems. Eng Optim 43(5):507–523
79. Bekdaş G, Nigdeli SM (2013) Optimization of T-shaped RC flexural members for different compressive strengths of concrete. Int J Mech 7:109–119
80. Shaqfa M, Orbán Z (2019) Modified parameter-setting-free harmony search (PSFHS) algorithm for optimizing the design of reinforced concrete beams. Struct Multidiscipl Optim 60(3):999–1019
81. de Medeiros GF, Kripka M (2012) Harmony search algorithm applied to the optimization of reinforced concrete columns
82. Bekdas G, Nigdeli SM (2014) The optimization of slender reinforced concrete columns. Proc Appl Math Mech 14(1):183–184
83. Nigdeli SM, Bekdas G, Kim S, Geem ZW (2015) A novel harmony search based optimization of reinforced concrete biaxially loaded columns. Struct Eng Mech 54(6):1097–1109
84. Medeiros GF, Kripka M (2017) Modified harmony search and its application to cost minimization of RC columns. Adv Comput Des 2(1):1–13
85. Nigdeli SM, Bekdaş G (2014) Optimization of reinforced concrete shear walls using harmony search. In: 11th international congress on advances in civil engineering, Istanbul, Turkey. Academic
86. Akin A, Saka MP (2015) Harmony search algorithm based optimum detailed design of reinforced concrete plane frames subject to ACI 318-05 provisions. Comput Struct 147:79–95
87. Bekdaş G, Nigdeli SM (2017) Modified harmony search for optimization of reinforced concrete frames. In: 3rd international conference on the harmony search algorithm (ICHSA 2017), Bilbao, Spain
88. Boscardin JT, Yepes V, Kripka M (2019) Optimization of reinforced concrete building frames with automated grouping of columns. Autom Constr 104:331–340

89. Bekdas G (2014) Optimum design of axially symmetric cylindrical reinforced concrete walls. Struct Eng Mech 51(3):361–375. https://doi.org/10.12989/sem.2014.51.3.361
90. Bekdaş G (2015) Harmony search algorithm approach for optimum design of post-tensioned axially symmetric cylindrical reinforced concrete walls. J Optim Theory Appl 164(1):342–358
91. Bekdaş G, Nigdeli SM (2018) Optimum reduction of flexural effect of axially symmetric cylindrical walls with post-tensioning forces. KSCE J Civil Eng 1–8
92. Bekdaş G (2018) New improved metaheuristic approaches for optimum design of posttensioned axially symmetric cylindrical reinforced concrete walls. Struct Des Tall Spec Build 27(7)
93. Molina-Moreno F, García-Segura T, Martí JV, Yepes V (2017) Optimization of buttressed earth-retaining walls using hybrid harmony search algorithms. Eng Struct 134:205–216
94. Manahiloh KN, Nejad MM, Momeni MS (2015) Optimization of design parameters and cost of geosynthetic-reinforced earth walls using harmony search algorithm. Int J Geosynth Ground Eng 1(2):15
95. Bekdaş G, Nigdeli SM (2015) Multi-objective optimization of reinforced concrete footings using harmony search. In: 23rd international conference on multiple criteria decision making (MCDM 2015), Hamburg, Germany. Academic
96. Bekdaş G, Arama ZA, Kayabekir AE, Geem ZW (2020) Optimal design of cantilever soldier pile retaining walls embedded in frictional soils with harmony search algorithm. Appl Sci 10(9):3232
97. Arama ZA, Kayabekir AE, Bekdaş G, Geem ZW (2020) CO_2 and cost optimization of reinforced concrete cantilever soldier piles: a parametric study with harmony search algorithm. Sustainability 12(15):5906
98. Kayabekir AE, Arama ZA, Bekdaş G, Nigdeli SM, Geem ZW (2020) Eco-friendly design of reinforced concrete retaining walls: multi-objective optimization with harmony search applications. Sustainability 12(15):6087
99. Aydogdu I, Akin A (2015) Biogeography based CO_2 and cost optimization of RC cantilever retaining walls. In: 17th international conference on structural engineering, pp 1480–1485
100. Toklu YC, Bekdaş G, Geem ZW (2020) Harmony search optimization of nozzle movement for additive manufacturing of concrete structures and concrete elements. Appl Sci 10(12):4413
101. García-Segura T, Yepes V, Alcalá J, Pérez-López E (2015) Hybrid harmony search for sustainable design of post-tensioned concrete box-girder pedestrian bridges. Eng Struct 92:112–122
102. Bekdaş G, Nigdeli SM (2011) Estimating optimum parameters of tuned mass dampers using harmony search. Eng Struct 33(9):2716–2723
103. Bekdaş G, Nigdeli SM (2013) Mass ratio factor for optimum tuned mass damper strategies. Int J Mech Sci 71:68–84
104. Nigdeli SM, Bekdas G (2013) Optimum tuned mass damper design for preventing brittle fracture of RC buildings. Smart Struct Syst 12(2):137–155
105. Nigdeli SM, Bekdaş G (2014) Optimization of TMDs for different objectives. In: An international conference on engineering and applied sciences optimization, Kos Island, Greece, pp 4–6
106. Bekdaş G, Nigdeli SM (2012) Preventing the pounding of adjacent buildings with harmony search optimized tuned mass damper. In: 3rd European conference of civil engineering, pp 2–4
107. Nigdeli SM, Bekdas G (2013) The effect of impulsive motions on optimum tuned mass damper parameters. In: 11th international conference on vibration problems, pp 2–10
108. Nigdeli SM, Bekdas G (2014) Optimum tuned mass damper approaches for adjacent structures. Earthq Struct 7(6):1071–1091
109. Nigdeli SM, Bekdaş G (2017) Optimum tuned mass damper design in frequency domain for structures. KSCE J Civil Eng 21(3):912–922
110. Bekdaş G, Nigdeli SM (2017) Metaheuristic based optimization of tuned mass dampers under earthquake excitation by considering soil-structure interaction. Soil Dyn Earthq Eng 92:443–461

111. Nigdeli SM, Bekdaş G, Yang XS (2017). Optimum tuning of mass dampers by using a hybrid method using harmony search and flower pollination algorithm. In: International conference on harmony search algorithm, February. Springer, Singapore, pp 222–231
112. Zhang HY, Zhang LJ (2017) Tuned mass damper system of high-rise intake towers optimized by improved harmony search algorithm. Eng Struct 138:270–282
113. Keshtegar B, Etedali S (2018) Nonlinear mathematical modeling and optimum design of tuned mass dampers using adaptive dynamic harmony search algorithm. Struct Control Health Monit 25(7):e2163
114. Bekdaş G, Kayabekir AE, Nigdeli SM, Toklu YC (2019) Tranfer function amplitude minimization for structures with tuned mass dampers considering soil-structure interaction. Soil Dyn Earthq Eng 116:552–562
115. Amini F, Ghaderi P (2013) Hybridization of harmony search and ant colony optimization for optimal locating of structural dampers. Appl Soft Comput 13(5):2272–2280
116. Nigdeli SM, Bekdaş G, Alhan C (2014) Optimization of seismic isolation systems via harmony search. Eng Optim 46(11):1553–1569
117. Kayabekir AE, Bekdaş G, Nigdeli SM, Geem ZW (2020) Optimum design of PID controlled active tuned mass damper via modified harmony search. Appl Sci 10(8):2976
118. de Almeida FS (2019) Optimization of laminated composite structures using harmony search algorithm. Compos Struct 221:110852
119. Kaveh A, Abadi ASM (2010) Cost optimization of a composite floor system using an improved harmony search algorithm. J Constr Steel Res 66(5):664–669
120. Cakiroglu C, Bekdaş G, Kim S, Geem ZW (2020) Optimisation of shear and lateral-torsional buckling of steel plate girders using meta-heuristic algorithms. Appl Sci 10(10):3639
121. Cakiroglu C, Bekdaş G, Geem ZW (2020) Harmony search optimisation of dispersed laminated composite plates. Materials 13(12):2862
122. Toklu YC (2004) Nonlinear analysis of trusses through energy minimization. Comput Struct 82(20–21):1581–1589
123. Toklu YC, Bekdaş G, Temur R (2013) Analysis of trusses by total potential optimization method coupled with harmony search. Struct Eng Mech 45(2):183–199
124. Toklu YC, Temür R, Bekdaş G (2015) Computation of nonunique solutions for trusses undergoing large deflections. Int J Comput Methods 12(03):1550022
125. Temür R, Bekdaş G, Toklu YC (2017) Total potential energy minimization method in structural analysis considering material nonlinearity. Chall J Struct Mech 3:129–133
126. Toklu YC, Bekdaş G, Temür R (2017) Analysis of cable structures through energy minimization. Struct Eng Mech 62(6):749–758
127. Toklu YC, Uzun F (2016) Analysis of tensegric structures by total potential optimization using metaheuristic algorithms. J Aerosp Eng 29(5):04016023
128. Kayabekir AE, Toklu YC, Bekdaş G, Nigdeli SM, Yücel M, Geem ZW (2020) A novel hybrid harmony search approach for the analysis of plane stress systems via total potential optimization. Appl Sci 10(7):2301
129. García-Segura T, Yepes V, Alcalá J (2017) Computer-support tool to optimize bridges automatically. Int J Comput Methods Exp Meas 5(2):171–178
130. Gholizadeh R, Amiri GG, Mohebi B (2010) An alternative approach to a harmony search algorithm for a construction site layout problem. Can J Civ Eng 37(12):1560–1571
131. Kayhan AH, Korkmaz KA, Irfanoglu A (2011) Selecting and scaling real ground motion records using harmony search algorithm. Soil Dyn Earthq Eng 31(7):941–953
132. Gold S, Krishnamurty S (1997) Trade-offs in robust engineering design. In: Proceedings of the ASME design engineering technical conferences, Sept 1997

Chapter 3
Teaching Learning Based Optimum Design of Transmission Tower Structures

Musa Artar and Ayse T. Daloglu

3.1 Introduction

In recent years, metaheuristic search techniques are highly preferred for optimum designs of civil engineering structures. Many different metaheuristic methods have been developed to carry out the solutions of discrete design problems. Some of these methods are Genetic Algorithm (GA), Harmony Search (HS) algorithm, Tabu Search (TS) algorithm, Simulated Annealing (SA) algorithm, Ant Colony Optimization (ACO), Artificial Bee Colony (ABC) algorithm, Particle Swarm Optimization (PSO), Biogeography-based Optimization (BBO), Firefly Algorithm (FA), Teaching Learning Based Optimization (TLBO), Jaya Algorithm (JA).

Saka [1] researched optimum design of pin-jointed steel structures with practical applications. Hasançebi and Erbatur [2] focused on another fundamental study on structural optimization. Lee and Geem [3] achieved a highly effective algorithm technique in structural optimum design. Değertekin et al. [4] and Kaveh and Talatahari [5] investigated comparative optimum solutions with different algorithms in structural design problems. Rao et al. [6] developed another new algorithm technique called TLBO. Talatahari et al. [7] studied on structural optimization in truss problems. Değertekin [8] focused on optimal solutions of nonlinear structures via ABC algorithm. Aydoğdu and Saka [9], Rafiee et al. [11], Saka and Geem [12], Hadidi and Rafiee [14], Carbas [17] studied optimum designs of steel frame structures. Kaveh and Talatahari [10] investigated a hybrid algorithm on structural problems. Talatahari et al. [13] investigated design optimization of tower structures via Firefly Algorithm.

M. Artar (✉)
Department of Civil Engineering, Bayburt University, Bayburt 69000, Turkey
e-mail: martar@bayburt.edu.tr

A. T. Daloglu
Department of Civil Engineering, Karadeniz Technical University, Trabzon 61000, Turkey
e-mail: aysed@ktu.edu.tr

S. Carbas et al. (eds.), *Nature-Inspired Metaheuristic Algorithms for Engineering Optimization Applications*, Springer Tracts in Nature-Inspired Computing,
https://doi.org/10.1007/978-981-33-6773-9_3

Dede and Togan [15], Artar [19] used TLBO algorithm for design problems of steel structures. Değertekin and Geem [16] studied metaheuristic optimization in structural engineering. Daloglu et al. [18] included soil effect in design problems. Rao [20] developed a new algorithm technique called Jaya for structural design problems. Aydogdu et al. [21], Carbas [22], Aydogdu [23], Dede [24], Topal et al. [25], Shallan et al. [28], Dede et al. [29], Kaveh et al. [31], Kalemci and Ikizler [32] researched different structural design problems using metaheuristic algorithm methods. Grzywinski et al. [26], Artar and Daloglu [27], Atmaca et al. [30] used Jaya algorithm for optimum design problems.

Teaching Learning Based Optimization (TLBO) is one of successfully applied metaheuristic search methods developed in recent years [6]. This novel algorithm technique has been tested with different studies [15, 19, 25] in literature for the optimum solutions of structural engineering problems. In this study, optimum designs of a 47-, 72- and 244-bar transmission towers taken from the literature are carried out by using a prepared list including 127 single-angle (*L*) profiles taken from American Institute of Steel Construction (AISC). The optimum solutions are determined with generations of 20 individuals. In the third problem, a population of 30 individuals is also run. A computer program is coded in MATLAB [33] interacting with SAP2000[34]-OAPI to obtain optimum results. Although the predetermined values are used as stress constraints in the 47- and 72-bar towers, the stress constraints taken from AISC-ASD [35] specifications are applied on 244-bar tower. The obtained minimum weight results are compared with the literature results found by different metaheuristic methods. The results prove that this novel search technique is very useful and convenient for multi-element structural problems.

3.2 Optimum Design Problem

The discrete optimum design problem of the truss structure is determined as below,

$$\min W = \sum_{k=1}^{ng} A_k \sum_{i=1}^{nk} \rho_i L_i \tag{3.1}$$

where W is the weight of the truss structure, A is cross-sectional area, ρ is the density, L length of member, i is member number, k group number, ng and nk total numbers.

The displacement and stress constraints applied on the transmission tower are taken from AISC-ASD [35] specifications. These constraints are shown as below,

$$g_j(x) = \frac{\delta_{jl}}{\delta_{ju}} - 1 \le 0 \quad j = 1, \ldots, n \tag{3.2}$$

$$g_m(x) = \frac{\sigma_m}{\sigma_{m,\text{all}}} - 1 \le 0 \quad m = 1, \ldots, ne \tag{3.3}$$

where δ_{jl} is displacement of jth degree of freedom under load case l,δ_{ju} is upper bound, σ_m and $\sigma_{m,\text{all}}$ are the calculated and allowable axial stresses of mth truss member, respectively.

The axial stresses of the truss members are calculated as follows (AISC-ASD [35]),

- for tension members;

$$\sigma_{t,\text{all}} = 0.6F_y \tag{3.4}$$

- for compression members;

$$\lambda_m = \frac{K_m L_m}{r_m} \quad m = 1, \ldots, ne \tag{3.5}$$

$$C_C = \sqrt{\frac{2\pi^2 E}{F_y}} \tag{3.6}$$

- for inelastic buckling ($\lambda_m < C_c$);

$$\sigma_{c,\text{all}} = \frac{\left[1 - \frac{\lambda_m^2}{2C_c^2}\right] F_y}{\frac{5}{3} + \frac{3\lambda_m}{8C_c} - \frac{(\lambda_m^3)}{8C_C^3}} \tag{3.7}$$

- for elastic buckling ($\lambda_m \geq C_c$);

$$\sigma_{c,\text{all}} = \frac{12\pi^2 E}{23\lambda_m^2} \tag{3.8}$$

where F_y is yield stress, r_m is minimum radius of gyration, $K_m(K = 1)$ is the effective length factor, λ_m is the slenderness ratio and C_c is critical slenderness ratio parameter.

$$g_i(\text{x}) > 0 \rightarrow c_i = g_i(\text{x}) \tag{3.9}$$

$$g_i(\text{x}) \leq 0 \rightarrow c_i = 0 \tag{3.10}$$

- The penalized objective function $\varphi(x)$ is determined as,

$$\varphi(x) = W(x)\left(1 + P\sum_{i=1}^{s} c_i\right) \tag{3.11}$$

where P *is* a penalty constant and c_i is constraint violations.

3.3 Teaching Learning Based Optimization (TLBO)

This novel search algorithm technique was developed by Rao et al. [6]. The optimization method consists of two basic phases which are teaching and learning. Teaching and learning processes between teacher and students in the class are applied on structural optimization problems to carry out optimum solutions. The teacher sharing his/her knowledge with the students is called as teaching phase. On the other hand, students in the class sharing information among themselves are named as learning phase. The main purpose of the two phases is to approach the best results in the optimization. The first class (population) is randomly created when starting optimization. Through TLBO technique, optimum global solutions can be obtained practically after many iterations. Important parameters of TLBO are listed below.

$$\text{class(population)} = \begin{bmatrix} x_1^1 & x_2^1 & \ldots & x_{n-1}^1 & x_n^1 \\ x_1^2 & x_2^2 & \ldots & x_{n-1}^2 & x_n^2 \\ \ldots & \ldots & \ldots & \ldots & \ldots \\ x_1^{S-1} & x_2^{S-1} & \ldots & x_{n-1}^{S-1} & x_n^{S-1} \\ x_1^S & x_2^S & \ldots & x_{n-1}^S & x_n^S \end{bmatrix} \begin{matrix} \to & f(x^1) \\ \to & f(x^2) \\ \to & \ldots \\ \to & f(x^{S-1}) \\ \to & f(x^S) \end{matrix} \tag{3.12}$$

where S and n indicate the numbers of row and column, respectively. In other words, S shows the number of students in the class and n gives the number of design variables. Each student in the class (population) gives a design solution and $f(x^{1,2,\ldots,S})$ show unconstrained objective function values of students (design solutions) in the class (population).

- In the teaching phase: The individual (student) having best solution in the class is determined as teacher. The other students are updated as,

$$x^{\text{new},i} = x^i + r(x_{\text{teacher}} - T_F x_{\text{mean}}) \tag{3.13}$$

$$x_{\text{mean}} = (\text{mean}(x_1)\ldots\text{mean}(x_S)) \tag{3.14}$$

where $x^{\text{new},i}$ shows the new student (design solution), x^i is the current student, r is a random number in the range [0,1], T_F is a teaching factor (either 1 or 2), x_{mean} is the mean of the class. If the new student ($x^{\text{new},i}$) gives a better solution than the current student, the design solution (student) replaces the current solution (student).

- In the learning phase: In order to get a better solution, information is shared among students in the class. The formulations given below are quite similar to those above.

$$\begin{aligned} &\text{if } f(x^i) < f(x^j) \Rightarrow x^{\text{new},i} = x^i + r(x^i - x^j) \\ &\text{if } f(x^i) > f(x^j) \Rightarrow x^{\text{new},i} = x^i + r(x^j - x^i) \end{aligned} \quad (3.15)$$

As mentioned above, if the new solution ($x^{\text{new},i}$) offers a better solution, it replaces the current student. All kinds of structural problems can be easily designed and analysed in SAP2000 software. In MATLAB programming, the preferred algorithm method is practically coded and conducted. Thanks to OAPI application, algorithm information (cross section, etc.) from MATLAB to SAP2000, static and dynamic analysis results from SAP2000 to MATLAB are easily transferred and optimum solutions are successfully carried out. Teaching learning based optimization procedures including MATLAB-SAP2000 OAPI (Open Application Programming Interface) interaction used in this study is shown in Fig. 3.1.

3.4 Design Examples

In this study, three different problems such as 47-member plane truss tower, 72- and 244-member space truss tower studied with different algorithm methods in literature studies are examined to test the optimum design performance of TLBO method in multi-element structures.

3.4.1 47-Member Plane Transmission Tower

Figure 3.2 shows a 47-member plane transmission tower that was studied with different algorithm methods in literature by Vanderplaats and Moses [36], Felix [37], Hansen and Vanderplaats [38], Hasancebi and Ulusoy [39], Kaveh et al. [40].

All the members are collected into 27 groups. In this study, optimum design of the structure is carried out by using a prepared list including 127 single-angle (*L*) profiles taken from AISC. The cross-sectional areas in the profile list varied between 3.12 and 107.74 cm^2. The loading information including different load cases is given in Table 3.1. The material properties used in the problem are modulus of elasticity (E) = 206,850 MPa and density (ρ) = 0.0816 N/cm^3. As the stress constraints, allowable tensile and compressive stress values used in the analysis are 137.90 MPa

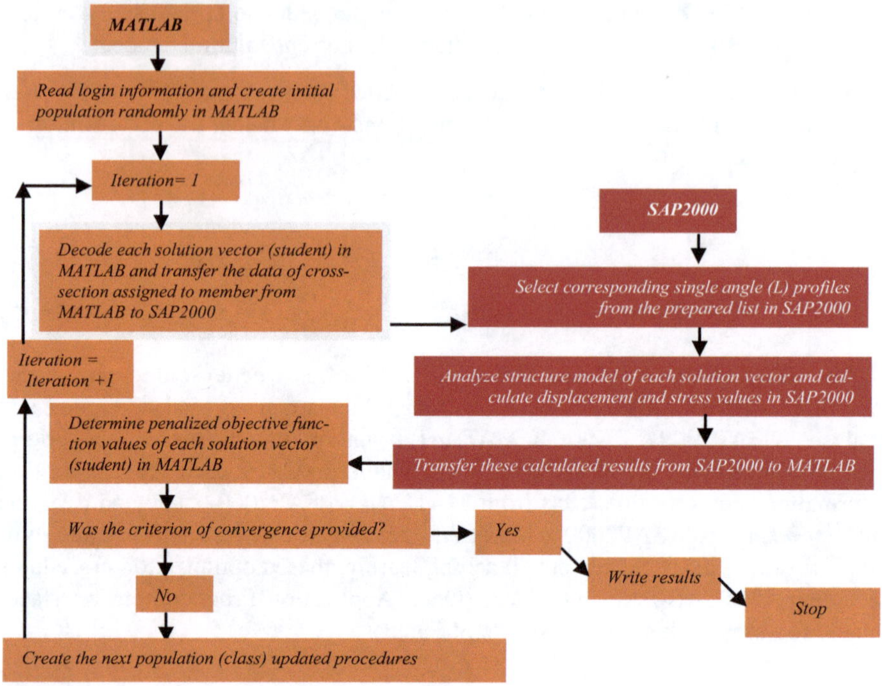

Fig. 3.1 Flowchart of TLBO algorithm procedures including MATLAB-SAP2000 OAPI interaction

and 103.425 MPa, respectively. The obtained cross-sectional areas and minimum weight value are compared with results provided in literature in Table 3.2. Also, the design history of the solution obtained in this study is presented in Fig. 3.3.

It is shown from Table 3.2 that the cross-sectional areas and minimum weight determined in this study are quite compatible with the literature results. The minimum weight value (8.29 kN) calculated is about 1% lighter than the result of Vanderplaats and Moses [36], about 2% lighter than the result of Felix [37], but 0.6% heavier than the result of Hansen and Vanderplaats [38], about 1% heavier than the results of Hasancebi and Ulusoy [39] and also Kaveh et al. [40]. The reason for this is that while the minimum cross-sectional area of 127 angle sections available in AISC used in this study is 3.12 cm^2, studies in literature were conducted with using smaller cross-sectional areas such as 0.64, 0.90, 0.95, 1.65 cm^2 etc. as shown in Table 3.2. In addition, it can be said that number of iterations to obtain optimum results is considerable less with TLBO. Optimum solution is obtained after 150 iterations for this example as shown in Fig. 3.3.

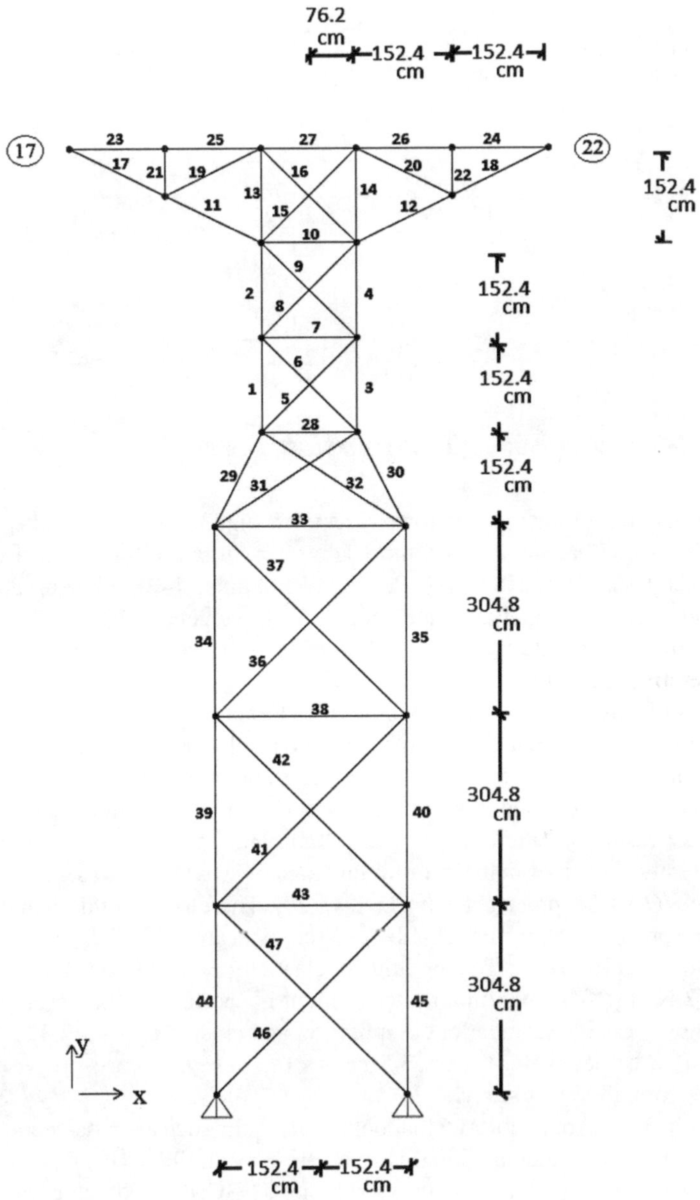

Fig. 3.2 47-member plane transmission tower

Table 3.1 Loading information of 47-member plane transmission tower

Joint	*Fx* (kN)	*Fy* (kN)
Load case 1		
17	26.72	−62.35
22	26.72	−62.35
Load case 2		
17	26.72	−62.35
22	0	0
Load case 3		
17	0	0
22	26.72	−62.35

3.4.2 72-Member Space Transmission Tower

A 72-member space transmission tower as shown Fig. 3.4 is the second space truss example to test the proposed algorithm taken form literature for comparison of the optimum solutions. Venkayya [41], Perez and Behdinan [42], Xicheng and Guixu [43], Erbatur et al. [44], Gellatly and Berke [45], Kaveh et al. [40] studied this truss structure before by using genetic algorithm, particle swarm optimization algorithm, parallel iterative algorithm.

The bars of truss tower are grouped in 16 different design variables as shown in Table 3.3. Loading information used in the problem solution is given in Table 3.4. The cross-sectional areas obtained, minimum weights and maximum displacement value are presented in Table 3.5 along with results obtained in other studies. Considering the available results in literature shown in Table 3.5, it is shown that many cross sections are much smaller than the minimum area of cross section (3.12 cm^2) of 127 single angles (*L*) in the preselected list in this study. Therefore, for this example, three solid square profiles having very small cross-sectional areas (0.65, 0.8 and 1.00 cm^2) are added to this profile list. The modulus of elasticity *E* is 68,950 MPa, and density ρ is 0.0272 N/cm^3. The maximum displacement is restricted to 0.635 cm in *x*- and *y*-directions. Stress constraints for compression and tension are—172.375 MPa and 172.375 MPa, respectively. Figure 3.5 shows the design history of the optimum solution performed in this study.

As it is observed from Table 3.5 that optimum results such as cross-sectional areas and minimum weight calculated in this study are very similar to the results obtained from the literature. As in the previous example, TLBO method achieves optimum results after very little iteration (after 250 iterations as seen in Fig. 3.5). As shown in Table 3.5 that the minimum weight value (1.76 kN) is same with the result of Gellatly and Berke [45]. However, this value is approximately 2.3–4% heavier than the other results of literature studies [40–44] given in Table 3.5. The reason is that optimum solutions are obtained with 127 angles available in AISC in this study. In addition to this, it can be said that if more different cross-sectional areas are added in addition to the three solid sections to the existing profile list, it can be said that the minimum

Table 3.2 Minimum weights of 47-element tower

	Vanderplaats and Moses [36]	Felix [37]	Hansen and Vanderplaats [38]	Hasancebi and Ulusoy [39]	Kaveh et al. [40]	This study (cm^2)
A3	17.78	17.34	15.37	16.89	16.94	15.68
A4	16.26	15.69	14.92	15.81	15.68	14.84
A5	4.89	4.64	5.21	4.32	4.30	4.61
A7	–	1.33	0.64	0.64	0.64	3.12
A8	4.25	5.97	5.46	5.78	5.76	5.22
A10	11.43	6.86	7.30	7.75	7.45	7.03
A12	12.32	10.73	11.24	11.68	11.59	10.90
A14	4.13	4.38	4.25	4.19	4.09	4.61
A15	6.73	6.73	5.46	5.08	5.03	6.84
A18	10.76	8.95	7.87	7.94	8.01	7.68
A20	1.02	1.65	2.10	1.84	1.83	3.12
A22	3.87	5.14	7.75	6.99	6.92	4.61
A24	8.32	6.73	5.91	5.97	6.01	9.42
A26	8.38	6.67	5.46	6.41	6.34	11.16
A27	6.67	5.21	4.38	11.37	10.80	9.29
A28	3.30	1.91	0.95	0.95	0.90	3.12
A30	18.67	17.59	15.62	16.77	16.81	16.52
A31	4.64	4.19	5.72	4.89	4.97	6.84
A33	–	1.33	0.64	0.64	0.64	3.12
A35	20.07	18.42	17.40	18.16	18.13	17.10
A36	6.16	1.71	5.84	5.91	5.72	7.03
A38	1.14	8.95	0.64	0.64	0.64	3.12
A40	22.04	21.78	18.67	19.56	19.55	18.52
A41	6.48	6.29	7.18	6.60	6.77	9.29
A43	–	1.08	0.64	0.64	0.64	3.12
A45	23.56	23.18	19.81	20.13	20.47	19.36
A46	5.91	6.41	6.99	7.56	7.54	9.29
Weight (kN)	8.37	8.48	8.24	8.21	8.21	8.29

weight may decrease accordingly. Moreover, it is understood from Table 3.5 that maximum displacement (0.63 cm) found in this study is very close to the limit value of (0.635 cm). This shows that the displacement constraints play a very effective role in the optimum design of 72-element space tower.

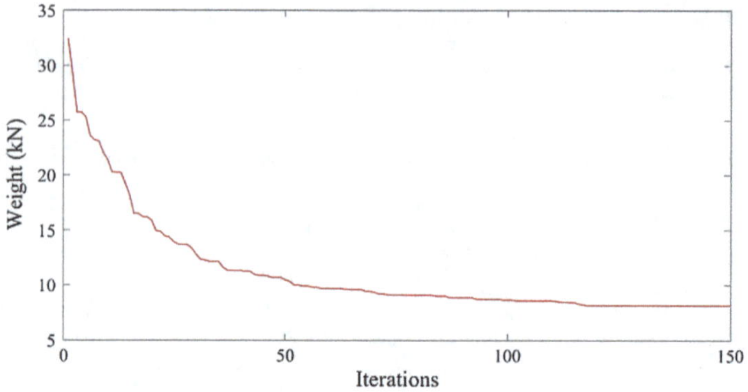

Fig. 3.3 Design history of 47-bar plane tower

Fig. 3.4 72-member space transmission tower

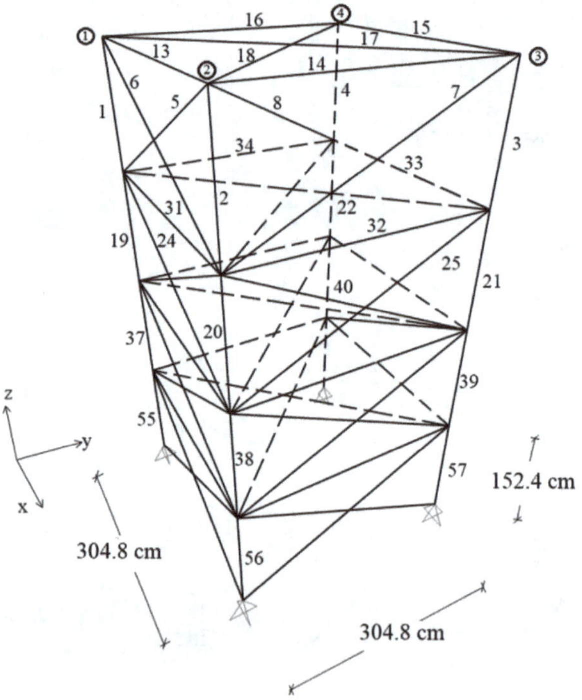

3.4.3 244-Member Space Transmission Tower

In order to show the robustness of the novel algorithm, optimum design of a 244-bar transmission tower structure presented in Fig. 3.6 is taken from studies Kaveh and Talatahari [5], Talatahari et al. [13], Talatahari et al. [7]. In this study, the 244 members of the truss structure are collected into 32 groups as shown in Fig. 3.6. A similar

Table 3.3 Member groups

Group	Members
A1	1–4
A2	5–12
A3	13–16
A4	17, 18
A5	19–22
A6	23–30
A7	31–34
A8	35, 36
A9	37–40
A10	41–48
A11	49–52
A12	53, 54
A13	55–58
A14	59–66
A15	67–70
A16	71, 72

Table 3.4 Loading information of 72-member space transmission tower

Loading condition	Joint	*X* (kN)	*Y* (kN)	*Z* (kN)
1	1	22.2685	22.2685	−22.2685
2	1	0	0	−22.2685
	2	0	0	−22.2685
	3	0	0	−22.2685
	4	0	0	−22.2685

example was studied by Saka[1]. Kaveh and Talatahari [5] investigated minimum weight of the transmission tower by using Heuristic Particle Swarm Ant Colony Optimization (HPSACO). Talatahari et al. [13] and Talatahari et al. [7] used Firefly Algorithm (FA) and chaotic imperialist competitive algorithm (CICA), respectively.

In this study, besides using TLBO algorithm, minimum weight design of the truss structure is carried out by using a prepared list including 127 single-angle (*L*) profiles taken from American Institute of Steel Construction (AISC). Moreover, despite the larger cross-sectional requirement, a double angle (125.81 cm^2) from AISC is added to the list. In this respect, this study is different from the literature studies. Therefore, it is expected at the beginning of the study that the minimum weight to be calculated in this study may be slightly higher than the values given in reference studies. The loads for the 244-bar transmission tower are taken as specified in reference studies by Kaveh and Talatahari [5], Talatahari et al. [13], Talatahari et al. [7]. The modulus

Table 3.5 Minimum weights of 72-element space tower

Group	Venkayya [41]	Perez and Behdinan [42]	Xicheng and Guixu [43]	Erbatur et al. [44]	Gellatly and Berke [45]	Kaveh et al. [40]	This study
A1	1.04	1.04	1.01	1.00	0.96	1.01	0.65
A2	3.59	3.29	3.46	3.45	4.99	3.53	3.12
A3	2.43	3.20	2.65	3.10	2.93	2.63	3.12
A4	3.26	3.63	3.68	3.35	2.20	3.70	5.22
A5	3.94	3.32	3.28	2.97	3.56	3.35	3.12
A6	3.43	3.53	3.37	3.42	3.93	3.34	3.12
A7	0.65	0.65	0.65	0.77	0.65	0.65	0.65
A8	0.65	0.71	0.65	1.06	0.65	0.65	3.12
A9	8.04	8.44	8.30	7.45	6.60	8.21	7.68
A10	3.38	3.35	3.33	3.77	3.50	3.27	4.61
A11	0.65	0.65	0.65	0.65	0.65	0.65	1.00
A12	0.65	0.65	0.65	0.65	0.65	0.65	0.65
A13	11.73	11.24	12.29	11.32	9.44	12.17	10.07
A14	3.38	3.35	3.34	3.26	3.36	3.32	3.12
A15	0.65	0.65	0.65	0.68	0.65	0.65	0.65
A16	0.65	0.65	0.65	1.00	0.65	0.65	0.65
Weight (kN)	1.70	1.70	1.70	1.72	1.76	1.69	1.76
Max. disp. (cm)							0.63

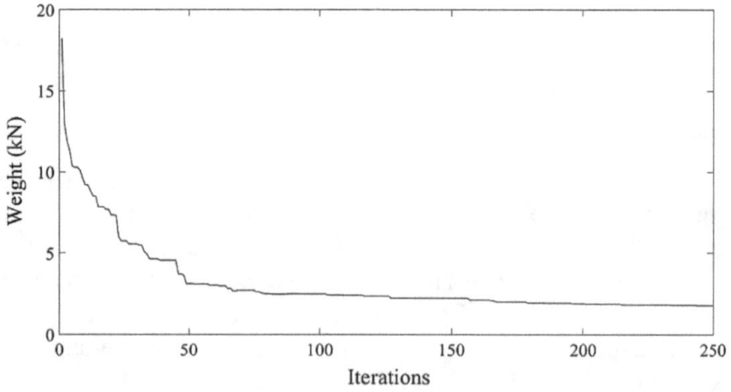

Fig. 3.5 Design history of 72-bar space tower

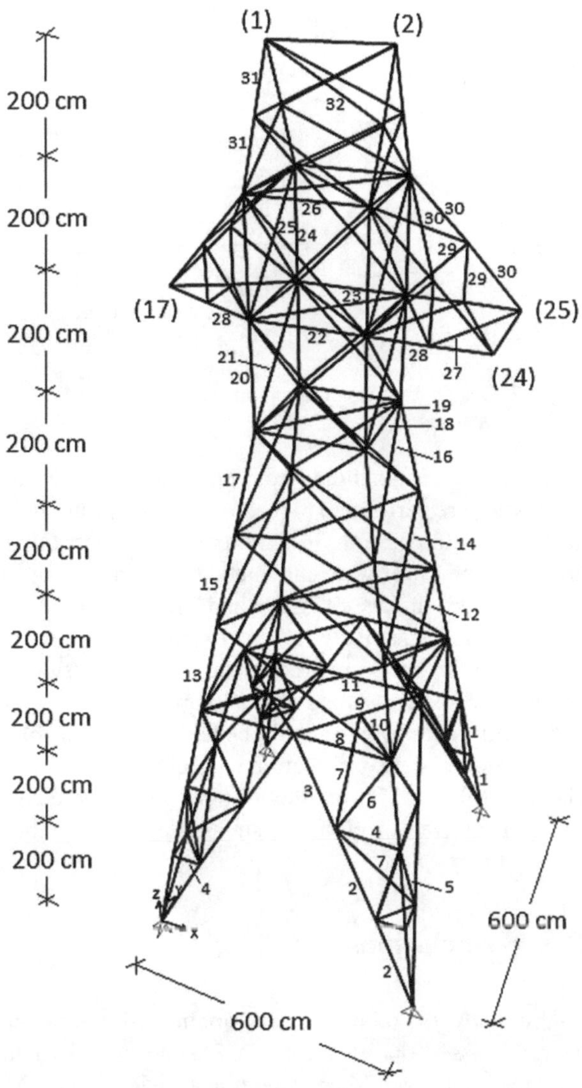

Fig. 3.6 A 244-bar transmission tower structure

of elasticity is 210,000 MPa, and material density is 2767.990 kg/m^3. The allowable stress is 140 MPa for tensile stress, and AISC-ASD [35] (Allowable Stress Design) specifications are used. Also, the displacements are restricted as stated in reference studies. Moreover, optimum solutions of the structure are investigated separately with generations of 20 and 30 individuals in this study.

Design histories of the optimum solutions for the generations of 20 and 30 individuals are presented in Fig. 3.7. Both solutions are determined with 800 iterations. Therefore, 16,000 and 24,000 analyses are carried out with 20 and 30 individuals, respectively.

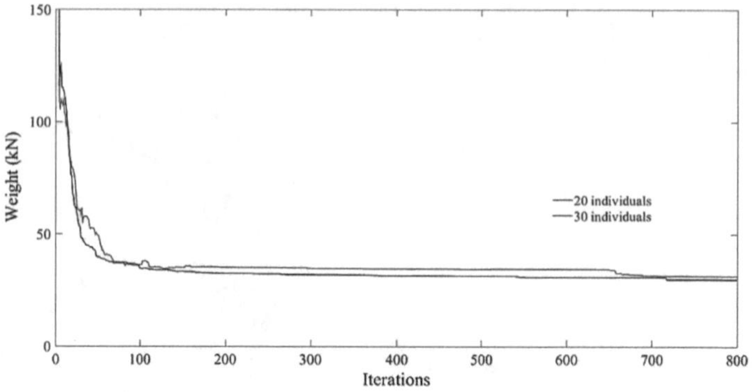

Fig. 3.7 Design histories of optimum solutions

In the reference studies from literature [5, 7, 13], approximately 10,000–150,000 analyses were performed to reach different optimum solutions via different algorithm techniques. In addition, in the literature studies, the minimum weight values of the transmission tower were calculated between 23.6 and 26 kN. In this study, minimum weights are found as 29.76 kN and 31.19 kN by using the generations of 30 and 20 individuals, respectively. In this example, the minimum structure weight is found to be quite higher compared to the literature results. For this reason, it is seen that the section list consisting of 127 L profiles is not sufficient for this problem, and many different profiles should be added to the profile list. Moreover, the maximum displacement in the y-direction is calculated as 14.9 mm. This value is very close to the limit value of 15 mm. Accordingly, it is observed that the displacement constraints are quite active as well as the stress constraints in the optimum solution.

3.5 Conclusions

In this study, the robustness and applicability of the novel algorithm called teaching learning based optimization (TLBO) are investigated via three different literature examples such as 47-bar plane transmission tower, 72- and 244-bar space transmission towers. In this study, optimum designs of these transmission towers are carried out by using a prepared list including 127 single-angle (*L*) profiles taken from American Institute of Steel Construction (AISC). When the optimum solutions reached are compared with the results provided in literature, it is shown that the results obtained for the first two examples are quite similar. The minimum weight is slightly heavier in the third example depending on the section list used in the analyses. In addition, it has been observed that the TLBO algorithm gives practically reasonable results after less number of iterations compare to other algorithms. Finally, the results prove that TLBO algorithm technique provides a practical and successful optimum design to multi-element structural problems with real profiles.

References

1. Saka MP (1990) Optimum design of pin jointed steel structures with practical applications. J Struct Eng ASCE 116:2599–2620
2. Hasançebi O, Erbatur F (2002) On efficient use of simulated annealing in complex structural optimization problems. Acta Mech 157:27–50
3. Lee KS, Geem ZW (2004) A new structural optimization method based on the harmony search algorithm. Comput Struct 82:781–798
4. Değertekin SO, Saka MP, Hayalioglu MS (2008) Optimal load and resistance factor design of geometrically nonlinear steel space frames via tabu search and genetic algorithm. Eng Struct 30:197–205
5. Kaveh A, Talatahari S (2009) Particle swarm optimizer, ant colony strategy and harmony search scheme hybridized for optimization of truss structures. Comput Struct 87:267–283
6. Rao RV, Savsani VJ, Vakharia DP (2011) Teaching-learning-based optimization: a novel method for constrained mechanical design optimization problems. Comput Aided Des 43:303–315
7. Talatahari S, Kaveh A, Sheikholeslami R (2012) Chaotic imperialist competitive algorithm for optimum design of truss structures. Struct Multidisc Optim 46:355–367
8. Degertekin SO (2012) Optimum design of geometrically non-linear steel frames using artificial bee colony algorithm. Steel Compos Struct 12:505–522
9. Aydoğdu I, Saka MP (2012) Ant colony optimization of irregular steel frames including elemental warping effect. Adv Eng Softw 44:150–169
10. Kaveh A, Talatahari S (2012) A hybrid CSS and PSO algorithm for optimal design of structures. Struct Eng Mech 42:783–797
11. Rafiee A, Talatahari S, Hadidi A (2013) Optimum design of steel frames with semi-rigid connections using Big Bang-Big Crunch method. Steel Compos Struct 14:431–451
12. Saka MP, Geem ZW (2013) Mathematical and Metaheuristic applications in design optimization of steel frame structures: an extensive review. Math Problems Eng 2013:1–33
13. Talatahari S, Gandomi AH, Yun GJ (2014) Optimum design of tower structures using firefly algorithm. Struct Des Tall Spec 23:350–361
14. Hadidi A, Rafiee A (2014) Harmony search based, improved Particle Swarm Optimizer for minimum cost design of semi-rigid steel frames. Struct Eng Mech 50:323–347
15. Dede T, Togan V (2015) A teaching learning based optimization for truss structures with frequency constraints. Struct Eng Mech 53:833–845
16. Degertekın SO, Geem ZW (2016) Metaheuristic optimization in structural engineering. Model Optim Sci Tech 7:75–93
17. Carbas S (2016) Design optimization of steel frames using an enhanced firefly algorithm. Eng Optim 48:2007–2025
18. Daloglu AT, Artar M, Ozgan K, Karakas AI (2016) Optimum design of steel space frames including soil-structure interaction. Struct Multidiscip O 54:117–131
19. Artar M (2016) Optimum design of braced steel frames via teaching learning based optimization. Steel Compos Struct 22:733–744
20. Rao RV (2016) Jaya: A simple and new optimization algorithm for solving constrained and unconstrained optimization problems. Int J Ind Eng Comput 7:19–34
21. Aydogdu I, Carbas S, Akin A (2017) Effect of levy flight on the discrete optimum design of steel skeletal structures using metaheuristics. Steel Compos Struct 24:93–112
22. Carbas S (2017) Optimum structural design of spatial steel frames via biogeography-based optimization. Neural Comput Appl 28:1525–1539
23. Aydogdu I (2017) Cost optimization of reinforced concrete cantilever retaining walls under seismic loading using a biogeography-based optimization algorithm with Levy flights. Eng Optim 49:381–400
24. Dede T (2018) Jaya algorithm to solve single objective size optimization problem for steel grillage structures. Steel Compos Struct 26:163–170

25. Topal U, Vo-Duy T, Dede T, Nazarimofrad E (2018) Buckling load optimization of laminated plates resting on Pasternak foundation using TLBO. Struct Eng Mech 67:617–628
26. Grzywinski M, Dede T, Ozdemir YI (2019) Optimization of the braced dome structures by using Jaya algorithm with frequency constraints. Steel Compos Struct 30:47–55
27. Artar M, Daloglu AT (2019) Optimum design of steel space truss towers under seismic effect using Jaya algorithm. Struct Eng Mech 71:1–12
28. Shallan O, Maaly HM, Sagiroglu M, Hamdy O (2019) Design optimization of semi-rigid space steel frames with semi-rigid bases using biogeography-based optimization and genetic algorithms. Struct Eng Mech 70:221–231
29. Dede T, Grzywinski M, Selejdak J (2020) Continuous size optimization of large-scale dome structures with dynamic constraints. Struct Eng Mech 73:397–405
30. Atmaca B, Dede T, Grzywinski M (2020) Optimization of cables size and prestressing force for a single pylon cable-stayed bridge with Jaya algorithm. Steel Compos Struct 34:853–862
31. Kaveh A, Kabir MZ, Bohlool M (2020) Optimum design of three-dimensional steel frames with prismatic and non-prismatic elements. Eng Comput 36:1011–1027
32. Kalemci EN, Ikizler SB (2020) Rao-3 algorithm for the weight optimization of reinforced concrete cantilever retaining Wall. Geomech Eng 20:527–536
33. MATLAB (2009) The language of technical computing. The Mathworks Inc., Natick, MA, USA
34. SAP2000 (2008) Integrated finite elements analysis and design of structures, computers and structures, Inc, Berkeley, CA, USA
35. AISC—ASD (1989) Manual of steel construction: allowable stress design. American Institute of Steel Construction, Chicago, IL, USA
36. Vanderplaats GN, Moses F (1972) Automated design of trusses for optimum geometry. J Struct Div ASCE 98:671–690
37. Felix JE (1981) Shape optimization of trusses subject to strength displacement and frequency constraints. Master thesis, Naval Postgraduate School
38. Hansen SR, Vanderplaats GN (1990) Approximation method for configuration optimization of trusses. AIAA J 28:161–168
39. Hasancebi O, Ulusoy AF (2005) Discrete and continuous structural optimization using evolution strategies. In: Topping BHV (ed) Proceedings of the eight international conference on the application of artificial intelligence to civil, structural and environmental engineering stirling: civil-comp press paper 23
40. Kaveh A, Gholipour Y, Rahami H (2008) Optimal design of transmission towers using genetic algorithm and neural networks. Int J Space Struct 23:1–19
41. Venkayya VB (1971) Design of optimum structures. Comput Struct 1:265–309
42. Perez RE, Behdinan K (2007) Particle swarm approach for structural design optimization. Comput Struct 85:1579–1588
43. Xicheng W, Guixu M (1992) A parallel iterative algorithm for structural optimization. Comput Methods Appl Mech Eng 96:25–32
44. Erbatur F, Hasancebi O, Tutuncu I, Kilic H (2000) Optimal design of planar and space structures with genetic algorithms. Comput Struct 75:209–224
45. Gellatly R, Berke L (1971) Optimal structural design. Technical Report AFFDL-TR-70-165. Air Force Flight Dynamics Laboratory (AFFDL)

Chapter 4
Modified Artificial Bee Colony Algorithm for Sizing Optimization of Truss Structures

Sadik Ozgur Degertekin, Luciano Lamberti, and Mehmet Sedat Hayalioglu

4.1 Introduction

Metaheuristic algorithms have been implemented for solving any kind of design problems in the last three decades. The main philosophy of metaheuristic search algorithms is to mimic a natural phenomenon. The name of each algorithm is indicative of its underlying principle. Particle swarm optimization (PSO) [1], ant colony optimization (ACO) [2], firefly algorithm (FFA) [3], eagle strategy (ES) [4], bat algorithm (BA) [5], dolphin echolocation [6], cuckoo search (CS) [7], crow search [8], bird swarm algorithm (BSO) [9], mouth brooding fish algorithm (MBF) [10], butterfly optimization algorithm (BOA) [11] and squirrel search algorithm (SSA) [12] are among the most popular colony-based methods.

Social sciences and human activities are another important source of inspiration for metaheuristic algorithms although less popular than swarm intelligence, for example tabu search (TS) [13], harmony search (HS) [14], teaching–learning-based optimization (TLBO) [15], mine blast algorithm (MBA) [16] and search group algorithm (SGA) [17].

Astronomy and physics are another interesting field of inspiration for metaheuristic optimization. Big bang-big crunch (BBBC) [18], gravitational search algorithm (GSA) [19], charged system search (CSS) [20], ray optimization (RO) [21], colliding bodies optimization (CBO) [22], water cycle algorithm (WCA) [23], flower pollination algorithm (FPA) [24], water evaporation optimization (WEO) [25], thermal exchange optimization (TEO) [26] and cyclical parthenogenesis algorithm

S. O. Degertekin (✉) · M. S. Hayalioglu
Department of Civil Engineering, Dicle University, 21280 Diyarbakir, Turkey
e-mail: sozgur@dicle.edu.tr

L. Lamberti
Dipartimento di Meccanica, Matematica e Management, Politecnico di Bari, 70125 Bari, Italy

S. Carbas et al. (eds.), *Nature-Inspired Metaheuristic Algorithms for Engineering Optimization Applications*, Springer Tracts in Nature-Inspired Computing,
https://doi.org/10.1007/978-981-33-6773-9_4

(CPA) [27] are the well-known optimization algorithms in these topics. Comprehensive reviews of the applications of metaheuristic optimization algorithms to structural design problems with critical assessment of the relative merits of different methods can be found in [28–32].

Artificial bee colony algorithm (ABC) is a swarm intelligence technique developed quite recently. ABC simulates the intelligent behaviour of honeybees. The ABC was originally proposed by Karaboga [33] for numerical optimization and later applied to function optimization [34], multimodal problems [35], design of digital filters [36], minimum spanning tree problem [37], economic load dispatch optimization [38], multi-objective design optimization of laminated composite components [39], design optimization of mechanical draft counter flow wet-cooling tower [40], economic optimization of shell and tube heat exchangers [41], capacitated vehicle routing problem [42], optimal reactive power flow [43], engineering design optimization [44], machine scheduling [45], travelling salesman problem [46], economic lot scheduling [47], spherical roller bearing design [48], multi-objective optimisation [49], optimization of multiple hydropower reservoir operation [50], design water quality monitoring networks in river basins [51], multiple traffic sign detection [52], quadratic assignment problem [53], mechanical engineering design problems [54], trim loss optimization in paper production [55].

Improved/modified/hybrid ABC has been developed for any kind of design optimization problems, such as modified ABC for constrained optimization problems [56], hybrid ABC for flexible job shop scheduling problems [57], improved ABC for global optimization [58], modified ABC for function optimization [59], modified ABC for real-parameter optimization [60], hybrid ABC for job shop scheduling [61], hybrid ABC for robust optimal design and manufacturing [62], hybrid ABC and particle swarm search for global optimization [63], modified ABC inspired by grenade explosion method [64], modified ABC with self-adaptive extended memory [65], hybrid artificial bee colony algorithm [66], enhanced ABC via differential evolution [67], improved ABC for robust design of power system stabilizers [68], improved ABC for constrained optimization problems [69], improved ABC for reliability optimization problems [70], improved ABC for energy-efficient clustering in wireless sensor networks [71], improved ABC for multi-objective distributed unrelated parallel machine scheduling [72].

ABC and its variants have been also implemented successfully in the field of structural design. For instance, optimization of truss structures [73], optimum design of geometrically nonlinear steel frames [74], optimum design of reinforced concrete beam [75], discrete optimization of trusses [76], identification of structural models [77], slope stability analysis [78], optimum cost design of RC columns [79], design optimization of RC continuous beams [80], optimal placement of viscous dampers in planar building frames [81], topology optimization for dynamic stiffness problems [82], construction site layout planning [83], topology optimization for nonlinear structural problems [84], optimal placement of steel diagonal braces [85], structural damage detection [86], design optimization of real-world steel space frames [87], structural damage detection [88], design optimization of RC shallow tunnels [89], minimum cost design of reinforced concrete frames [90], damage identification of hinged joints for simply supported slab bridges [91] and assessment of risks in tunnel constructions [92].

Various metaheuristic algorithms have been used for sizing optimization of truss structures. Just focussing on the last decade for the sake of brevity, ABC with adaptive penalty function (ABC-AP) [73], self-adaptive harmony search (SAHS) [93], teaching–learning-based optimization (TLBO) [94], modified teaching–learning-based optimization (MTLBO) [95], hybrid particle swarm and swallow swarm optimization (HPSSO) [96], flower pollination algorithm (FPA) [24], water evaporation optimization (WEO) [25], heat transfer search (HTS) [97] and chaotic enhanced colliding bodies (Chaotic EBC) [98].

The literature survey reveals that the ABC is very efficient in structural optimization problems and other engineering applications. However, it is observed that the convergence capability of the ABC is poor respect to the other metaheuristic algorithms. This study aims to develop a modified artificial bee colony algorithm (MABC) for enhancing robustness of standard ABC. Four classical structural optimization problems previously solved by standard ABC and other state-of-the-art metaheuristic algorithms are compared to MABC. The results achieved by the proposed approach are compared with those of the ABC and other methods in terms of optimum weight, convergence capability and several statistical indices.

The rest of the article consists of the following sections: Sect. 4.2 presents the structural optimization problem formulation. The standard ABC is described in Sect. 4.3. The proposed MABC and its differences from ABC are described in Sect. 4.4. Section 4.5 presents the MABC implemented for sizing optimization of truss structures. Test problems and design results are given in Sect. 4.6. Finally, Sect. 4.7 presents the main findings of this study.

4.2 Formulation of the Truss Optimization Problem

The optimization problem of a truss structure can be given as:

$$\begin{aligned}
&\text{Find } X = [x_1, x_2, \ldots, x_{\text{ng}}] \\
&\text{to minimize } W(X) = \sum_{i=1}^{nm} \rho_i x_i L_i \\
&\text{subject to } \sigma_i^c \le \sigma_i \le \sigma_i^t \quad i = 1, 2, \ldots, nm \\
&d_{\min} \le d_j \le d_{\max},\ j = 1, 2, \ldots, nn \\
&x_k^l \le x_k \le x_k^u \quad k = 1, 2, \ldots, \text{ng}
\end{aligned} \tag{4.1}$$

where X is the design vector, x_i is the cross-sectional area corresponding to the k-th variable, ng is the number of optimization variables (i.e. number of bar groups), x_k^l and x_k^u, respectively, are the lower and upper limits of cross-sectional areas, $W(X)$

denotes the weight of the truss, ρ_i and L_i, respectively, state the material density and the length of the i-th member. σ_i is the member stresses for each element i, and d_j is the displacements for each node j. σ_i^c and σ_i^t are the allowable compression and tension stresses for the i-th member. d_{min} and d_{max} are the allowable displacements for node j. nm is the number of elements, and nn is the number of nodes of the truss.

The penalized objective function used in this study is defined as follows:

$$W_\text{p}(X) = W(X) \times \phi_\text{p} \tag{4.2}$$

where $\boldsymbol{\varphi_\text{p}}$ is the penalty function and expressed as:

$$\phi_p = (1 + \nu)^\varepsilon \tag{4.3}$$

where the penalty function exponent ε is set equal to 2. ν is the sum of stress and displacement penalties, defined as:

$$\nu = \sum_{i=1}^{nm} \nu_i^\sigma + \sum_{j=1}^{\text{ndof}} \nu_j^d \tag{4.4}$$

The penalty ν_i^σ for the i-th member stress constraint and the displacement constraint penalty ν_j^d for the j-th node are respectively defined as:

$$\nu_i^\sigma = 0 \quad \text{if} \quad \sigma_i^c \le \sigma_i \le \sigma_i^t \tag{4.5}$$

$$\nu_i^\sigma = \frac{\left|\sigma_i - \sigma_i^{c,t}\right|}{\left|\sigma_i^{c,t}\right|} \quad \text{if} \quad \sigma_i < \sigma_i^c \quad \text{or} \quad \sigma_i^t < \sigma_i \tag{4.6}$$

$$\nu_j^d = 0 \quad \text{if} \quad d_j^{\text{min}} \le d_j \le d_j^{\text{max}} \tag{4.7}$$

$$\nu_j^d = \frac{\left|d_j - d_j^{\text{max,min}}\right|}{\left|d_j^{\text{max,min}}\right|} \quad \text{if} \quad d_j \le d_j^{\text{min}} \quad \text{or} \quad d_j^{\text{max}} \le d_j \tag{4.8}$$

4.3 Artificial Bee Colony Algorithm (ABC)

ABC is a colony-based heuristic search method originally developed by Karaboga [33]. The method simulates the honeybee swarm intelligent behaviour. Every bee colony includes employed bees, onlooker bees and scout bees. Employed bees count by half colony, while onlooker bees count by the other half. If an employed bee or onlooker bee cannot find a better food source for a while, they turn into a scout bee. Employed bees visit a food source close to those stored in their memory and bring collected nectar to the hive. Food sources are ranked based on employed bee dancing, which is more intensive for richer sources. Onlooker bees observe these dances and choose a food source based on its amount of nectar. The richer food sources are thus visited by more bees. Scout bees randomly search new food sources near the hive. An employed bee becomes a scout bee when its source is entirely exploited [35]. Employed and onlooker bees use available food sources, while scout bees investigate new sources.

The ABC includes three phases. In the first phase, an initial bee colony is randomly generated. A set of food sources is randomly selected by the bees, and their nectar amounts are determined. Source x_i (i = 1, 2, …, ntb) is denoted by a vector with *ng* components. ntb and ng, respectively, denote the number of food sources and the number of design variables. The food sources in the colony are sorted according to their nectar amounts. The colony is divided in employed bees and onlooker bees, each of which count by half of the colony.

In the second phase, an employed bee in the position *i* goes to a new food source by using following equation:

$$\begin{aligned} x_{ij}^{\text{new}} &= x_{ij}^{\text{old}} + \varphi_{ij}(x_{ij}^{\text{old}} - x_{kj})i, \\ k &= \{1, 2, ..., \text{neb}\}, \ j = \{1, 2, ..., \text{ng}\} \end{aligned} \tag{4.9}$$

where $i \neq k$ are randomly selected indexes, *neb* are the employed bees, *j* is the currently perturbed optimization variable, ϕ_{ij} is a random number in [−1,1]. Then, the employed bee observes the nectar amount of the new food source. If the new source is richer than the previous one, the bee moves to the new one. As soon as employed bees completed their investigation, they partake information with onlooker bees. The onlooker bees take this information and choose a food source based on probability as [34]:

$$p_i = \frac{\text{obj}_i}{\sum_{n=1}^{\text{neb}} \text{obj}_n} \tag{4.10}$$

where obj_i is the nectar amount of the *i-th* employed bee. It can be seen that probability is proportional to the food source nectar amount in the position *i*. After that, the onlooker bee varies the selected employed bee's position as:

$$x_{lj}^{\text{new}} = x_{lj}^{\text{old}} + \varphi_{lj}(x_{lj}^{\text{old}} - x_{ij}) \quad l = \{1, 2, \ldots \text{nob}\}, i = \{1, 2, \ldots, \text{neb}\} \tag{4.11}$$

where *l* and *i* are randomly selected indexes, *j* is the currently perturbed optimization variable, φ_{lj} is a random number in the interval [−1,1] and *nob* is the number of onlooker bees. If the new source is richer in nectar than the previous one, it is selected by the onlooker bee as its new position.

In the last phase, if a bee could not increase its nectar amount after a predetermined number of trials (lv), its food source is removed and the bee turns into a scout bee. This provides a new search space to the ABC. A new food source is explored by scout bees when it (x_{ij}) is removed. That is [38]:

$$x_{ij} = x_{j\,\min} + \text{rand}(0, 1)(x_{j\,\max} - x_{j\,\min}) \quad j = \{1, 2, \ldots, ng\} \tag{4.12}$$

where $x_{j\,\min}$ and $x_{j\,\max}$ denote the bounds of the *j-th* variable currently perturbed. An iteration is completed upon performing the three stages. The ABC is terminated when the maximum number of iterations is performed. The ABC has three control parameters: total number of bees (ntb), equal to the sum of employed bees (neb) and onlooker bees (nob), the predetermined limit number of analyses (lv), and the maximum number of iterations ($\text{it}_{\max}$).

Selection processes performed by the ABC consist of four steps: (*i*) *Global probabilistic selection*, where probability value given in Eq. (4.10) is used by onlooker bees to discover promising search areas. (ii) *Local probabilistic selection* performed by employed and onlooker bees to find a new food source near the current food source using Eqs. (4.9) and (4.11), respectively. (iii) *Greedy selection* performed by employed and onlooker bees; they forget the previous food source and memorize the new one if this is richer in nectar. (iv) *Random exploration stage* performed by scout bees as shown in Eq. (4.12).

The analogy between ABC and sizing optimization of truss structures could be made as follows: an employed bee/onlooker bee indicate a possible truss design. Hence, the total number of bees in the colony correspond to population size. The nectar amount denotes the weight of the truss structure.

4.4 Modified Artificial Bee Colony Algorithm (MABC)

Although the ABC was successfully applied to optimization of truss structures and the optimum designs obtained by the ABC are as good as or better than other algorithms, convergence speed was often very low [73]. For more complicated design problems with a large number of design variables, the ABC requires excessive structural analyses to obtain the optimum design. For instance, the ABC founds the optimal weight of the 200-bar planar truss (29 design variables and 1200 nonlinear stress constraints) after 1450×10^3 structural analyses in [73].

In this study, the MABC developed to improve convergence speed of standard ABC. The proposed MABC diverges from standard ABC for four perspectives: (i) standard ABC determines the new food source by perturbing a randomly selected variable whereas MABC obtains the new food source by perturbing all components of old source with Eqs. (4.9) and (4.11). (ii) A novel strategy was implemented the MABC to reduce computational cost. Structural analysis and hence penalty evaluation are done only if the weight of new design is lighter than previous design

$(W(X^{new}) < W(X^{pre}))$. The case $W(X^{new}) > W(X^{pre})$ requires a new constraint evaluation only if X^{pre} is infeasible. If X^{pre} is feasible and the new design is heavier than X^{pre}, it is not logical to perturb design along a non-descent direction. The converge capability is aimed to be improved by using the first two strategies. (iii) Even if the best food source is improved after lv searches, the employed bee located at this source does not turn into a scout bee, thus preserving the best source.

The last difference between standard ABC and proposed MABC regards the termination of algorithm. In standard ABC, search steps are repeated until the limit number of iterations is completed. However, the best value of it_{max} is usually determined only after exhaustive trial executions for each design example. The maximum iteration number is not specified as a constant value by the user in MABC, thus avoiding the aforementioned drawback. The optimization process is automatically terminated by MABC as soon as the ratio between the standard deviation of penalized functions evaluated for the designs of the population and the average penalized functions becomes smaller than the convergence limit fixed by the user. In other words, the search process is terminated automatically when population includes almost same designs. That is, when the following equation is satisfied [99]:

$$\frac{\text{SD}\left[W_P\left(X^1\right), W_P\left(X^2\right), \ldots, W_P\left(X^{\text{ntb}}\right)\right]}{\sum_{i=1}^{\text{ntb}} W_P\left(X^1\right)/\text{ntb}} \leq \varepsilon_{\text{conv}} \tag{4.13}$$

where $W_p(X^i)$ is the penalized objective function value of the i-th design in the colony, ϵ_{conv} is the convergence limit set equal to 10^{-15} in this study, consistently with the double precision limit of floating operations.

4.5 Truss Sizing Optimization with the MABC

The proposed MABC for truss sizing optimization includes the following steps:

Step 1 Set ntb and lv. The iteration counter is set as it = 0.

Step 2 Each design variable in a truss design is generated as follows:

$$x_i^j = x_{i\,\min}^j + \text{rand}(0, 1)(x_{i\,\max}^j - x_{i\,\max}^j) \tag{4.14}$$

where i is the design variable (i = 1, 2, …, ng), j is the truss design in the colony (j = 1, 2, …, ntb). Truss designs are stored in a matrix:

$$X_i^j = \begin{bmatrix} x_1^1 & x_2^1 & .. & x_{ng}^1 \\ x_1^2 & x_2^2 & .. & x_{ng}^2 \\ . & . & .. & . \\ . & . & .. & . \\ x_1^{\text{ntb}} & x_2^{\text{ntb}} & .. & x_{ng}^{\text{ntb}} \end{bmatrix}$$

$$i = 1, 2, \ldots, ng, j = 1, 2, \ldots, \text{ntb} \quad (4.15)$$

where each row corresponds to a truss design. The first half of designs are employed bees (neb), while the rest are appointed as onlooker bees (nob).

Step 3 Perform structural analyses for the truss designs in the initial colony. Calculate the penalized objective functions $W(X)$ using Eqs. (4.1)–(4.8).

Step 4 it = it + 1.

Step 5 Select a truss design randomly with the value of $W(X^r)$ objective function (r = 1, 2, …, neb) from the first half of the colony. Modify the selected truss design for generating a new one. The new design (X^{new}) is obtained by perturbing all variables with Eq. (4.9). If $W(X^{\text{new}}) < W(X^{\text{pre}})$ or the case $W(X^{\text{new}}) > W(X^{\text{pre}})$ only if the previous design X^{pre} is infeasible, perform structural analysis for the new truss design and compute penalized objective function $W_{\text{p}}(X^{\text{new}})$, go to Step 6, otherwise go to Step 7.

Step 6 If $W_{\text{p}}(X^{\text{new}}) < W_{\text{p}}(X^{\text{pre}})$, X^{new} is included in the colony; otherwise, X^{pre} is not changed.

Step 7 Steps 5 and 6 are repeated until the first half of the colony is analysed.

Step 8 An individual in the first half of the colony is extracted with a probability p_i. Equation (4.10) is suitable for maximization problems. Since this study minimizes weight of truss structures, Eq. (4.10) is modified as follows:

$$p_i = \frac{1/W(X^i)}{\sum_{i=1}^{\text{neb}} 1/W(X^i)} \quad i = 1, 2, \ldots, \text{neb} \quad (4.16)$$

Step 9 Obtain the truss design with Eq. (4.9). If $W(X^{\text{new}}) < W(X^{\text{pre}})$ or the case $W(X^{\text{new}}) > W(X^{\text{pre}})$ only if the previous design X^{pre} is infeasible, perform structural analysis for the new truss design and compute penalized objective function $W_{\text{p}}(X^{\text{new}})$ go to Step 10, otherwise go to Step 11.

Step 10 If $W_{\text{p}}(X^{\text{new}}) < W_{\text{p}}(X^{\text{pre}})$,$X^{\text{new}}$ is included in the colony; otherwise, X^{pre} is not changed.

Step 11 Repeat steps 9 and 10 until the rest of the colony is analysed.

Step 12 If a truss design different from the current best individual is not improved after a predetermined number of analyses (lv), abandon this design. Generate a new truss design with Eq. (4.10). Record the feasible design with the lowest objective function value and set it as the current optimum.

Step 13 Terminate the MABC if Eq. (4.13) is satisfied. Set the current best record as the final optimum design. Conversely, execute Step 4.

4.6 Design Examples

The new MABC algorithm is tested in four classical truss optimization problems. The results obtained by the MABC are compared with those of ABC-AP [73], SAHS

[93], TLBO [94], MTLBO [95], HPSSO [96], FPA [24], WEO [25], HTS [97] and chaotic EBC [98]. Although lots of metaheuristic optimization methods have been used in the field of sizing optimization of truss structures, the methods compared in this study to the proposed algorithm present the best solutions for the test problems solved here.

Sensitivity analysis is performed to determine the best setting of internal parameters. Due to the stochastic nature of MABC, 20 independent runs are carried out for each problem starting from randomly generated populations. The best solution found by MABC and the number of structural analyses (NSA) are reported. Furthermore, the mean optimized weight and the standard deviation (SD) on optimized weight computed over the 20 independent runs are presented. Constraint violation tolerances (CV) evaluated at the optimum designs are presented as well.

The results of sensitivity analysis done for determining the most appropriate values of ntb and lv in the MABC are presented in Table 4.1. It can be seen that the values of 30 for ntb and 150 structural analyses for lv lead to have the minimum weight truss designs. Therefore, these values are selected for all design examples.

Table 4.1 Sensitivity analysis for determining the best parameter setting of MABC: planar 10-bar truss (case 1) and 25-bar spatial truss

Cases	MABC tuning parameters		10-bar truss (case 1)	25-bar truss
	ntb	lv	Weight (lb)	Weight (lb)
1	20	100	5105.30	563.13
2	20	150	5084.30	563.25
3	20	200	5081.80	563.35
4	20	250	5081.80	563.33
5	30	100	5061.49	545.22
6	**30**	**150**	**5060.88**	**545.11**
7	30	200	5061.30	545.24
8	30	250	5061.49	545.32
9	40	100	5062.36	545.57
10	40	150	5062.22	545.47
11	40	200	5062.11	545.47
12	40	250	5062.11	545.58
13	50	100	5062.37	545.24
14	50	150	5061.62	545.41
15	50	200	5061.62	545.43
16	50	250	5061.49	545.47
17	60	100	5064.27	545.32
18	60	150	5064.15	545.25
19	60	200	5063.60	545.24
20	60	250	5064.28	545.24

4.6.1 *Planar Ten-Bar Truss*

The first design example regards the planar ten-bar truss structure shown in Fig. 4.1. The Young's modulus and mass density of truss members are 10^4 ksi (1 ksi = 6.895 MPa) and 0.1 lb/in^3, respectively. The allowable element stress is 25 ksi in both tension and compression. All nodal displacements must be less than ± 2 in. Each element denoted a design variable, with a minimum gage of 0.1 in^2. Two load cases are as: Case 1, $P_1 = 100$ kips and $P_2 = 0$ and Case 2, $P_1 = 150$ kips and $P_2 = 50$ kips.

Tables 4.2 and 4.3 compare optimization results obtained by the MABC with those of other algorithms for Case 1 and Case 2, respectively. The proposed algorithm found better designs than SAHS, TLBO, HPSSO, WEO and the same optimum of ABC-AP. However, MABC is the second fastest method among all methods presented in Tables 4.2 and 4.3. MABC significantly reduced the computational effort in comparison with ABC-AP: it found the same optimized designs after only 11,040 (Case 1) and 12,360 (Case 2) structural analyses versus 500,000 analyses. Although MABC required more structural analyses than SAHS, it could find feasible intermediate designs lighter than the optimized weights of SAHS [93] after only 6602 and 6861 structural analyses for cases 1 and 2, respectively.

Figures 4.2 and 4.3 compare the convergence curves of the MABC and the other methods. It found nearly optimum designs after approximately 5000 structural analyses in both cases.

Fig. 4.1 Planar ten-bar truss

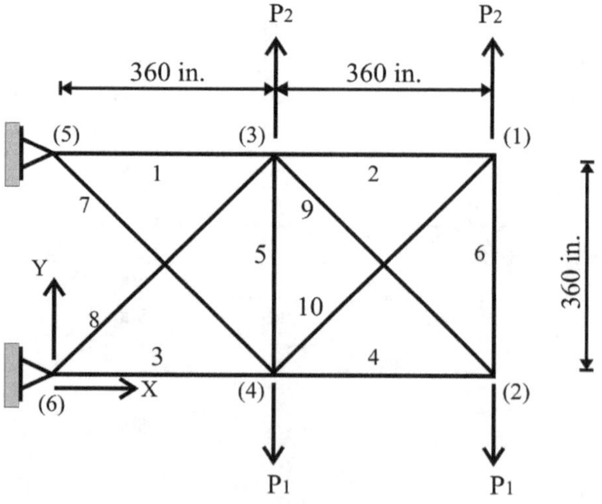

Table 4.2 Optimized designs for the 10-bar truss problem (case 1)

Design variables A_i (in^2)	ABC-AP [73]	SAHS [93]	TLBO [94]	HPSSO [96]	WEO [25]	Chaotic EBC [98]	MABC
A_1	30.5480	30.394	30.4286	30.5838	30.5755	NR	30.5480
A_2	0.1000	0.1000	0.1000	0.1000	0.1000	NR	0.10000
A_3	23.1800	23.0980	23.2436	23.1510	23.3368	NR	23.1800
A_4	15.2180	15.4910	15.3677	15.2056	15.1497	NR	15.2180
A_5	0.1000	0.1000	0.1000	0.1000	0.1000	NR	0.1000
A_6	0.5510	0.5290	0.5751	0.5488	0.5276	NR	0.55100
A_7	7.4630	7.4880	7.4404	7.46532	7.4458	NR	7.46300
A_8	21.0580	21.1890	20.9665	21.0643	20.9892	NR	21.0580
A_9	21.5010	21.3420	21.5330	21.5293	21.5236	NR	21.5010
A_{10}	0.1000	0.1000	0.1000	0.1000	0.1000	NR	0.1000
Weight (lb)	5060.880	5061.42	5060.96	5060.86	5060.99	5064.488	5060.88 [5061.42][a]
Mean weight (lb)	NA	5061.95	5062.08	5062.28	5062.09	5112.60	5060.99
SD (lb)	NA	0.71	0.79	4.325	2.05	NR	0.85
CV (%)	None	None	None	None	None	*	None
NSA	500,000	7081	16,872	14,118	19,540	20,000	11,040

[a]MABC found a feasible intermediate design lighter than the optimized weight of SAHS [93] after only 6602 structural analyses

*CV (%) could not be calculated because design variables are not reported

4.6.2 *Spatial Twenty-Five Bar Truss*

Figure 4.4 shows the spatial 25-bar spatial tower selected as second test problem. The Young's modulus and density of truss members are the same as in the previous design example. The structure must carry the loads listed in Table 4.4. Variable linking and corresponding allowable stress values could be found in the previous study of authors [94]. All nodal displacements are below 0.35 in. The minimum cross-sectional area is 0.01 in^2 for all bars.

Table 4.5 compares the results obtained in this study with the literature. It appears that MABC found very competitive designs with SAHS and TLBO. Although MABC required more structural analyses than SAHS and FPA, it could find feasible intermediate designs lighter than the optimized weight of SAHS and FPA after only 8962 and 8210 structural analyses, respectively. The convergence capability of MABC is obviously better than that of ABC-ABC, 10,640 versus 300,000 analyses. Moreover, MABC could find the same optimized weight obtained by ABC-AP after only 8015 structural analyses.

Table 4.3 Optimized designs for the 10-bar truss problem (case 2)

Design variables A_i (in^2)	ABC-AP [73]	SAHS [93]	TLBO [94]	HPSSO [96]	WEO [25]	Chaotic EBC [98]	MABC
A_1	23.4692	23.5250	23.5240	23.52377	23.5804	NR	23.4432
A_2	0.1005	0.1000	0.1000	0.10000	0.1003	NR	0.1000
A_3	25.2393	25.4290	25.4410	25.3686	25.1582	NR	25.3718
A_4	14.3540	14.4880	14.4790	14.3779	14.1801	NR	14.1360
A_5	0.1001	0.1000	0.1000	0.10000	0.1002	NR	0.1000
A_6	1.9701	1.9920	1.9950	1.9697	1.9708	NR	1.9699
A_7	12.4128	12.3520	12.3340	12.3678	12.4511	NR	12.4335
A_8	12.8925	12.6980	12.6890	12.7972	12.9349	NR	13.0173
A_9	20.3343	20.3410	20.3540	20.3257	20.3595	NR	20.2717
A_{10}	0.1000	0.1000	0.1000	0.10000	0.1001	NR	0.1000
Weight (lb)	4677.077	4678.84	4678.31	4676.95	4677.31	4680.232	4677.26 [4678.84][b]
Mean weight (lb)	NR[a]	4680.08	4680.12	4677.38	4679.06	4694.344	4677.97
SD (lb)	NR	1.89	1.016	0.463	2.07	NR	0.96
CV (%)	None	None	None	0.00142	None	*	None
NSA	500,000	7267	14,857	14,406	19,890	20,000	12,360

[a]Not reported
[b]MABC found a feasible intermediate design lighter than the optimized weight of SAHS [93] after only 6861 structural analyses.
*CV (%) could not be calculated because design variables are not available

The convergence curves are given in Fig. 4.5. It is clear that the MABC shows as good convergence performance as other algorithms.

4.6.3 Spatial Seventy-Two Bar Truss

The third example regards the spatial 72-bar truss of Fig. 4.6. Material properties are the same of previous examples. The cross-sectional areas of elements (sizing variables) are categorized into 16 groups (see Table 4.6). The structure must carry the loads indicated in [94]. The allowable element stress is 25 ksi (the same in tension and compression), while top nodes cannot displace by more than 0.25 in. The minimum area gage is 0.1 in^2 for Case 1 and 0.01 in^2 for Case 2.

Tables 4.6 and 4.7 compare results obtained by MABC with those reported in the literature. Both tables show that the MABC obtained the best design since the lighter design found by the FPA for Case 1 violated the design constraints. The optimum design obtained by MABC in Case 2 not only is lighter than the one of ABC-AP but also was obtained after only 10,520 structural analyses vs. the 400,000 analyses of

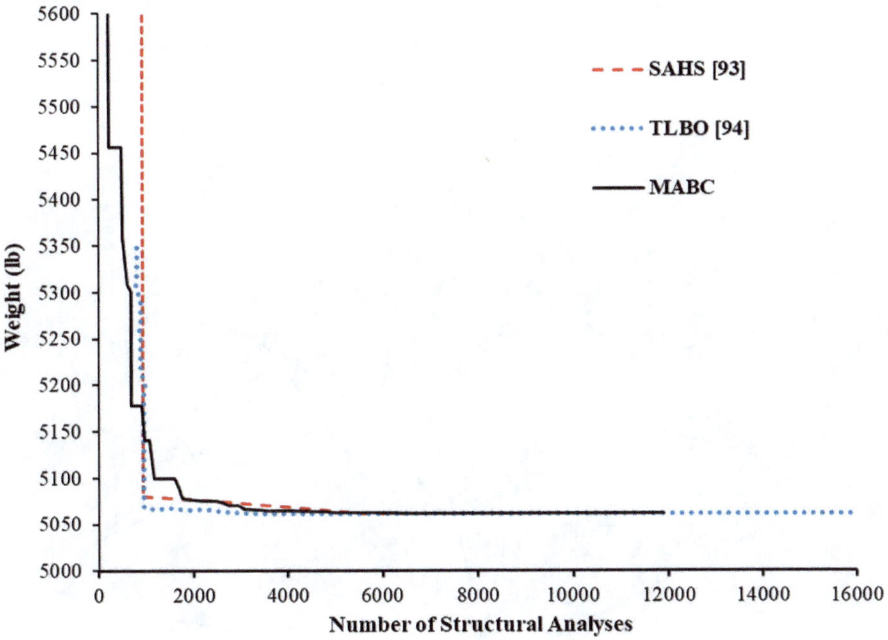

Fig. 4.2 Convergence curves for case 1 of the 10-bar truss problem

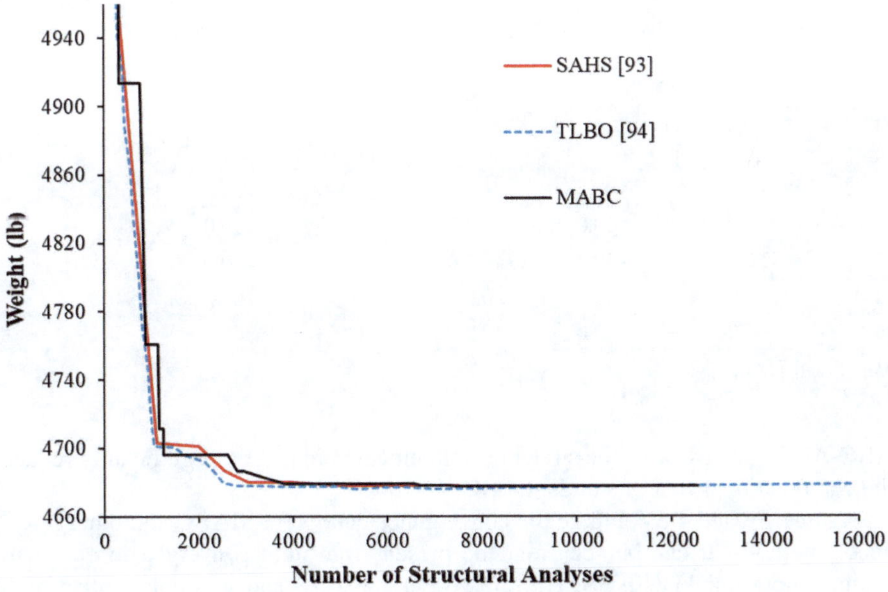

Fig. 4.3 Convergence curves for case 2 of the 10-bar truss problem

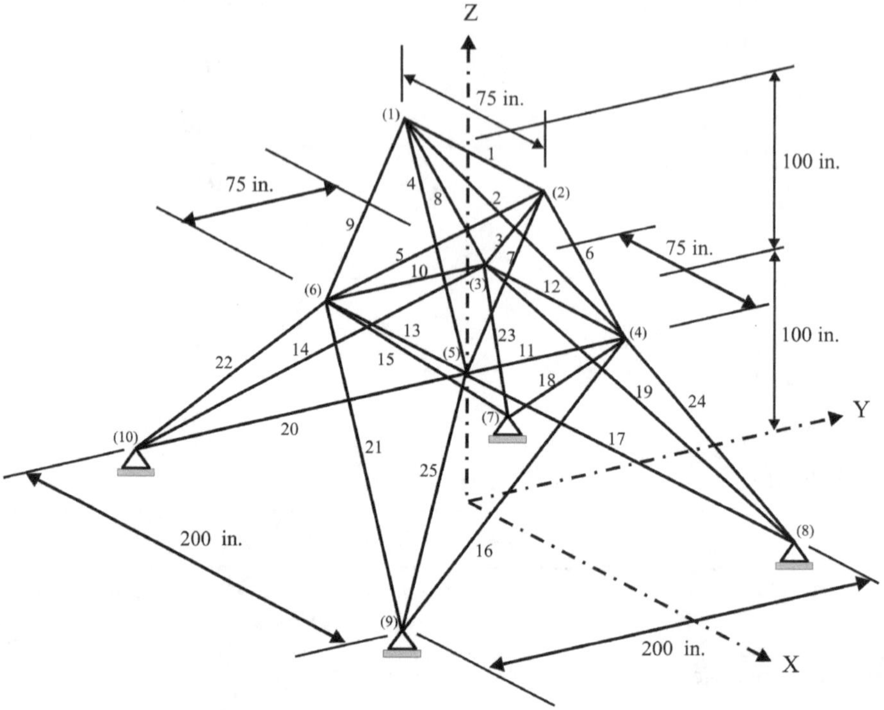

Fig. 4.4 Spatial twenty-five bar truss

Table 4.4 Loads of the spatial 25-bar tower

Node	Condition 1			Condition 2		
	F_x	F_y	F_z	F_x	F_y	F_z
1	0.0	20.0	−5.0	1.0	10.0	−5.0
2	0.0	−20.0	−5.0	0	10.0	−5.0
3	0.0	0.0	0.0	0.5	0	0
6	0.0	0.0	0.0	0.5	0	0

Note Loads are in kips

ABC-AP. In other words, the strategies implemented in MABC significantly reduce the computational cost of standard ABC.

Figures 4.7 and 4.8 compare the convergence curves for MABC and other referenced methods. It can be seen that the present algorithm converged to optimum designs more quickly SAHS, HPSSO, WEO and HTS and was competitive with TLBO.

Table 4.5 Optimized designs for the 25-bar truss problem

Design variables A_i (in^2)	ABC-AP [73]	SAHS [93]	TLBO [94]	FPA [24]	WEO [25]	MABC
A_1	0.0110	0.0100	0.0100	0.0100	0.0100	0.0100
A_2–A_5	1.9790	2.0740	2.0712	1.8308	1.9184	2.0700
A_6–A_9	3.0030	2.9610	2.9570	3.1834	3.0023	2.9590
A_{10}–A_{11}	0.0100	0.0100	0.0100	0.0100	0.0100	0.0100
A_{12}–A_{13}	0.0100	0.0100	0.0100	0.0100	0.0100	0.0100
A_{14}–A_{17}	0.6900	0.6910	0.6891	0.7017	0.6827	0.6880
A_{18}–A_{21}	1.6790	1.6170	1.6209	1.7266	1.6778	1.6210
A_{22}–A_{25}	2.6520	2.6740	2.6768	2.5713	2.6612	2.6780
Weight (lb)	545.193	545.12	545.09	545.159	545.166	545.11 [545.193][a] [545.12][b] [545.159][c]
Mean weight (lb)	NR	545.94	545.41	545.730	545.226	545.42
SD (lb)	NR	0.91	0.42	0.59	0.083	0.23
CV (%)	None	None	None	0.138	None	None
NSA	300,000	9051	15,318	8149	19,750	10,640

[a]MABC found a feasible intermediate design lighter than the optimized weight of ABC-AP [73] after only 8015 structural analyses

[b]MABC found a feasible intermediate design lighter than the optimized weight of SAHS [93] after only 8962 structural analyses.

[c]MABC found a feasible intermediate design lighter than the optimized weight of FPA [24] after only 8210 structural analyses.

4.6.4 Planar Two-Hundred Bar Truss

The optimum design of the planar two-hundred truss structure shown in Fig. 4.9 was searched in the last example. The Young's modulus of the material is 30,000 ksi, mass density is 0.283 lb/in^3. The allowable element stress is 10 ksi (the same in tension and compression). The optimization problem does not include displacement constraints. The structure includes 29 element groups (see Table 4.8). The minimum cross-sectional area of bars is 0.1 in^2. The structure must carry the three independent loading conditions indicated in Ref. [94].

Table 4.9 compares the optimum designs found by MABC, ABC-AP and other referenced methods. Once again, MABC is most efficient method in terms of optimized weight. Although MABC required more structural analyses than FPA, it found feasible intermediate designs lighter than the optimized weight of FPA after only 9740 structural analyses.

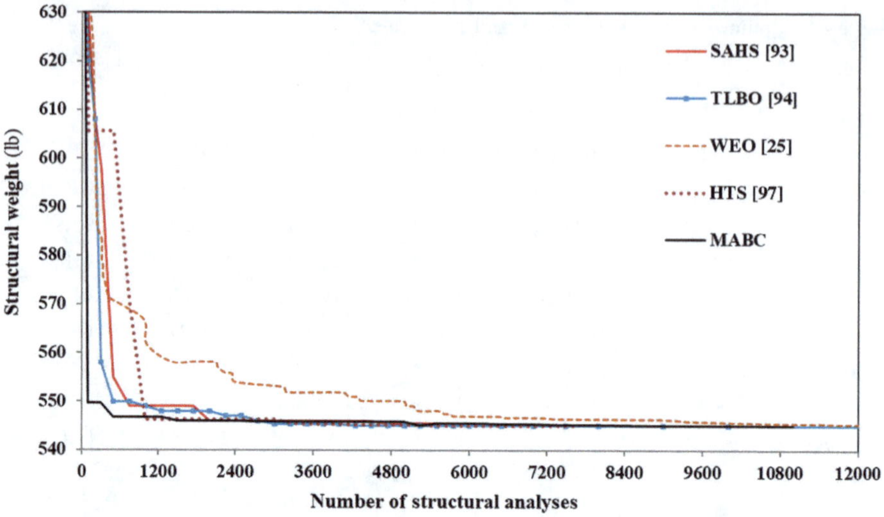

Fig. 4.5 Convergence curves for the 25-bar truss problem

Figure 4.10 compares the convergence curves of the MABC and other methods considered in this study. It is apparent that the MABC has a very good performance with respect to other methods.

4.7 Concluding Remarks

This study presented a modified variant of ABC (denoted as MABC) for truss sizing optimization. The design examples discussed in this study demonstrate that the MABC outperforms other state-of-the-art metaheuristic methods. MABC is powered by the implementation of elitist strategies in the search process. The trial designs which do not increase the quality of population are immediately rejected in MABC. This allows reducing the number of structural analyses required in the optimization process. Therefore, the computational cost of MABC is significantly lower than that of ABC-AP in all cases and better than other methods in almost all cases. Remarkably, standard deviation on optimized weight over 20 independent runs was always less than 0.5% of average optimized weight for all design examples. This proves that the MABC always converges to the global optimum or a nearly global optimum design and it is insensitive to initial population.

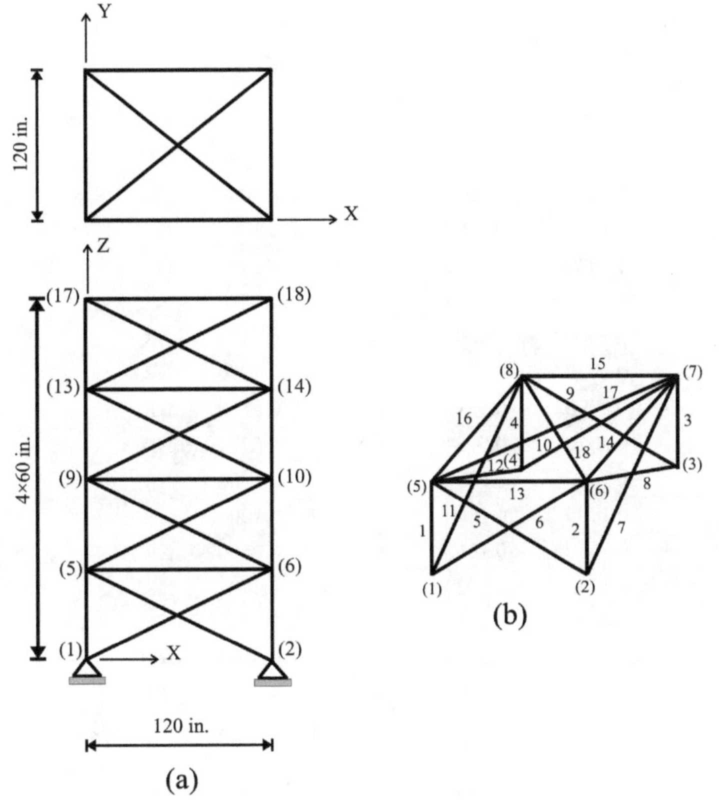

Fig. 4.6 Spatial 72-bar truss

Table 4.6 Optimized designs for the 72-bar truss problem (case 1)

Design variables A_i (in^2)	SAHS [93]	TLBO [94]	FPA [24]	HTS [97]	Chaotic EBC [98]	MABC
A_1–A_4	1.8600	1.8807	1.8758	1.9001	NR	1.8794
A_5–A_{12}	0.5210	0.5142	0.5160	0.5131	NR	0.5141
A_{13}–A_{16}	0.1000	0.1000	0.1000	0.1000	NR	0.1000
A_{17}–A_{18}	0.1000	0.1000	0.1000	0.1000	NR	0.1000
A_{19}–A_{22}	1.2930	1.2711	1.2993	1.2456	NR	1.2719
A_{23}–A_{30}	0.5110	0.5151	0.5246	0.5080	NR	0.5138
A_{31}–A_{34}	0.1000	0.1000	0.1001	0.1000	NR	0.1000
A_{35}–A_{36}	0.1000	0.1000	0.1000	0.1000	NR	0.1000
A_{37}–A_{40}	0.4990	0.5317	0.4971	0.5550	NR	0.5249
A_{41}–A_{48}	0.5010	0.5134	0.5089	0.5227	NR	0.5179
A_{49}–A_{52}	0.1000	0.1000	0.1000	0.1000	NR	0.1000
A_{53}–A_{54}	0.1000	0.1000	0.1000	0.1000	NR	0.1000
A_{55}–A_{58}	0.1680	0.1565	0.1575	0.1566	NR	0.1566
A_{59}–A_{66}	0.5840	0.5429	0.5329	0.5407	NR	0.5419
A_{67}–A_{70}	0.4330	0.4081	0.4089	0.4084	NR	0.4085
A_{71}–A_{72}	0.5200	0.5733	0.5731	0.5669	NR	0.5705
Weight (lb)	380.62	379.632	379.095	379.73	379.7181	379.62
Mean weight (lb)	382.42	379.759	379.534	382.26	380.8552	379.75
SD (lb)	1.38	0.149	0.272	1.94	NR	0.15
CV (%)	None	None	0.2039	None	*	None
NSA	13,742	21,542	9029	13,166	20,000	10,160

*CV (%) could not be calculated because design variables are not reported

Table 4.7 Optimized designs for the 72-bar truss problem (case 2)

Design variables A_i (in^2)	ABC-AP [73]	SAHS [93]	TLBO [94]	FFA [24]	HPSSO [96]	WEO [25]	HTS [97]	MABC
A_1–A_4	1.8907	1.8890	1.8929	1.8945	1.8932	1.8618	1.9000	1.8929
A_5–A_{12}	0.5166	0.5200	0.5160	0.5179	0.5111	0.5206	0.5238	0.5186
A_{13}–A_{16}	0.0100	0.0100	0.0100	0.0100	0.0100	0.0105	0.0100	0.0100
A_{17}–A_{18}	0.0100	0.0100	0.0100	0.0100	0.0100	0.0100	0.0100	0.0100
A_{19}–A_{22}	1.2968	1.2890	1.2917	1.2906	1.2912	1.2455	1.3039	1.2998
A_{23}–A_{30}	0.5191	0.5240	0.5175	0.5170	0.5151	0.5177	0.5073	0.5146
A_{31}–A_{34}	0.0100	0.0100	0.0100	0.0100	0.0100	0.0101	0.0100	0.0100
A_{35}–A_{36}	0.0101	0.0100	0.0100	0.0100	0.0100	0.0100	0.0100	0.0100
A_{37}–A_{40}	0.5208	0.5390	0.5229	0.5213	0.5360	0.5327	0.5194	0.5223
A_{41}–A_{48}	0.5178	0.5190	0.5192	0.5204	0.5211	0.5109	0.5169	0.5184
A_{49}–A_{52}	0.0100	0.0150	0.0100	0.0100	0.0100	0.0100	0.0100	0.0100
A_{53}–A_{54}	0.1048	0.1050	0.0997	0.1045	0.1108	0.1205	0.1067	0.0980
A_{55}–A_{58}	0.1675	0.1670	0.1679	0.1675	0.1666	0.1655	0.1674	0.1682
A_{59}–A_{66}	0.5346	0.5320	0.5359	0.5328	0.5339	0.5397	0.5330	0.5357
A_{67}–A_{70}	0.4443	0.4250	0.4456	0.4455	0.4537	0.4554	0.4536	0.4460
A_{71}–A_{72}	0.5803	0.5790	0.5818	0.5804	0.5745	0.5995	0.5786	0.5817
Weight (lb)	363.8392	364.0	363.841	363.833	363.8581	363.9827	363.885	363.826
Average weight (lb)	N/A	366.57	364.42	363.26	364.065	364.3536	366.03	364.19
SD (lb)	N/A	2.02	0.49	0.369	0.305	0.2188	2.29	0.41
CV (%)	None	None	None	None	None	None	None	None
NSA	400,000	12,852	17,954	39,404	13,086	19,860	13,592	10,520

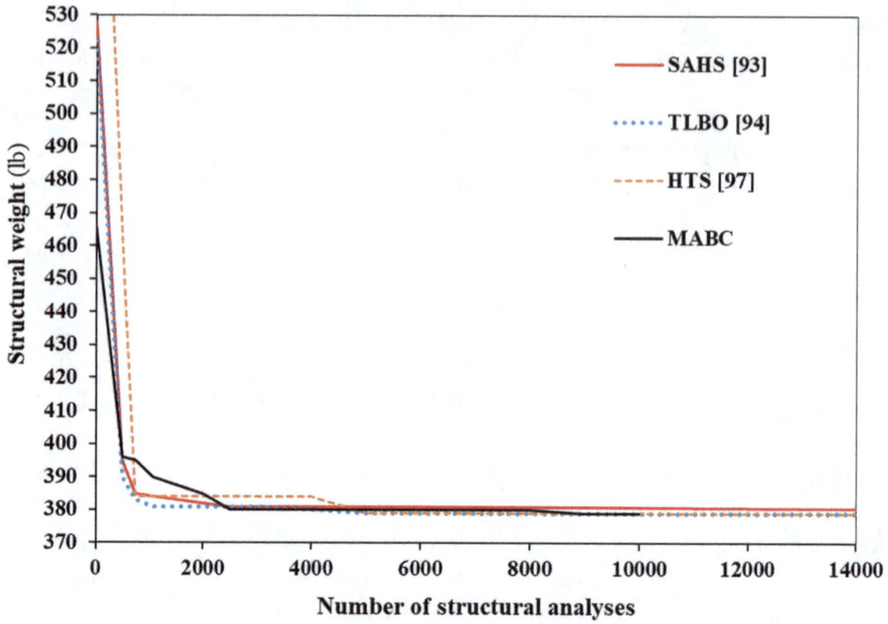

Fig. 4.7 Convergence curves for the 72-bar truss problem (case 1)

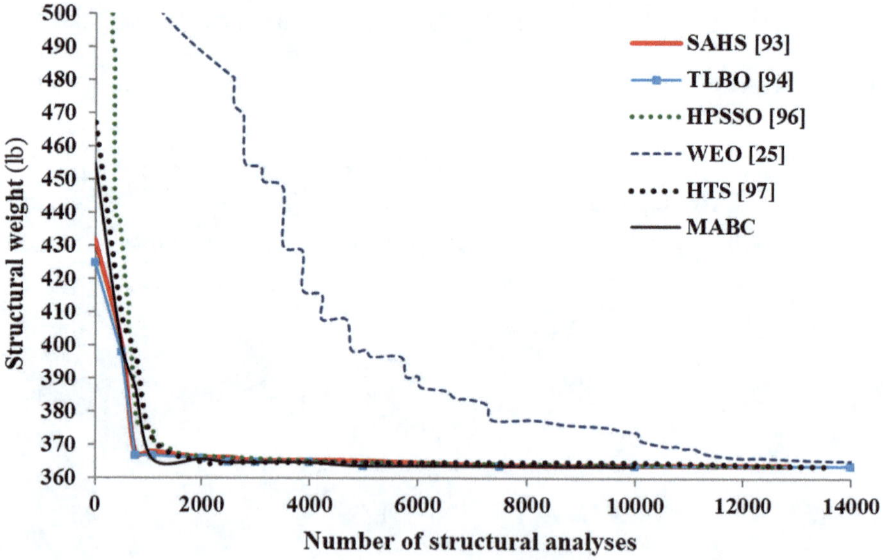

Fig. 4.8 Convergence curves for the 72-bar truss problem (case 2)

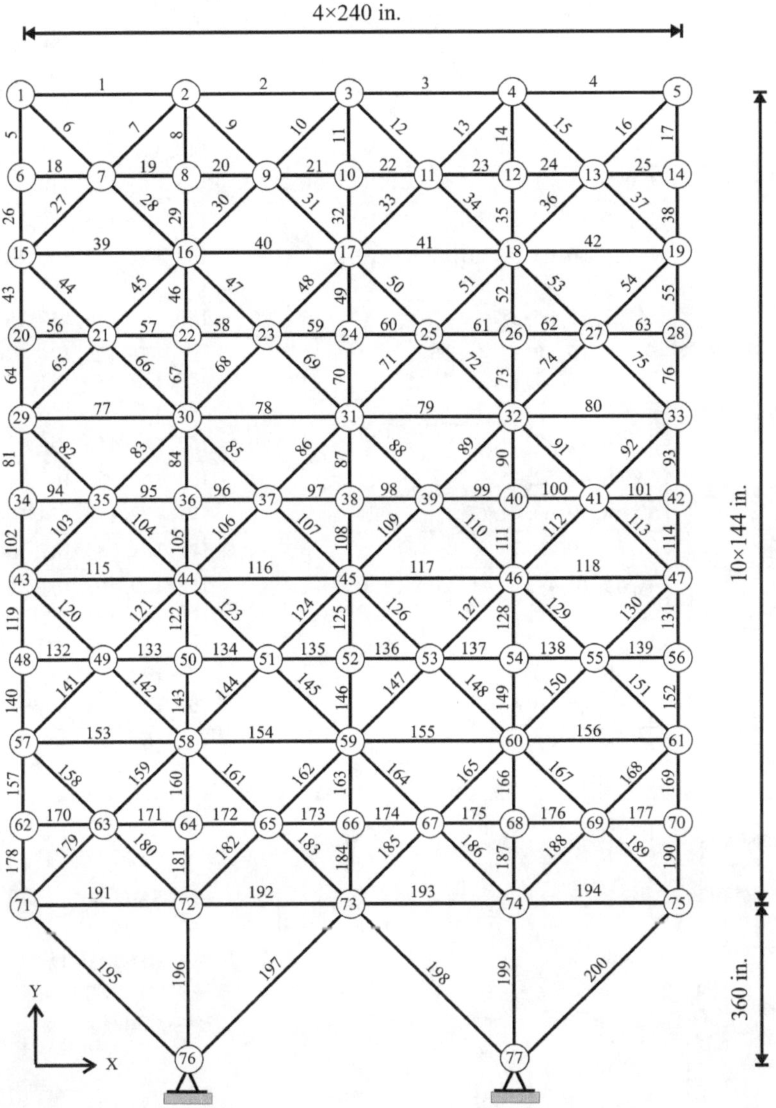

Fig. 4.9 Planar 200-bar truss

Table 4.8 Design variables for the planar 200-bar truss problem

Design variables	Member number	Design variables	Member number
1	1, 2, 3, 4	16	82, 83, 85, 86, 88, 89, 91, 92, 103, 104, 106, 107, 109, 110, 112, 113
2	5, 8, 11, 14, 17	17	115, 116, 117, 118
3	19, 20, 21, 22, 23, 24	18	119, 122, 125, 128, 131
4	18, 25, 56, 63, 94,101,132, 139, 170, 177	19	133, 134, 135, 136, 137, 138
5	26, 29, 32, 35, 38	20	140, 143, 146, 149, 152
6	6, 7, 9, 10, 12, 13, 15, 16, 27, 28, 30, 31, 33, 34, 36, 37	21	120, 121, 123, 124, 126, 127, 129, 130, 141, 142, 144, 145, 147, 148, 150, 151
7	39, 40, 41, 42	22	153, 154, 155, 156
8	43, 46, 49, 52, 55	23	157, 160, 163, 166, 169
9	57, 58, 59, 60, 61, 62	24	171, 172, 173, 174, 175, 176
10	64, 67, 70, 73, 76	25	178, 181, 184, 187, 190
11	44, 45, 47, 48, 50, 51, 53, 54, 65, 66, 68, 69, 71, 72, 74, 75	26	158, 159, 161, 162, 164, 165, 167, 168, 179, 180, 182, 183, 185, 186, 188, 189
12	77, 78, 79, 80	27	191, 192, 193, 194
13	81, 84, 87, 90, 93	28	195, 197, 198, 200
14	95, 96, 97, 98, 99, 100	29	196, 199
15	102, 105, 108, 111, 114		

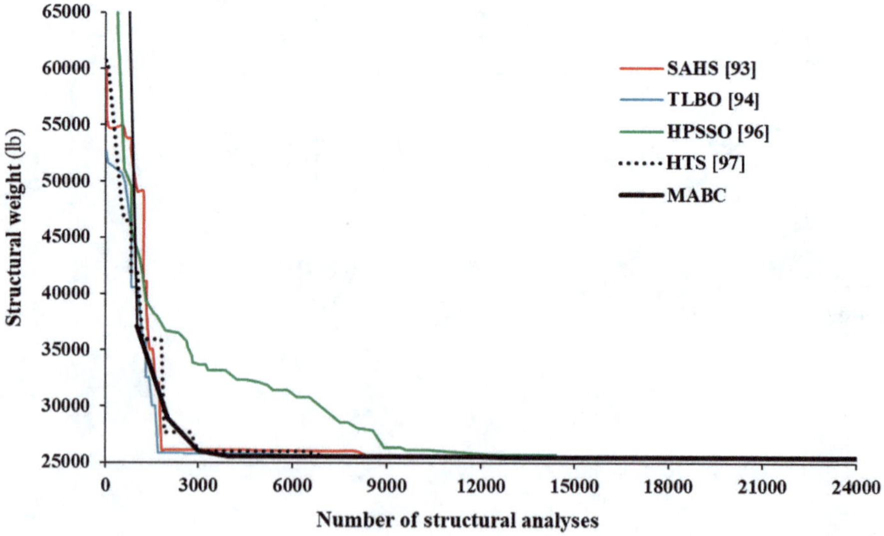

Fig. 4.10 Convergence curves for the 200-bar truss problem

Table 4.9 Optimized designs for the 200-bar truss problem

Design variables A_i (in^2)	ABC-AP [73]	SAHS [93]	TLBO	HPSSO	FPA [24]	WEO [25]	HTS [97]	MABC
1	0.1039	0.1540	0.1460	0.1213	0.1425	0.1144	0.1510	0.1480
2	0.9463	0.9410	0.9410	0.9426	0.9637	0.9443	0.9480	0.9400
3	0.1037	0.1000	0.1000	0.1220	0.1005	0.1310	0.1010	0.1000
4	0.1126	0.1000	0.1010	0.1000	0.1000	0.1016	0.1030	0.1000
5	1.9520	1.9420	1.9410	2.0143	1.9514	2.0353	1.9470	1.9400
6	0.2930	0.3010	0.2960	0.2800	0.2957	0.3126	0.2970	0.2990
7	0.1064	0.1000	0.1000	0.1589	0.1156	0.1679	0.1000	0.1050
8	3.1249	3.1080	3.1210	3.0666	3.1133	3.1541	3.1150	3.1090
9	0.1077	0.1000	0.1000	0.1002	0.1006	0.1003	0.1020	0.1000
10	4.1286	4.1060	4.1730	4.0418	4.1100	4.1005	4.2020	4.1110
11	0.4250	0.4090	0.4010	0.4142	0.4165	0.4350	0.4040	0.4060
12	0.1046	0.1910	0.1810	0.4852	0.1843	0.1148	0.1710	0.1930
13	5.4803	5.4280	5.4230	5.4196	5.4567	5.3823	5.4300	5.4490
14	0.1060	0.1000	0.1000	0.1000	0.1000	0.1607	0.1010	0.1000
15	6.4853	6.4270	6.4220	6.3749	6.4559	6.4152	6.4260	6.4390
16	0.5600	0.5810	0.5710	0.6813	0.5800	0.5629	0.5680	0.5730
17	0.1825	0.1510	0.1560	0.1576	0.1547	0.4010	0.1590	0.1340
18	8.0445	7.9730	7.9580	8.1447	8.0132	7.9735	7.9570	7.9930
19	0.1026	0.1000	0.1000	0.1000	0.1000	0.1092	0.1000	0.1000
20	9.0334	8.9740	8.9580	9.0920	9.0135	9.0155	8.9570	8.9800
21	0.7844	0.7190	0.7200	0.7462	0.7391	0.8628	0.7210	0.7080

(continued)

Table 4.9 (continued)

Design variables A_i (in^2)	ABC-AP [73]	SAHS [93]	TLBO	HPSSO	FPA [24]	WEO [25]	HTS [97]	MABC
22	0.7506	0.4220	0.4780	0.2114	0.7870	0.2220	0.4820	0.4210
23	11.3057	10.892	10.8970	10.9587	11.1795	11.0254	10.9010	10.8710
24	0.2208	0.1000	0.1000	0.1000	0.1462	0.1397	0.1030	0.1000
25	12.2730	11.887	11.8970	11.9832	12.1799	12.0340	11.9420	11.8730
26	1.4055	1.0400	1.0800	0.9241	1.3424	1.0043	1.0780	1.0370
27	5.1600	6.6460	6.4620	6.7676	5.4844	6.5762	6.4560	6.6890
28	9.9930	10.804	10.7990	10.9639	10.1372	10.7265	10.8190	10.8250
29	14.70144	13.870	13.9220	13.8186	14.5262	13.9666	13.9180	13.8470
Weight (lb)	25,533.79	25,491.9	25,488.15	25,698.85	25,521.81	25,674.83	25,517.31	25,485.78 [25533.79][a] [25491.9][b] [25521.81][c]
Mean weight (lb)	NR	25,610.2	25,533.14	28,386.72	25,543.51	26,613.45	25,565.33	25,560.12
SD (lb)	NR	141.85	27.44	2403	18.13	702.80	114.33	63.95
CV (%)	13.136	None	None	None	0.169	None	None	None
NSA	1,450,000	19,670	28,059	14,406	10,685	19,410	19,661	24,280

[a]MABC found a feasible intermediate design lighter than the optimized weight of ABC-AP [73] after only 8533 structural analyses
[b]MABC found a feasible intermediate design lighter than the optimized weight of SAHS [93] after only 18,800 structural analyses.
[c]MABC found a feasible intermediate design lighter than the optimized weight of FPA [24] after only 9740 structural analyses.

References

1. Eberhart RC, Kennedy J (1995) A new optimizer using particle swarm theory. In: Proceedings of the sixth international symposium on micro machine and human science, Nagoya, Japan
2. Dorigo M, Maniezzo V, Colorni A (1996) The ant system: optimization by a colony of cooperating agents. IEEE Trans Syst Man Cybern B 26(1):29–41
3. Yang XS (2010) Engineering optimization: an introduction with metaheuristic applications. Wiley
4. Yang XS, Deb S (2010) Eagle strategy using Levy walk and firefly algorithms for stochastic optimization. In: Cruz C, González JR et al (eds) Nature inspired cooperative strategies for optimization (NICSO2010), studies in computational intelligence, vol 284. Springer, pp 101–111
5. Yang XS, Gandomi AH (2012) Bat algorithm: a novel approach for global engineering optimization. Eng Comput 29(5):464–483
6. Kaveh A, Farhoudi N (2013) A new optimization method: dolphin echolocation. Adv Eng Softw 59:53–70
7. Gandomi AH, Yang XS, Alavi AH (2013) Cuckoo search algorithm: a metaheuristic approach to solve structural optimization problems. Eng Comput 29(1):17–35
8. Askarzadeh A (2016) A novel metaheuristic method for solving constrained engineering optimization problems: crow search algorithm. Comput Struct 169(2016):1–12
9. Xian-Bing Meng XZ, Gao LL, Liu Yu, Zhang H (2016) A new bio-inspired optimisation algorithm: bird swarm algorithm. J Exp Theor Artif Intell 28:673–687
10. Ehsan J, Mohammad C (2018) Tackling global optimization problems with a novel algorithm-mouth brooding fish algorithm. Appl Soft Comput 62:987–1002
11. Sankalap A, Satvir S (2019) Butterfly optimization algorithm: a novel approach for global optimization. Soft Comput 23:715–734
12. Mohit J, Vijander S, Asha R (2019) A novel nature-inspired algorithm for optimization: squirrel search algorithm. Swarm Evol Comput 44:148–175
13. Glover F, Laguna M (1997) Tabu search. Kluwer Academic Publishers, Boston (MA), USA
14. Geem ZW, Kim JH, Loganathan G (2001) A new heuristic optimization algorithm: harmony search. Simulation 76(2):60–68
15. Rao RV, Savsani VJ, Vakharia DP (2011) Teaching-learning-based optimization: a novel method for constrained mechanical design optimization problems. Comput Aided Design 43(3):303–315
16. Sadollah A, Bahreininejad A, Eskandar H, Hamdi M (2012) Mine blast algorithm for optimization of truss structures with discrete variables. Comput Struct 102–103:49–63
17. Gonçalves MS, Lopez RH, Miguel LFF (2015) Search group algorithm: a new metaheuristic method for the optimization of truss structures. Comput Struct 153:165–184
18. Erol OK, Eksin I (2006) A new optimization method: big bang–big crunch. Adv Eng Softw 37(2):106–111
19. Rashedi E, Nezamabadi-Pour H, Saryazdi S (2009) GSA: a gravitational search algorithm. Inform Sciences 179(13):2232–2248
20. Kaveh A, Talatahari S (2010) A novel heuristic optimization method: charged system search. Acta Mech 213(3–4):267–289
21. Kaveh A, Khayat Azad M (2012) A new meta-heuristic method: ray optimization. Comput Struct 112–113:283–294
22. Kaveh A, Mahdavi VR (2014) Colliding bodies optimization: a novel meta-heuristic method. Comput Struct 139:18–27
23. Eskandar H, Sadollah A, Bahreininejad A, Hamdi M (2012) Water cycle algorithm–a novel metaheuristic optimization method for solving constrained engineering optimization problems. Comput Struct 110:151–166
24. Bekdaş G, Nigdeli SM, Yang XS (2015) Sizing optimization of truss structures using flower pollination algorithm. Appl Soft Comput 37:322–331

25. Kaveh A, Bakhshpoori T (2016) A new metaheuristic for continuous structural optimization: water evaporation optimization. Struct Multidiscip Optimiz 54(1):23–43
26. Kaveh A, Dadras A (2017) A novel meta-heuristic optimization algorithm: thermal exchange optimization. Adv Eng Softw 110:69–84
27. Kaveh A, Zolghadr A (2017) Cyclical parthenogenesis algorithm for guided modal strain energy based structural damage detection. Appl Soft Comput 57:250–264
28. Lamberti L, Pappalettere C (2011) Metaheuristic design optimization of skeletal structures: a review. Comput Technol Rev 4(1):1–32
29. Saka MP, Dogan E (2012) Recent developments in metaheuristic algorithms: a review. Comput Technol Rev 5:31–78
30. Kaveh A (2014) Advances in metaheuristic algorithms for optimal design of structures. Springer International Publishing, Switzerland, pp 9–40
31. Kaveh A (2017) Applications of metaheuristic optimization algorithms in civil engineering. Springer International Publishing, Switzerland
32. Kaveh A, Ghazaan MI (2018) Meta-heuristic algorithms for optimal design of real-size structures. Springer International Publishing, Switzerland
33. Karaboga D (2005) An idea based on honey bee swarm for numerical optimization, technical report-TR06, Erciyes University Engineering Faculty Computer Engineering Department
34. Karaboga D, Basturk B (2007) A powerful and efficient algorithm for numerical function optimization: artificial bee colony (ABC) algorithm. J Global Optim 39(3):459–471
35. Karaboga D, Basturk B (2008) On the performance of artificial bee colony (ABC) algorithm. Appl Soft Comput 8(1):687–697
36. Karaboga N (2009) A new design method based on artificial bee colony algorithm for digital IIR filters. J Frankl Ins 346(4):328–348
37. Singh A (2009) An artificial bee colony algorithm for the leaf-constrained minimum spanning tree problem. Appl Soft Comput 9(2):625–631
38. Hemamalini S, Simon SP (2010) Artificial bee colony algorithm for economic load dispatch problem with non-smooth cost functions. Electr Power Compos Syst 38(7):786–803
39. Omkar SN, Senthilnath J, Khandelwal R, Naik GN, Gopalakrishnan S (2011) Artificial bee colony (ABC) for multi-objective design optimization of composite structures. Appl Soft Comput 11(1):489–499
40. Rao RV, Patel VK (2011) Optimization of mechanical draft counter flow wet-cooling tower using artificial bee colony algorithm. Energ Convers Manage 52(7):2611–2622
41. Şahin AŞ, Kılıç B, Kılıç U (2011) Design and economic optimization of shell and tube heat exchangers using artificial bee colony (ABC) algorithm. Energ Convers Manage 52(11):3356–3362
42. Szeto WY, Wu Y, Ho SC (2011) An artificial bee colony algorithm for the capacitated vehicle routing problem. Eur J Oper Res 215(1):126–135
43. Ayan K, Kılıç U (2012) Artificial bee colony algorithm solution for optimal reactive power flow. Appl Soft Comput 12(5):1477–1482
44. Akay B, Karaboga D (2012) Artificial bee colony algorithm for large-scale problems and engineering design optimization. J Intell Manuf 23(4):1001–1014
45. Lin SW, Ying KC (2013) Increasing the total net revenue for single machine order acceptance and scheduling problems using an artificial bee colony algorithm. J Oper Res Soc 64(2):293–311
46. Kıran MS, İşcan H, Gündüz M (2013) The analysis of discrete artificial bee colony algorithm with neighborhood operator on traveling salesman problem. Neural Comput Appl 23(1):9–21
47. Bulut O, Tasgetiren MF (2014) An artificial bee colony algorithm for the economic lot scheduling problem. Int J Prod Res 52(4):1150–1170
48. Tiwar R, Waghole V (2015) Optimization of spherical roller bearing design using artificial bee colony algorithm and grid search method. Int J Comput Meth Eng Sci Mech 16(4):221–233
49. Luo J, Liu Q, Yang Y, Li X, Chen MR, Cao W (2017) An artificial bee colony algorithm for multi-objective optimisation. Appl Soft Comput 50:235–251

50. Choong SM, El-Shafie A, Mohtar WW (2017) Optimisation of multiple hydropower reservoir operation using artificial bee colony algorithm. Water Resour Manage 31(4):1397–1411
51. Pérez CJ, Vega-Rodríguez MA, Reder K, Flörke M (2017) A multi-objective artificial bee colony-based optimization approach to design water quality monitoring networks in river basins. J Clean Prod 166:579–589
52. Banharnsakun A (2018) Multiple traffic sign detection based on the artificial bee colony method. Evolv Syst 9(3):255–264
53. Dokeroglu T, Sevinc E, Cosar A (2019) Artificial bee colony optimization for the quadratic assignment problem. Appl Soft Comput 76:595–606
54. Sharma TK, Abraham A (2020) Artificial bee colony with enhanced food locations for solving mechanical engineering design problems. J Amb Intel Human Comput 11(1):267–290
55. Fairee S, Khompatraporn C, Sirinaovakul B, Prom-On S (2020) Trim loss optimization in paper production using reinforcement artificial bee colony. IEEE Access 8:130647–130660
56. Karaboga D, Akay B (2011) A modified artificial bee colony (ABC) algorithm for constrained optimization problems. Appl Soft Comput 11(3):3021–3031
57. Li JQ, Xie SX, Pan QK, Wang S (2011) A hybrid artificial bee colony algorithm for flexible job shop scheduling problems. Int J Comput Commun Control 6(2):286–296
58. Gao W, Liu S (2011) Improved artificial bee colony algorithm for global optimization. Inform Process Lett 111(17):871–882
59. Gao WF, Liu SY (2012) A modified artificial bee colony algorithm. Comput Oper Res 39(3):687–697
60. Akay B, Karaboga D (2012) A modified artificial bee colony algorithm for real-parameter optimization. Inform Sci 192:120–142
61. Zhang R, Song S, Wu C (2013) A hybrid artificial bee colony algorithm for the job shop scheduling problem. Int J Prod Econ 141(1):167–178
62. Yildiz AR (2013) A new hybrid artificial bee colony algorithm for robust optimal design and manufacturing. Appl Soft Comput 13(5):2906–2912
63. Chun-Feng W, Kui L, Pei-Ping S (2014) Hybrid artificial bee colony algorithm and particle swarm search for global optimization. Math Probl Eng 832949
64. Zhang C, Zheng J, Zhou Y (2015) Two modified artificial bee colony algorithms inspired by grenade explosion method. Neurocomputing 151:1198–1207
65. Mao M, Duan Q (2016) Modified artificial bee colony algorithm with self-adaptive extended memory. Cybernet Syst 47(7):585–601
66. Ma L, Zhu Y, Zhang D, Niu B (2016) A hybrid approach to artificial bee colony algorithm. Neural Comput Appl 27(2):387–409
67. Gao WF, Huang LL, Wang J, Liu SY, Qin CD (2016) Enhanced artificial bee colony algorithm through differential evolution. Appl Soft Comput 48:137–150
68. Guesmi T, Alshammari BM (2017) An improved artificial bee colony algorithm for robust design of power system stabilizers. Eng Comput 34(7):2131–2153
69. Liang Y, Wan Z, Fang D (2017) An improved artificial bee colony algorithm for solving constrained optimization problems. Int J Mach Learn Cyber 8(3):739–754
70. Ghambari S, Rahati A (2018) An improved artificial bee colony algorithm and its application to reliability optimization problems. Appl Soft Comput 62:736–767
71. Mann PS, Singh S (2019) Improved artificial bee colony metaheuristic for energy-efficient clustering in wireless sensor networks. Artif Intell Rev 51(3):329–354
72. Lei D, Yuan Y, Cai J (2020) An improved artificial bee colony for multi-objective distributed unrelated parallel machine scheduling. Int J Prod Res:1–13
73. Sonmez M (2011) Artificial bee colony algorithm for optimization of truss structures. Appl Soft Comput 11(2):2406–2418
74. Degertekin SO (2012) Optimum design of geometrically non-linear steel frames using artificial bee colony algorithm. Steel Compos Struct 12(6):505–522
75. Ozturk HT, Durmus A, Durmus A (2012) Optimum design of a reinforced concrete beam using artificial bee colony algorithm. Comput Conc 10(3):295–306

76. Fiouz AR, Obeydi M, Forouzani H, Keshavarz A (2012) Discrete optimization of trusses using an artificial bee colony (ABC) algorithm and the fly-back mechanism. Struct Eng Mech 44(4):501–519
77. Sun H, Luş H, Betti R (2013) Identification of structural models using a modified artificial bee colony algorithm. Comput Struct 116:59–74
78. Kang F, Li J, Ma Z (2013) An artificial bee colony algorithm for locating the critical slip surface in slope stability analysis. Eng Optimiz 45(2):207–223
79. Ozturk HT, Durmus A (2013) Optimum cost design of RC columns using artificial bee colony algorithm. Struct Eng Mech 45(5):643–654
80. Jahjouh MM, Arafa MH, Alqedra MA (2013) Artificial bee colony (ABC) algorithm in the design optimization of RC continuous beams. Struct Multidiscip Optimiz 6:963–979
81. Sonmez M, Aydin E, Karabork T (2013) Using an artificial bee colony algorithm for the optimal placement of viscous dampers in planar building frames. Struct Multidiscip Optimiz 48(2):395–409
82. Park JY, Han SY (2013) Application of artificial bee colony algorithm to topology optimization for dynamic stiffness problems. Comput Math Appl 66(10):1879–1891
83. Yahya M, Saka MP (2014) Construction site layout planning using multi-objective artificial bee colony algorithm with Levy flights. Automat Constr 38:14–29
84. Park JY, Han SY (2015) Topology optimization for nonlinear structural problems based on artificial bee colony algorithm. Int J Precis Eng Man 16(1):91–97
85. Aydin E, Sonmez M, Karabork T (2015) Optimal placement of elastic steel diagonal braces using artificial bee colony algorithm. Steel Compos Struct 19(2):349–368
86. Xu HJ, Ding ZH, Lu ZR, Liu JK (2015) Structural damage detection based on chaotic artificial bee colony algorithm. Struct Eng Mech 55(6):1223–1239
87. Aydoğdu İ, Akın A, Saka MP (2016) Design optimization of real-world steel space frames using artificial bee colony algorithm with Levy flight distribution. Adv Eng Softw 92:1–14
88. Xu HJ, Ding ZH, Lu ZR, Liu JK (2016) Structural damage detection using a modified artificial bee colony algorithm. CMES-Comp Model Eng 111:335–355
89. Ozturk HT, Turkeli E, Durmus A (2016) Optimum design of RC shallow tunnels in earthquake zones using artificial bee colony and genetic algorithms. Comput Conc 17(4):435–453
90. Tapao A, Cheerarot R (2017) Optimal parameters and performance of artificial bee colony algorithm for minimum cost design of reinforced concrete frames. Eng Struct 151:802–820
91. Liu H, He X, Jiao Y (2018) Damage identification algorithm of hinged joints for simply supported slab bridges based on modified hinge plate method and artificial bee colony algorithms. Algorithms 11(12):198
92. Mikaeil R, Beigmohammadi M, Bakhtavar E, Haghshenas SS (2019) Assessment of risks of tunneling project in Iran using artificial bee colony algorithm. SN Appl Sci 1(12):1711
93. Degertekin SO (2012) Improved harmony search algorithms for sizing optimization of truss structures. Comput Struct 92:229–241
94. Degertekin SO, Hayalioglu MS (2013) Sizing truss structures using teaching-learning-based optimization. Comput Struct 119:177–188
95. Camp CV, Farshchin M (2014) Design of space trusses using modified teaching–learning based optimization. Eng Struct 62:87–97
96. Kaveh A, Bakhshpoori T, Afshari E (2014) An efficient hybrid particle swarm and swallow swarm optimization algorithm. Comput Struct 143:40–59
97. Degertekin SO, Lamberti L, Hayalioglu MS (2017) Heat transfer search algorithm for sizing optimization of truss structures. Lat Am J Solids Stru 14(3):373–397
98. Kaveh A, Dadras A, Montazeran AH (2018) Chaotic enhanced colliding bodies algorithms for size optimization of truss structures. Acta Mech 229(7):2883–2907
99. Lamberti L, Pappalettere C (2009) An improved harmony-search algorithm for truss structure optimization. In: Topping BHV, Neves LFC, Barros RC (eds) Proceedings of the twelfth international conference civil, structural and environmental engineering computing. Civil-Comp Press, Stirlingshire (UK)

Chapter 5
Electrostatic Discharge Algorithm for Optimum Design of Real-Size Truss Structures

Ibrahim Aydogdu, Tevfik Oguz Ormecioglu, and Serdar Carbas

5.1 Introduction

A large number of different designs can be achieved in accordance with architectural and static criteria in structural engineering. Obtaining the safest and, concurrently, the most suitable design among these is the main aim of a structural engineer. However, in order to find the most suitable design from an infinite design space, it is necessary to make many iterative computations and spend a lot of time and effort. In this context, the accuracy of the design to be called optimal and acquiring it in a relatively short time may differ according to the knowledge, experience, and experience of the designer. In parallel with this, the growing population and industrialization around the world have created the need for a large number of various types of structures. So, the expeditious raise in the need for construction has increased the demand for designers to provide accurate and safe engineering solutions in a short time, and as a result, the importance of structural optimization has emerged. The optimization of the structures has been developed since human beings began to build up and it has come up to the present. Yet, the increase in design criteria and having large sizes of a structural system to be designed leaves the mathematical-based classical optimization methods helpless at the point of generating feasible designs. These and many other reasons led the designers to different analytical searches to get faster and more feasible results from conventional optimization studies. Consequently, it has been determined

I. Aydogdu (✉) · T. O. Ormecioglu
Department of Civil Engineering, Akdeniz University, Antalya, Turkey
e-mail: aydogdu@akdeniz.edu.tr

T. O. Ormecioglu
e-mail: tevfik.oguz@gmail.com

S. Carbas
Department of Civil Engineering, Karamanoglu Mehmetbey University, Karaman, Turkey
e-mail: scarbas@kmu.edu.tr

S. Carbas et al. (eds.), *Nature-Inspired Metaheuristic Algorithms for Engineering Optimization Applications*, Springer Tracts in Nature-Inspired Computing,
https://doi.org/10.1007/978-981-33-6773-9_5

that suitable solutions can be produced more quickly and effectively by modeling completely natural phenomena in the field of design engineering. Based on this logic, natural models such as the human brain, evolution theory, swarm intelligence, and colony characteristics are expressed mathematically and successfully applied in solving complex optimization problems [1, 2]. These kinds of methods are called as metaheuristic optimization techniques [3–5]. With the development of computer technologies over time, designs have been made more systematic with the use of various metaheuristic methods in the optimization of structural systems and human-induced errors have been minimized in obtaining optimum designs.

Common problems encountered in structural engineering optimization can be listed as optimum sizing of structures according to static and dynamic behavior, optimum shape and geometric design of structural systems, and optimum control of planned systems based on reliability [6]. The main objectives of structural engineering are safety, economy, and aesthetics, respectively. In a building design with a steel skeleton, the safest design can only be yielded by minimizing the design cost. This is possible if the steel structure to be designed has a minimum structural weight. This has become a design rule for accomplishing optimum designs of steel skeleton structures, such as steel trusses and/or steel frames, via metaheuristic algorithms [7–21]. However, steel skeleton structural systems having the least weight should be designed with sufficient strength against the loads acting on it. Only in this way, the lightest steel skeleton (truss and/or frame) carrier system design that can withstand the loads acting on it within the safety limits becomes the optimum design for that structure.

In this chapter, it is concentrated on obtaining optimal designs of real-size steel truss structures with electrostatic discharge algorithm (ESDA), which is a new generation metaheuristic optimization method inspired by a natural phenomenon where electrostatic discharge occurs toward a less electrically charged body from a more electrically charged body as a result of the interaction of two objects with different voltage potentials.

The design constraints are fulfilled from American Institute of Steel Construction-Load and Resistance Factor Design (AISC-LRFD) [22]. These constraints comprise the displacement and slenderness ratio restrictions and combined axial and bending strength requirements, as well. The ESDA-based design algorithm chooses the steel profiles for the truss members from the available tubular steel pipe sections listed in AISC-LRFD such that the weight of the structural steel truss systems is obtained as the minimum while satisfying the whole design constraints. The algorithmic performance of the ESDA is examined and proved utilizing two real-size steel truss structures treated as design examples. With this chapter, it is successfully demonstrated for the first time that ESDA can be applied for optimum design of real-size steel truss structure design problems.

5.2 Discrete Optimization Problem Formulation of Truss Structures

As mentioned before, the most important optimization criterion considered in the design problems of steel trusses is the weight of the structure, since it directly affects the construction cost. Therefore, it is important to size the steel truss structure to have the minimum cost or/and minimum weight. So, the objective function of those types of optimization problems is to achieve a design vector $\boldsymbol{X}$ as described in Eq. 5.1 constituted of sizing design variables ($I_1, I_2, I_3, \ldots, I_{nk}$) for a truss with *nm* number of elements.

$$X^T = [I_1, I_2, I_3, \ldots\ldots, I_{nk}] \tag{5.1}$$

So, the $\boldsymbol{X}$ can minimize the objective function of the optimization problem as shown in below-mentioned Eq. 5.2;

$$W = \sum_{i=1}^{nm} \rho_i L_i A_i \tag{5.2}$$

here, *W(X)* is the total weight of the truss structure, ρ_i is weight per unit volume, L_i is the length of the *ith* structural member, and A_i is the area of cross section of the same member. The design constraints are enforced as specified from AISC-LRFD [22] in the discrete optimization formulation of the steel truss design problem for obtaining the least structural weight.

$$\text{If } \frac{P_u}{\phi P_n} \geq 0.2 \text{ then } \frac{P_u}{\phi P_n} + \frac{8}{9}\left(\frac{M_{ux}}{\phi_b M_{nx}} + \frac{M_{uy}}{\phi_b M_{ny}}\right) \leq 1 \tag{5.3}$$

$$\text{If } \frac{P_u}{\phi P_n} < 0.2 \text{ then } \frac{P_u}{2\phi P_n} + \left(\frac{M_{ux}}{\phi_b M_{nx}} + \frac{M_{uy}}{\phi_b M_{ny}}\right) \leq 1 \tag{5.4}$$

In Eqs. (5.3) and (5.4), the strength requirements for a member subjected to both bending and axial force are presented. In these inequalities $\O_b$ is the resistance factor for flexure given as 0.90, $\O$ is the resistance factor for compression given as 0.85, M_{ux} is the required flexural strength relating to *x*-axis bending, M_{uy} is the required flexural strength related with the *y*-axis bending, and M_{nx} and M_{ny} are the nominal flexural strength related with *x*-axis and *y*-axis bendings, respectively. P_u is the required tensile or compressive strength, and P_n is the nominal tensile or compressive strength which is derived from Eqs. (5.5) and (5.6), respectively.

$$P_n = F_y A_g \tag{5.5}$$

here, F_y is the denominated yield stress and A_g is the gross cross-sectional area of the member.

$$P_n = F_{cr} A_g \tag{5.6}$$

here, F_{cr} is the critical flexural buckling stress of the member calculated as;

$$\text{If } \lambda_c \leq 1.5 \text{ then } F_{\text{cr}} = (0.658^{\lambda_c^2}) F_y \tag{5.7}$$

$$\text{If } \lambda_c > 1.5 \text{ then } F_{\text{cr}} = \left[0.877/\lambda_c^2\right] F_y \tag{5.8}$$

here, the λ_c is attained from Eq. (5.9) in which K is the effective length factor taken as 1, l is the length of a truss member, r is governing radius of gyration about the buckling axis, and E is the modulus of elasticity.

$$\lambda_c = \frac{Kl}{r\pi} \sqrt{\frac{F_y}{E}} \tag{5.9}$$

Additionally, the restriction on the slenderness ratio, λ_i, of the *ith* truss member under impact of tension and compression is denoted in Eqs. (5.10) and (5.11), respectively.

$$\text{If } \lambda_i = \frac{Kl}{r} \leq 300 \text{ for the members under impact of tension} \tag{5.10}$$

$$\text{If } \lambda_i = \frac{Kl}{r} \leq 200 \text{ for the members under impact of compression} \tag{5.11}$$

Furthermore, the displacement limitation can be taken into account through Eq. (5.12) as follows;

$$d_{j,k} \leq (d_{j,k})_{\text{allowable}} \quad (j = 1, 2, \ldots, N_j) \tag{5.12}$$

here, *dj,k* and (*dj,k*)allowable are the displacements calculated in the *kth* direction of the *jth* joint and its allowable value, respectively, and N_j is the total joints number.

5.2.1 Penalty Function and Penalized Objective Function

In this study, the optimum design problem defined up to this section comes to a feasible solution by using some constraints. However, since metaheuristic algorithms are the methods developed for the design of unconstrained optimization problems, at this stage, a design problem determined based on constraints should be transformed into an unconstrained problem. To do this, a function called penalty function is used, which calculates the degree of violation of constraints. Thus, by adding the penalty function to the objective function, it is brought into a form that includes the constraints. Normalized forms of constraints should be utilized in order to determine the penalty function based on the constraints. After the constraints are defined and normalized with Eqs. (5.2)–(5.12) given above, the calculation of the penalty function, C, which is required to transform the design optimization problem into an unconstrained one is initiated. Here, the penalty function (C) depending on the strength, slenderness ratio, and displacement constraints can be computed by Eqs. (5.13) and (5.14) as follows;

$$C = \sum_{i=1}^{nc} c_i \tag{5.13}$$

$$c_i = \begin{cases} 0 \text{ if } g_j \leq 0 \\ g_j \text{ if } g_j > 0 \end{cases} \quad j = 1, 2, \ldots, nc \tag{5.14}$$

here, g_j is the *jth* constraint function and *nc* is the total constraints number in the steel truss design optimization problem. Eventually, the following Eq. (5.15) is employed to add the calculated penalty function to the objective function [23].

$$W_p = W(1 + C)^{\Psi} \tag{5.15}$$

here, W is the value of objective function of optimum design problem given in Eq. (5.2), W_p is the penalized objective function, C is the total constraint violations value computed by adding the violation of each individual constraint as described above [Eqs. (5.13) and (5.14)], and ψ is coefficient of penalty determining the effect of the constraints on the feasible solution and its value is determined according to the problem. The value penalty coefficient is considered as 2.0 in this chapter.

5.3 Electrostatic Discharge Algorithm (ESDA)

5.3.1 Electrostatic Discharge (ESD)

All matter in nature is made up from atoms and an atom has a nucleus in its center. Atomic nucleus contains positively charged protons and uncharged neutrons. Besides, there are negatively charged electrons around it. The number of electrons is equal to the number of protons. But electrons can easily move from one atom to another. Consequently, the sudden transfer of electrical charges that is caused by the friction of two bodies is called electrostatic. In more technical terms, electrostatic occurs spontaneously because two materials, that are different or identical, conductive, or non-conductive, come into contact and then separate or create friction. Electron transfer occurs across the contact surface between bodies getting in touch with each other. The electrical characteristic of this boundary layer is different from those of both bodies which are in contact. If these two contacted bodies are separated from each other, the boundary layer disappears and as a result, while an electron excess (negative charge) occurs in one of them, an electron deficiency (positive charge) occurs in the other. As a matter of fact, these two separate charges attract each other, and they want to be discharged by making an arc (spark) through a conductive layer like air and balance the charge difference. This arc formation can be very dangerous in some environments [24]. The static electric, which has existed since the formation of nature, but its effects were understood toward the end of the 1980s, is a fact mankind encounter in almost every phase of daily life. Lightning can be given as the best example of static electricity in nature. As it is known, nature's largest static electricity discharge occurs during the approach of positive and negative charged clouds to each other. In daily life, people are, also, charged with static electric. This charge may be due to the clothes they put on and take off, the friction while walking on the carpet, getting on and off the vehicles, from the desk they work on, or sitting on a chair, and so forth. Static electric taken by people harms both their health and the electronic devices they use. Especially, it is inevitable that sensitive electronic materials (chips, mosfet, etc.) malfunctioning due to this static electric [25].

To say in short, electrostatic discharge (ESD) consists of electrical charge exchange between two objects with different voltage potentials. ESD is undoubtedly encountered in every area where static electric is available. In nature, electrostatics involves a wide range of physical phenomena, from the work of our heart to lightning and thunderbolt, or the interactions of charges within an atom.

5.3.2 Interpretation of the ESD Algorithm

The ESDA is one of the contemporary metaheuristic optimization methods developed by Bouchekara in 2019 [26] simulating the electrostatic discharge phenomena naturally exist in the nature. There are very limited study in the literature conducted taking the ESDA into account [27, 28]. In this respect, this chapter is the first in the literature where the ESDA is utilized to obtain optimum designs of real-size steel truss structures.

The interpretation of the ESDA is detailed with the steps of algorithm given below;

Step 1 At the beginning, in order to initiate the ESDA objects size standing for the population is randomly generated. This population mimic electrical instruments (objects) and it is the individuals population taking place in the search space of the algorithm. Every instrument is comprised of various ingredients that are analog to design variables. The search space position of an instrument directly affects the fitness value of it. If an object has better fitness value, this means that it is a feasible solution. Moreover, a counter assigned to each instrument calculates how many a time it has become a victim instrument. This step is also called initialization step.

Step 2 Then, the algorithm is iteratively executed for acquiring the optimal solution of the defined design problem until reaching the termination criteria which is maximum number of iterations. At every iteration, randomly selected three objects are sorted from best to worst in a descending order. According to rand1, which is a randomly generated uniform number, the objects are evaluated as to be regarded or not. If rand1 is greater than 0.5, it is assumed that ESD comes into existence through only two objects, else whole three objects are regarded. In the case, only two objects are considered, the object having better position is enumerated as Object 1, and the other one becomes Object 2 and this worse positioned object moves in direction to Object 1 via Eq. (5.16).

$$x_{2new} = x_2 + 2\beta_1(x_1 - x_2) \tag{5.16}$$

here, x_{2new} denotes the novel position of the Object 2 which is worse positioned instrument. While the x_1 stands for the former position of Object 1, the x_2 presents the previous position of Object 2. A randomly generated number, whose normal distribution with mean parameter is 0.7 and standard deviation parameter is 0.2, symbolized as β_1.

It is assumed that the ESD constitutes regarding only two objects, if the Object 2 comes closer to the Object 1. So, the Object 2 comes in sight as victim instrument. This is conceived to be equivalent to the direct ESD event.

The whole three objects are regarded, provided that rand1 is generated as smaller than 0.5. If this is the case, it is presumed that the third object, which has the worst fitness value, heads over to the other two objects by way of Eq. 5.17.

$$x_{3new} = x_3 + 2\beta_2(x_1 - x_3) + 2\beta_3(x_2 - x_3) \tag{5.17}$$

here, each of β_2 and β_3 are also the randomly generated numbers, whose normal distributions with mean parameters are 0.7 and standard deviation parameters are 0.2.

It is supposed that ESD takes place if Object 3 comes closer to Object 1 and Object 2. So, the Object 3 is treated as victim instrument. This is due to fact that, because the Object 3 is in vicinity of the ESD incident, it turns into a victim instrument and receives the electromagnetic zones by virtue of discharge. This is conceived to be equivalent to the indirect ESD event. When an object or a victim instrument is exposed to an ESD incident, the counter of those is increased one.

Step 3 The boundaries are controlled whether there are objects exterior the search space, or not. If any object identified exterior, they are carried back interior of the search space.

Step 4 The objects are controlled one after another. If an instrument is exposed to ESD greater than three times, it is assumed that this instrument has been demolished, and it has to be switched by a novel object is generated randomly in the search space. On the other hand, if an instrument is exposed to ESD incident less than or equal to three times, a number rand2 is generated randomly. Then, it is observed whether the rand2 is smaller than 0.2. If it is the case, one item of the objects is presumed to be demolished and it has to be exchanged, else it is guaranteed that all items are in secure.

Step 5 The newly generated population is summed up to the existing one and all kept in the archive. When the population of this total archive is assorted in an order of ascending objective function values, the foremost objects are chosen as the population of the following iteration.

The pseudocode showing the operational steps of ESDA is presented in below Algorithm 5.1 [26, 27].

Algorithm 5.1 Pseudocode of ESDA

Assigning; population size (*ObjectsSize*), lower bound (*LB*) and upper bound (*UB*) of the design variables, and maximum number of iterations ($max_{iteration}$)
Initialization
while (iter < $max_{iteration}$)
  for *i*=1 : round(ObjectsSize/3)
    choose randomly generated three objects
    rand1 = random number generated between [0, 1]
    if *(rand1>0.5)*
      $x_{2new} = x_2 + 2\beta_1(x_1 - x_2)$
      $x_{2\,ESDcounter} = x_{2\,ESDcounter} + 1$
    else
      $x_{3new} = x_3 + 2\beta_2(x_1 - x_3) + 2\beta_3(x_2 - x_3)$
      $x_{3\,ESDcounter} = x_{3\,ESDcounter} + 1$
    end if
  end for
  control that if there are objects exterior the search space
  for $i_1 = 1$: ObjectsSize
    if ($x_{i_1\,ESDcounter} \leq 3$)
      for $i_2 = 1$: ObjectsSize
        rand2 = random number generated between [0, 1]
        if *(rand2<0.2)*
        one item has to be exchanged
        end if
      end for
    else
    the instrument has to be exchanged
      $x_{i_1} = LB + rand(UB - LB)$
    end if
  end for
choose the foremost of the objects as the population of the following iteration
end while

Table 5.1 Search parameter analysis results of the ESDA

Objects size	20	40	70	100
The minimum weights (kN)	27.658	27.349	27.349	27.349
Average weights (kN)	28.260	28.344	27.349	27.359

5.3.3 Determination of Search Parameters of ESDA

Performances of the metaheuristic optimization methods can vary for different problem types utilizing the usage of random computations. Therefore, the search parameters of the metaheuristics should be rearranged for new problems. In this chapter, the most convenient search parameters of the ESDA method are investigated since it is applied to the steel truss optimization problem for the first time. ESDA has many search parameters such as objects size (population size), arc size, μ, and σ. Trying combinations of all search parameters would significantly increase the computation time and make it difficult to find the balance in the algorithm. Therefore, only the population size, which is directly related to the problem size, is taken as a variable in this study. For the issue, the 160-bar steel truss pyramid structure, which is the first design example of the study, is solved by taking four different population sizes as 20, 40, 70, and 100, respectively, in the ESDA. In each test, 10 different analyzes have been performed by taking different seed values. Average and the best values of the tests are given in Table 5.1. According to this table, the population size is selected 70 for both design examples.

5.4 Design Examples

Two real-size pin-jointed truss structures are used to test the performance of the ESDA optimization technique. In the design examples, the design variables are selected from a database of 37 pipe sections made of steel. The mechanical properties of the steel material are modulus of elasticity (E) = 200,000 MPa, yield stress for tension (F_y) = 248 MPa, and unit weight (ρq) = 76.82 kN/m^3. The Sap2000 Open Application Programming Interface (OAPI) [29] is integrated to the ESDA-based optimization algorithm for analyzing and designing the steel truss structures to verify optimum results with Sap2000 [30].

5.4.1 160-Bar Steel Truss Pyramid

The 160-bar steel truss pyramid structure which was previously optimized by many researchers is considered as the first example in the study [16, 31, 32]. The truss has 55 joints and 160 tubular steel members which are collected in 7 member groups. The base dimensions and the height of the structure are 16 m × 16 m and 8 m, respectively. The plan and 3D views of the steel truss pyramid structure are shown in Fig. 5.1. Side view and group definition are depicted in Fig. 5.2. The 8.53 kN vertical loads in gravity direction are applied to all unsupported joints. The displacement limit of the truss structure is taken as 0.0445 m.

To statistically validate ESDA method, 160-bar steel truss pyramid structure is optimized 10 times using different seed values. Statistical information of the optimum

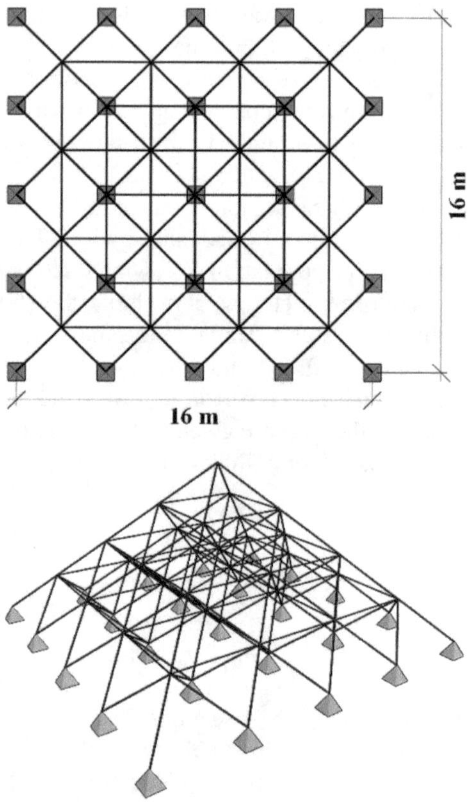

Fig. 5.1 Plan and 3D views of the first example

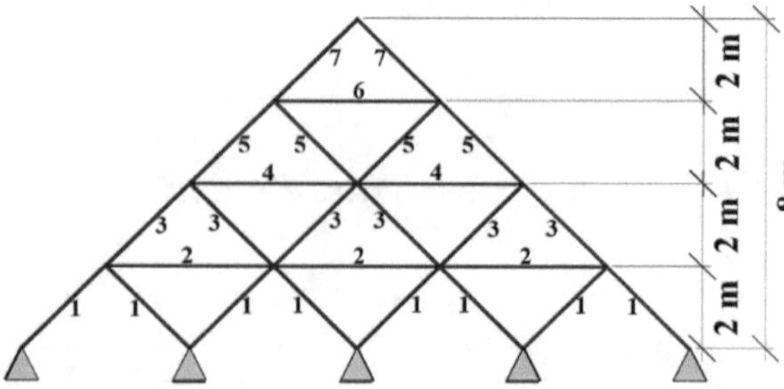

Fig. 5.2 Side view and group definition of the first example

Table 5.2 Statistical validation of ESDA for the first design example

The minimum weight (kN)	The average weight (kN)	Standard deviation	$Q1$	$Q3$	Median	The maximum weight (kN)
27.349	27.349	0	27.349	27.349	27.349	27.349

Table 5.3 Optimum design sections and critical constraint values of the first design example

Group ID	ACO [16]	HS [16]	PSO [16]	SBB-BC [31]	MBB-BC [31]	ESDA
1	P2	P2	P2	P2	P2	P2
2	P1.25	P1.25	P1.25	P1.25	P1.25	P1.25
3	P2	P2	P2	P2	P2	P2
4	P1.25	P1.25	P1.25	P1.25	P1.25	P1.25
5	P2	P2	P2	P2	P2	P2
6	P1.25	P1.25	P1.25	P2	P1.25	P1.25
7	P2	P2	P2	P2	P2	P2
Weight (kN)	27.349	27.349	27.349	27.665	27.349	27.349

solutions obtained by ESDA is given in Table 5.2. According to this table, the ESDA accomplished the optimum design weight of 27.349 kN in all analyses and displayed very stable convergence performance behavior in this design example. The design details of the optimal solution achieved by the ESDA method are compared to previously reported literature results in Table 5.3. Similar to the ant colony optimization (ACO) [16], particle swarm optimizer (PSO) [16], harmony search (HS) [16], and modified big bang-big crunch (MBB-BC) [31] methods, the ESDA produces an identical optimal design weight as 27.349 kN. The standard big bang–big crunch (SBB-BC) method found slightly heavy design whose weight is 27.665 kN. Moreover, the search histories of the optimal designs for all optimization algorithms are plotted in Fig. 5.3. This figure proves that the ESDA display quite smooth convergence rate against the MBB-BC which exhibits the best convergence rate.

5.4.2 1032-Bar Double-Layer Steel Truss Roof Structure

The second design example is a double-layer steel truss roof structure that is particularly created for this study and it is examined firstly in this chapter. The truss roof structure has 283 joints and 1032 tubular steel members which are collected in 24 member groups. 3-D and plan views of the structure are given in Fig. 5.4. The upper layer dimensions, lower layer dimensions, and the height of the structure, respectively, are 21.6 m × 28.8 m, 24 m × 31.2 m, and 2.4 m. Eight different regions are generated for the structural elements (the lower layer, the upper layer, and the diagonal elements) in each layer of the truss and the structural member grouping

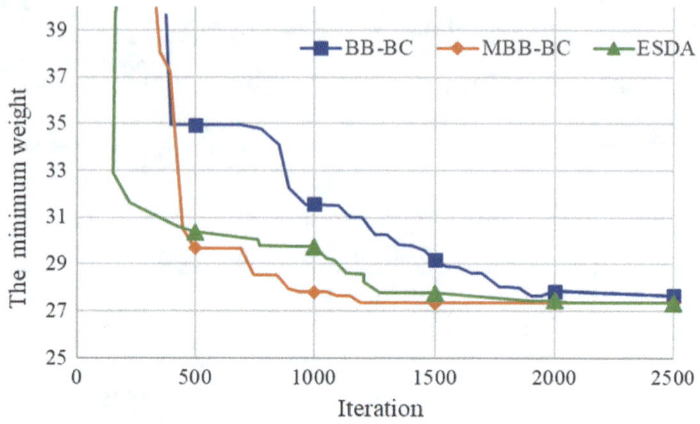

Fig. 5.3 Search histories of the optimization methods for the first design example

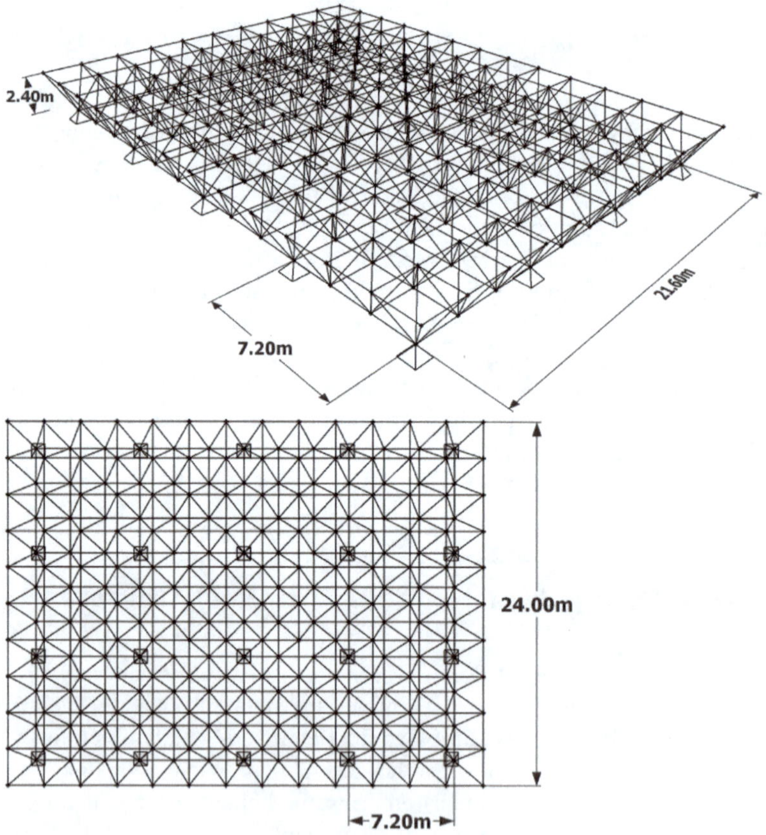

Fig. 5.4 Plan and 3D views of the second example

is designed according to these regions. The regions created for the layers are given in Fig. 5.5. The structural member grouping details are given in Table 5.4. The wind, snow, dead, and live loads act on the structure, and according to these load types, the structure is designed by a total of seven load combinations. The maximum displacement of the joints in this truss roof structure is limited to 0.018 m in each direction.

In this design example, artificial bee colony (ABC) and biogeography-based optimization (BBO) methods are used for attaining the optimum solution of the 1032-bar double-layer steel truss roof structure in addition to the ESDA method in order to test the algorithmic performance of which efficiently. Similar to the first design example, for each optimization method, the 1032-bar truss roof structure is optimized 10 times using different seed values. The statistical information on the optimum solutions obtained by ABC, BBO, and ESDA methods is given in Table 5.5. According to this table, the BBO algorithm yields the lowest minimum and average weight values. The ABC algorithm evinces the second-best performance. The ESDA presents the worst performance in this design example. The difference between the minimum design weights obtained by the ESDA and BBO algorithms is 60.65% that can be evaluated as very high. Design details of the optimal designs obtained by the proposed optimization methods are presented in Table 5.6. This table represents that the opti-

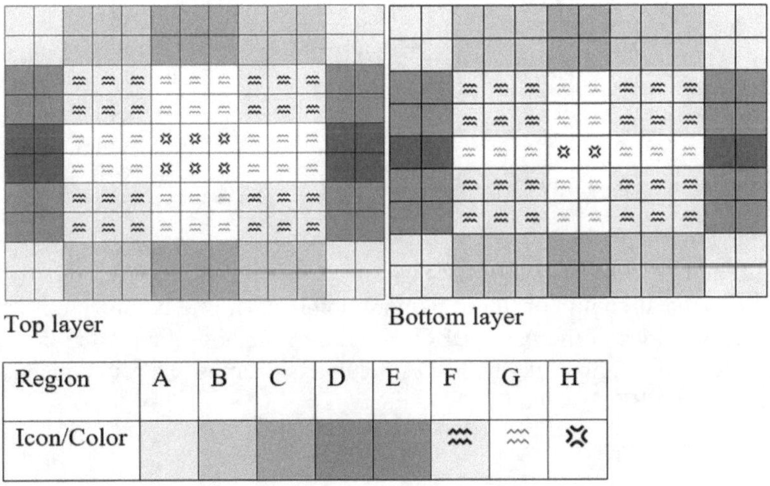

Fig. 5.5 Group regions of the second design example

Table 5.4 Group details of the second design example

Group regions	*A*	*B*	*C*	*D*	*E*	*F*	*G*	*H*
Top layer member groups	1	4	7	10	13	16	19	22
Diagonal members groups	2	5	8	11	14	17	20	23
Bottom layer members groups	3	6	9	12	15	18	21	24

Table 5.5 Statistical validation of ESDA for the second design example

The method	The minimum weight (kN)	The average weight (kN)	Standard deviation	$Q1$	$Q3$	Median	The maximum weight (kN)
ABC	143.427	151.363	4673	147.242	154.769	153.110	156.938
BBO	101.501	117.078	14.645	104.230	129.917	116.515	140.311
ESDA	162.562	175.543	9.208	168.727	180.866	174.329	191.995

Table 5.6 Optimum design sections and critical constraint values of the second design example

Group ID	ABC	BBO	ESDA	Group ID	ABC	BBO	ESDA
1	P1.25	P1.25	XP1.5	13	XP2	P1.25	XP2
2	XP1.5	P1.25	XP1.5	14	XP1.5	P1.5	XP1.5
3	XP1.5	P1.5	P1.5	15	XP2	P1.5	XP2
4	P1.25	P1.25	XP1.5	16	P2	P1.25	XP1.5
5	P1.25	P1.25	XP1.25	17	XP1.5	P1.25	XP1.5
6	P1.5	P1.5	P2	18	P2	P1.5	XP2
7	XP1.5	P1.25	XP2	19	XP1.5	P1.25	P1.25
8	XP2	P1.25	XP2	20	P2	P1.25	XP1.5
9	XP2	P1.5	XP2	21	XP1.5	P1.5	XP2
10	P1.5	P1.25	P1.5	22	P1.25	P1.25	P2
11	P1.25	P1.25	XP2	23	XP1.5	P1.5	P1.5
12	XP1.5	P1.5	XP2	24	XP2.5	P1.5	XP2
Weight (kN)	143,427	101,501	162,562				

mization methods use steel tubular sections close to each other in their optimum designs and the strength constraint is more dominant in the optimization process. The search histories of the ABC, BBO, and ESDA methods are plotted in Fig. 5.6. It is seen from this figure that the ESDA method converges in a rough manner than the ABC and the BBO methods.

5.5 Conclusions

In this chapter, it is aimed to test the performance of the ESDA method in the design optimization problems of real-size steel truss systems, which can be categorized as one of the challenging structural optimization problems. For this purpose, a 160-bar steel truss pyramid structure, which is one of the benchmark design examples of the structural optimization problem of truss systems, and a 1032-bar double-layer steel truss roof structure, specially designed and investigated for this chapter, are treated as

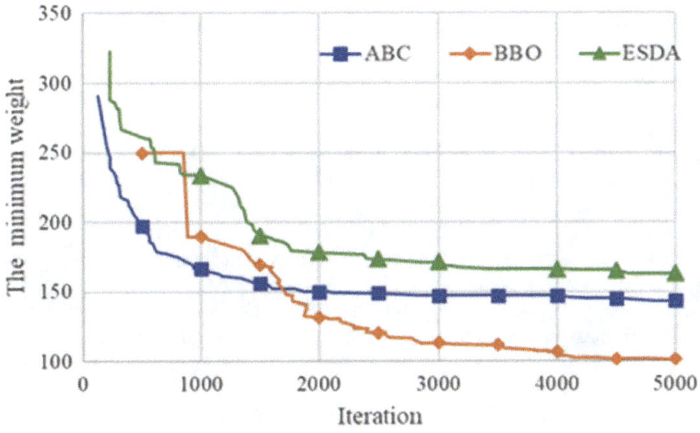

Fig. 5.6 Search histories of the optimization methods for the second design example

design examples for this chapter. In the 160-bar steel truss pyramid structure example, the ESDA method yielded the optimal design in all 10 different analyzes and showed a quite stable and successful algorithmic performance. However, in the solution of the 1032-bar double-layer steel truss roof structure, which is a quite complicated design example in terms of size and difficulty, unfortunately the ESDA did not perform better than the ABC and BBO methods which are selected for comparison.

When ESDA was examined in terms of convergence speed, it was seen that this method has a lower convergence speed rate compared to the reference methods, especially in yielding the solution of the 1032-bar double-layer steel truss roof structure design example.

In conclusion, although the ESDA method shows good performance in small-scale structural steel design problems, it needs some improvements and/or modifications to increase its algorithmic performance and convergence rate for acquiring the optimum designs of large-scale real-size steel truss structural problems. Thus, this chapter study is planned to be extended to improve the algorithmic capacity of the ESDA method as a future work.

References

1. Yang X (2010) Nature-inspired metaheuristic algorithms. Luniver press.https://doi.org/10.1016/B978-0-12-416743-8.00005-1
2. Yang XS (2014) Nature-inspired optimization algorithms. Elsevier Inc. https://doi.org/10.1016/C2013-0-01368-0
3. Saka MP, Carbas S, Aydogdu I, Akin A (2016) Use of swarm intelligence in structural steel design optimization. https://doi.org/10.1007/978-3-319-26245-1_3
4. Kaveh A (2016) Advances in metaheuristic algorithms for optimal design of structures, second edition. Springer International Publishing. https://doi.org/10.1007/978-3-319-46173-1
5. Yang X-S (2018) Optimization techniques and applications with examples. Wiley

6. Baldock R (2007) Structural optimisation in building design pratice: case-studies in topology optimisation of bracing systems
7. Aydogdu I, Carbas S, Akin A (2017) Effect of levy flight on the discrete optimum design of steel skeletal structures using metaheuristics. Steel Compos Struct 24. https://doi.org/10.12989/scs.2017.24.1.093
8. Hasançebi O, Çarbaş S, Doğan E, Erdal F, Saka MP (2010) Comparison of non-deterministic search techniques in the optimum design of real size steel frames. Comput Struct 88. https://doi.org/10.1016/j.compstruc.2010.06.006
9. Aydogdu I, Akin A, Saka MP (2016) Design optimization of real world steel space frames using artificial bee colony algorithm with Levy flight distribution. Adv Eng Softw 92:1–14. https://doi.org/10.1016/j.advengsoft.2015.10.013
10. Carbas S, Aydogdu I, Tokdemir T, Saka MP (2014) Design optimization of low-rise cold-formed steel frames with thin-walled sections using the artificial bee colony algorithm. In: Proceedings of the twelfth international conference on computational structures technology. Civil-Comp Press, Stirlingshire, Scotland
11. Aydogdu I, Akin A, Saka MP (2012) Optimum design of steel space frames by artificial bee colony algorithm. In: 10th international congress on advances in civil engineering, pp 1–10
12. Aydogdu I, Saka MP (2012) Ant colony optimization of irregular steel frames including elemental warping effect. Adv Eng Softw 4:150–169. https://doi.org/10.1016/j.advengsoft.2011.05.029
13. Saka MP, Aydogdu I (2020) Performance evaluation of artificial bee colony algorithm and its variants in the optimum design of steel skeletal structures. Asian J Civ Eng 1–19
14. Hasançebi O, Çarbaş S, Doğan E, Erdal F, Saka MP (2009) Performance evaluation of metaheuristic search techniques in the optimum design of real size pin jointed structures. Comput Struct 87:284–302. https://doi.org/10.1016/j.compstruc.2009.01.002
15. Hasançebi O, Çarbas S, Saka MP (2011) A reformulation of the ant colony optimization algorithm for large scale structural optimization. Civil-Comp Proc 97
16. Hasançebi O, Çarbaş S (2011) Ant colony search method in practical structural optimization. Iran University of Science and Technology
17. Çarbaş S (2017) Optimum structural design of spatial steel frames via biogeography-based optimization. Neural Comput Appl 28. https://doi.org/10.1007/s00521-015-2167-6
18. Carbas S (2020) Enhanced firefly algorithm for optimum steel construction design. In: Dey N (ed) Applications of firefly algorithm and its variants. Springer, Singapore, pp 119–146. https://doi.org/10.1007/978-981-15-0306-1_6
19. Carbas S, Aydogdu I (2021) Cuckoo search for optimum design of real-sized high-level steel frames.https://doi.org/10.1007/978-981-15-5163-5_6
20. Aydogdu I, Akin A (2015) Optimum design of steel space frames by hybrid teaching-learning based optimization and harmony search algorithms. In: The international conference on computing in civil and building engineering, pp. 1351–1358
21. Aydogdu I, Efe P, Yetkin M, Akin A (2017) Optimum design of steel space structures using social spider optimization algorithm with spider jump technique. Struct Eng Mech 62:259–272. https://doi.org/10.12989/sem.2017.62.3.259
22. AISC (1999) Load and resistance factor design specification for structural steel buildings. Chicago, Illinois
23. Coello Coello CA (2002) Theoretical and numerical constraint-handling techniques used with evolutionary algorithms: a survey of the state of the art. Comput Methods Appl Mech Eng 191:1245–1287. https://doi.org/10.1016/S0045-7825(01)00323-1
24. Mardiguian M (2009) Electrostatic discharge: understand, simulate and fix ESD problems. IEEE Press, Piscataway, NJ
25. Cheng KB, Ueng TH, Dixon G (2001) Electrostatic discharge properties of stainless steel/polyester woven fabrics. Text Res J 71:732–738. https://doi.org/10.1177/004051750107100813
26. Bouchekara HREH (2019) Electrostatic discharge algorithm: a novel nature-inspired optimisation algorithm and its application to worst-case tolerance analysis of an EMC filter. IET Sci Meas Technol 13:518–522. https://doi.org/10.1049/iet-smt.2018.5194

27. Khasanov M, Kamel S, Abdel-Mawgoud H (2019) Minimizing power loss and improving voltage stability in distribution system through optimal allocation of distributed generation using electrostatic discharge algorithm. In: 2019 21st international middle east power systems conference, MEPCON 2019—proceedings. Institute of Electrical and Electronics Engineers Inc., pp 354–359. https://doi.org/10.1109/MEPCON47431.2019.9007943
28. Nandy S, Roy PK (2020) Electrostatic discharge algorithm for economic load dispatch problems including renewable energy. In: 2020 national conference on emerging trends on sustainable technology and engineering applications, NCETSTEA 2020. Institute of Electrical and Electronics Engineers Inc. https://doi.org/10.1109/NCETSTEA48365.2020.9119957
29. OAPI—Technical knowledge base—Computers and Structures, Inc.
30. Structural software for analysis and design|SAP2000
31. Hasançebi O, Kazemzadeh Azad S (2014) Discrete size optimization of steel trusses using a refined big bang-big crunch algorithm. Eng Optim 46:61–83. https://doi.org/10.1080/0305215X.2012.748047
32. Tunca O, Aydogdu I, Omercioglu TO, Carbas S Grasshopper optimization algorithm based design of structures. Co-chair 170

Chapter 6
Solving of Distinct Engineering Optimization Problems Using Metaheuristic Algorithms

Ali Mortazavi and Vedat Togan

6.1 Introduction

Along with developing and rising the population in our world, optimum usage of the available resources has become one of the most crucial subjects for humanity. In this regard, the engineering systems (e.g., structural, mechanical, and industrial systems) play a great role in providing the required balance between resources and consumptions. Generally, engineering design can be defined as the art of obtaining a system that meets all expectations using minimum resources. Here, resource in a general term covers the different concepts like time, material, and human labor. Based on this definition, most engineering design processes can be converted to an optimization problem, and to solve them, a robust optimization technique is required [1].

Generally, the optimization methods are divided into deterministic and non-deterministic approaches. The deterministic-based techniques are gradient-based approaches which can find high accuracy (or even exact) solutions with fast convergence rates. These methods demand continuous objective functions and their gradient information to determine the search direction and step size of the iterative process, respectively. However, defining continuous objective function for several complex engineering problems is very difficult or even impossible. Also, in most cases, acquiring the gradient information of the objective function is a costly process. On

A. Mortazavi (✉)
Graduate School of Natural and Applied Sciences, Ege University, 3500 Izmir, Turkey
e-mail: ali.mortazavi.phd@gmail.com

V. Togan
Civil Engineering Department, Karadeniz Technical University, Trabzon, Turkey
e-mail: togan@ktu.edu.tr

S. Carbas et al. (eds.), *Nature-Inspired Metaheuristic Algorithms for Engineering Optimization Applications*, Springer Tracts in Nature-Inspired Computing,
https://doi.org/10.1007/978-981-33-6773-9_6

the other hand, the starting point of the algorithm is very important for these algorithms. If these algorithms are not initiated from a proper location(s) in the search domain, they can easily be trapped into local minima [2, 3].

Many engineering optimization problems have a non-convex search domain since to find a feasible solution, several side constraints and boundary conditions should be taken into account. Consequently, the gradient-based techniques cannot easily solve these types of problems. On the other hand, the non-deterministic approaches provide a group of alternative techniques to solve these kinds of problems [4]. They are gradient-free methods that do not demand any continuous and differentiable objective functions. One of the most important methods of this group is the meta-heuristic algorithms. These algorithms have simple mathematical formulations that are generally inspired by a natural phenomenon, physical rule, or social behavior. These methods can be categorized into four main groups based on their inspiration sources. These groups are given as following [5]:

- The human behavior-inspired methods: They simulate the humans' social behavior in the society. As examples of these methods quantum-inspired social evolution (QSE) [6] and teaching and learning-based optimization (TLBO) [7] can be mentioned.
- The swarm-inspired methods: They mimic the collective behavior of the birds and fish colonies to find food sources and avoid enemy attacks. Particle swarm optimization (PSO) [8], Drosophila food-search (dso) [9], grey wolf optimization (GWO) [10], and symbiotic organisms search (SOS) [11] are the examples of this type of optimization methods.
- The physical-rule-inspired methods: They models the physical principles in the nature. Gravitational search algorithm (GSA) [12], weighted superposition attraction (WSA) [13] and simulating annealing (SA) [14] are the examples for this kind of techniques.
- The evolutionary-inspired methods: They imitate the evolution process in the species in the nature. Genetic algorithm (GA) [15] differential evolution (DE) [16], and evolutionary optimization based on biological evolution in plants (BBO) [17] are the examples for this group of techniques.

In the current chapter, the butterfly optimization algorithm (BOA), firefly algorithm (FA) and Drosophila food-search algorithm (DSO) as the swarm-based methods, teaching and learning-based optimization (TLBO) as human-inspired, and interactive search algorithm (ISA) as a hybrid human-swarm-inspired methods are employed to solve different engineering optimization problems. These methods are well-established techniques that are selected based on their different working mechanisms. The utilized BOA, FA, and ISA calculate objective function once in each iteration, TLBO evaluates the objective function twice per each iteration (one in teaching and another in learning phase), DSO based on the improvement level in the results, equal or more than two times, calculates the objective function during each iteration. The detailed explanation for applied methods in the current chapter is provided in the following.

6.2 The Optimization Methods Employed in the Current Chapter

In the current section, the utilized optimization techniques and their features and formulation are explained. Applied algorithms are firefly algorithm (FA), teaching and learning-based optimization (TLBO) [7], interactive search algorithm (ISA) [18], interactive search algorithm (ISA) [18] and butterfly optimization algorithm (BOA) [19]. All of these algorithms are population-based methods that initiated from a random location in the search domain. During the optimization process, using a simple strategy, they iteratively conduct the population inside the search space until at least one of the stopping criteria is met. In each iteration, the population is perturbed to improve the previous solution candidates.

6.2.1 Firefly Algorithm (FA)

Firefly algorithm (FA) is an optimization method introduced by Yang [20]. This method simulates the mating behavior of the firefly insects. In this method, each solution candidate is named as a firefly. Each firefly in the flock is encouraged to move toward the other fireflies according to the intensity of the light they produce. In the mathematical model, the light intensify of each firefly is determined by its objective function value. In other words, the better objective function value equals with more intense light. All fireflies are assumed unisex; it means that they are attracted to each other just based on their light intensity. Also, the distance between a pair of fireflies affects their attraction level. In other words, closer insects are more attractive than distant ones. It should be noted that the landscape of the search domain, also, can affect their interest in each other (i.e., equivalent to a foggy day in which they cannot see each other well). Based on these descripts, the FA algorithm is mathematically formulated as follows:

$$\begin{aligned} \Delta \mathbf{x}_i &= \beta_0 e^{-\gamma r_{ij}^2}\left(\mathbf{x}_i^t - \mathbf{x}_j^t\right) + \alpha\left(rand - \frac{1}{2}\right) \\ \mathbf{x}_i^{t+1} &= \mathbf{x}_i^t + \Delta \mathbf{x}_i \end{aligned} \tag{6.1}$$

where $\mathbf{x}_i^{t+1}$ and $\mathbf{x}_i^t$, respectively, demonstrate the updated and current positions of the ith firefly, respectively; α is the random coefficient selected from interval of [0,1]; β_0 is the take as unity and r_{ij} is the Euclidean distance between a pair of ith and jth fireflies. For more clarity, the pseudo of the FA method is given in below (Table 6.1).

Table 6.1 Pseudo code of FA

```
Adjust internal parameters of the algorithm;
Produce n random firefly;
Determine the light intensity of each firefly (X_i) by calculating f(X_i)
while (any termination criteria is not met)
  for (current firefly X_i)
    for (each firefly X_j)
      the ith firefly moves towards ith firefly based on Eq. (6.1)
    end
  end
end
```

6.2.2 Teaching and Learning-Based Optimization (TLBO)

The teaching and learning-based optimization (TLBO) method is developed by Rao [7]. This method mimics the learning process in a classroom in which each student can learn in two ways, first from his/her teacher and second from her/his classmates. In this method, each solution candidate is called a student, while in each iteration the most knowledge member is selected as the teacher. It mathematically means that the member with the best objective function value is designated as the teacher. Based on this fact, the TLBO method consists of two phases called teaching phase and learning phase. In teaching phase, each student is conducted toward the best student in the classroom (i.e., teacher). In the learning phase, each student learns from another random student based on a pairwise knowledge sharing. Based on these explanations, the two phases of TLBO method are mathematically formulated as below.

Teaching phase:

$$\begin{aligned} &\mathbf{X}^{(\text{new},i)} = \mathbf{X}^i + r(\mathbf{X}_{\text{teacher}} - T_F\mathbf{X}_{\text{mean}}) \\ &\text{if } f\left(\mathbf{x}_i^{t+1}\right) < f\left(\mathbf{x}_i^t\right), \mathbf{x}_i^t = \mathbf{x}_i^{t+1} \\ &\text{if } f\left(\mathbf{x}_i^{t+1}\right) \geq \text{f}\left(\mathbf{x}_i^t\right), \mathbf{x}_i^t = \mathbf{x}_i^t \end{aligned}$$

Learning phase:

$$\begin{aligned} \mathbf{x}_i^{t+1} &= \mathbf{x}_i^t + r.\left(\mathbf{x}_i^t - \mathbf{x}_j^t\right) \text{ if } f\left(\mathbf{x}_i^t\right) \leq f\left(\mathbf{x}_j^t\right) \\ \mathbf{x}_i^{t+1} &= \mathbf{x}_i^t + r.\left(\mathbf{x}_j^t - \mathbf{x}_i^t\right) \text{ if } f\left(\mathbf{x}_i^t\right) > f\left(\mathbf{x}_j^t\right) \end{aligned} \tag{6.2}$$

where $\mathbf{x}_i^{t+1}$ and $\mathbf{x}_i^t$, respectively, demonstrate the updated and current positions of the ith student, respectively; r is a random coefficient selected from interval of [0,1]; $f(.)$ returns the objective function value for the current student. For more clarity, the pseudo of the TLBO method is given in below (Table 6.2).

Table 6.2 Pseudo-code of TLBO

Adjust internal parameters of the algorithm
Produce a class with n random students
Sort the class and choose the best student as the teacher
while *(any termination criteria is not met)*
  for *(each student in the class)*
    conduct the students toward the teacher position using Eq. (6.2)
    evaluate updated student calculating $f\left(X_i^{t+1}\right)$
    if *updated location of the current student is improved*
      hold the new one and replace with old student
    else
      reset it to its prior location
    end
    select an arbitrary student$\left(X_j^t\right)$ *which (i≠j)*
    if $f\left(\mathbf{x}_i^t\right) \leq f\left(\mathbf{x}_j^t\right)$
      the current student should move away from the selected student
    else $f\left(\mathbf{x}_i^t\right) > f\left(\mathbf{x}_j^t\right)$
      the current student should go toward the selected student
    end
    evaluate updated student calculating $f\left(X_i^{t+1}\right)$
    if *updated location of the current student is improved*
      hold the new one and replace with old student
    else
      reject the updated location
    end
  end
end

6.2.3 Drosophila Food-Search Optimization (DSO)

The Drosophila food-search optimization (DSO) is introduced by Das and Singh [9]. This method mimics the food-searching process of the Drosophila Melanogaster incest. In the proposed DSO, each solution candidate is called insect. This method employs two main search patterns to search the problems domain. First pattern is based on the neighborhood searching paradigm, and second pattern is based on the quadratic searching scheme. In the neighborhood search mode, each incest tries to search the vicinity of the other insects in the colony. In the quadratic approximation (QA) search, the current insect is replaced by a child one which is introduced using QA formulation. The DSO searching mechanism based on these three search patterns is mathematically formulated as follows.

Neighborhood search pattern:

Table 6.3 Pseudocode of DSO method

Adjust internal parameters of the algorithm
Produce a colony with p random insects
while *(any termination criteria is not met)*
Apply the tournament selection
for *(each insect in the colony)*
perform a neighborhood searching applying Eq. (6.3)
Calculate the objective function of the updated insect $f(\mathbf{x}_i^{t+1})$
if the improvement is lower than 1%
replace the current insect with its child using the QA formulation in Eq. (6.3)
end
end
end

$$U_{i,k} = \mathrm{x}_{i,k}^t + \left|\mathrm{x}_{r3,k}^t - \mathrm{x}_{r4,k}^t\right| \tag{6.3}$$

$$W_{i,k} = \mathrm{x}_{i,k}^t - \left|\mathrm{x}_{r3,k}^t - \mathrm{x}_{r4,k}^t\right| \text{for} k = r1 \text{ and } r2;$$
$$\text{for } j \neq r1 \text{ and } j \neq r2,\ U_{i,k} = \mathbf{x}_{i,j}^t \text{ and } W_{i,j} = \mathbf{x}_{i,j}^t$$

$$\mathbf{x}_i^{t+1} = \mathrm{Min}\left\{\left(\mathbf{x}_i^t\right), (\boldsymbol{U}_i), (\boldsymbol{W}_i)\right\} \text{for} \begin{cases} i = 1, 2, \ldots, P \\ j = 1, 2, \ldots, D \end{cases}$$

Quadratic Approximation search pattern:

$$\text{if}\left|f(\mathbf{x}_i^{t+1}) - f(\mathbf{x}_i^t)/f(\mathbf{x}_i^t)\right| < 1\%$$

$$\mathbf{x}_i^{t+1} = 0.5\frac{\left(R_2^2 - R_3^2\right)f(R_1) + \left(R_3^2 - R_1^2\right)f(R_2) + \left(R_1^2 - R_2^2\right)f(R_3)}{\left(R_2 - R_3\right)f(R_1) + \left(R_3 - R_1\right)f(R_2) + \left(R_1 - R_2\right)f(R_3)}$$

where $\mathbf{x}_i^{t+1}$ and $\mathbf{x}_i^t$, respectively, demonstrate the updated and current positions of the ith insect, respectively; P and D, respectively, show size of population and problem dimension; $r1$, $r2$, $r3$ and $r4$ are random integer numbers such that $r1, r2 \in [1, D]$ and $r3, r4 \in [1, P]$. R_1, R_2 and R_3 are three random selected insects from the swarm. For more clarity, the pseudo of the proposed DSO method is given below (Table 6.3).

6.2.4 *Interactive Search Algorithm (ISA)*

Interactive search algorithm (ISA) is introduced by Mortazavi et al. [18]. In this method, each solution candidate is called agent. This method applies two different search patterns called tracking and interacting phases. In the tracking phase, agents try to search the vicinity of some promising locations in the domain. These locations are spotted by the best agent ($\mathbf{X}^{G}$), the previous random location hold in swarm memory ($\mathbf{X}_j^P$), and weighted agent ($\mathbf{X}^{W}$). In the interacting phase, the agents are in interaction in pair. Based on these clarifications, the proposed ISA is mathematically formulated as follows.

$$\begin{aligned}
&\quad Tracking\ phase\\
&\quad \text{if } \tau \geq 0.3\\
&{}^{t+1}V_i = \omega \cdot {}^{t}V_i + \varphi_1 \odot \left({}^{t}\mathbf{X}_i^P - {}^{t}\mathbf{X}_i\right) + \varphi_2 \odot \left({}^{t}\mathbf{X}^G - {}^{t}\mathbf{X}_i^P\right) + \varphi_3 \odot \left({}^{t}\mathbf{X}^W - {}^{t}\mathbf{X}_i\right)\\
&\quad Interacting\ phase:\\
&\quad \text{if } \tau < 0.3\\
&{}^{t+1}V_i = \varphi_4 \odot \left({}^{t}\mathbf{X}_i - {}^{t}\mathbf{X}_j\right) \qquad \text{if} \qquad f(\mathbf{X}_i) \leq f(\mathbf{X}_j)\\
&{}^{t+1}V_i = \varphi_4 \odot \left({}^{t}\mathbf{X}_j - {}^{t}\mathbf{X}_i\right) \qquad \text{if} \qquad f(\mathbf{X}_i) > f(\mathbf{X}_j)\\
&\quad Updating\ formulation:\\
&{}^{t+1}\mathbf{X}_i = {}^{t}\mathbf{X}_i + {}^{t+1}\mathbf{V}_i
\end{aligned} \tag{6.4}$$

where ${}^{t+1}\mathbf{X}_i$ and ${}^{t}\mathbf{X}_i$, respectively, demonstrate the updated and current positions of the ith agent, respectively; the coefficients of φ_1, φ_2, φ_3 and φ_4 are random vectors which their components are uniformly selected from [0,1] interval; also, $\odot$ indicates the Hadamard multiplier; ω is the inertia coefficient, and it is set as 0.4 [18]; *f(.)*returns the objective function value for the current agent. For more clarity, the pseudo of the ISA method is given in below (Table 6.4).

Table 6.4 Pseudocode of ISA method

```
Adjust internal parameters of the algorithm
Produce a population with p random agents
while (any termination criteria is not met)
  for (each agent in the population)
     if τ ≥ 0.3
       perform the tracking search using Eq. (6.4)
     else
       perform the interacting search using Eq. (6.4)
     end
  end
end
```

6.2.5 Butterfly Optimization Algorithm (BOA)

The butterfly optimization algorithm (BOA) has been developed by Arora et al. [19]. This method mimics the mating and collective food-search behavior of the butterfly insect. In this method, each solution candidate is called a butterfly. This method uses two different search patterns for search the current problem's domain. In the nature, the behavior of these butterflies is complex, but to provide a mathematical model, some simplification assumptions are imposed. First, all butterflies are unisex and can be attracted each other just based on their fragrance intensities. Second, all butterflies are attracted from the best butterfly (i.e., the butterfly with the best objective function value). Third, the fragrance emanating from butterflies is affected by the landscape of the search space. Based on the given information the BOA method applies two different search patterns for global and local search as follows.

$$\begin{aligned}
&\text{If } p < 0.8 \\
&Global\ search \\
&\mathbf{X}_i^{t+1} = \mathbf{X}_i^t + \left(r^2 \times \mathbf{g}^* - \mathbf{X}_i^t\right) \times f_i \\
&\text{If } p \geq 0.8 \\
&Local\ search \\
&\mathbf{X}_i^{t+1} = \mathbf{X}_i^t + \left(r^2 \times \mathbf{X}_j^t - \mathbf{X}_k^t\right) \times f_i
\end{aligned} \tag{6.5}$$

where $\mathbf{x}_i^{t+1}$ and $\mathbf{x}_i^t$, respectively, demonstrate the updated and current positions of the ith butterfly, respectively; r is the Euclidean distance between tow butterflies; the best butterfly is shown by $\mathbf{g}^*$; k is the random integer such that $k \in [1, P]$, and P demonstrates the population size; the fragrance coefficient is shown by f and defined as $f = cI^a$ and $a = 0.1$ and $c = 0.01$; I is the fragrance intensity and is equivalent with the objective function of each butterfly. For more clarity, the pseudo of the BOA method is given in below (Table 6.5).

Table 6.5 Pseudocode of BOA method

```
Adjust internal parameters of the algorithm
Produce a colony with n random butterflies
while (any termination criteria is not met)
  for (each agent in the population)
    if p<0.8
      perform the global search using Eq. (6.5)
    else
      perform the interacting search using Eq. (6.5)
    end
  end
end
```

Table 6.6 Parameter setting for the applied algorithms

Algorithm	Parameter
FA [20]	$\alpha = 0.5, \beta_{\min} = 0.2, \gamma = 1$
TLBO [7]	TF = round[1 + rand(0, 1){2 − 1}]
DSO [9]	–
ISA [18]	$\tau = 0.3$
BOA [19]	$C = 0.01, a = 0.1, p = 0.8$

6.3 Numerical Examples

In the current section, a suit of numerical examples is solved using given optimization methods. The examples consist of four mathematical functions, two mechanical problems, a structural design problem and a time–cost trade of problem for project management. To prevent any premature convergence conditions, each problem is run for 30 times. The setting for applied methods is given in below.

6.3.1 Mathematical Functions

The current section is devoted to minimize four mathematical function using five selected metaheuristic techniques. The functions formulas are given in Table 6.6. To provide more insight into the search domain of the selected functions, their 2D surfaces are plotted in Fig. 6.1. The convergence history diagrams for optimization process are given in Fig. 6.2. Based on the given diagrams for Ackley, Levy, Rastrigin and Schaffer functions, BOA, DSO, ISA and TLBO provide the fasted and the most accurate solutions, respectively. Based on these data, each method can show high or low search performance depending on the problem under consideration.

6.3.2 Mechanical Problems

In the current section, the search performance of the selected metaheuristic methods is assessed on two mechanical problems. The manufacturing cost minimization of a pressure vessel and optimal designing of a gear system are considered. These problems are constrained optimization problems. The problems definitions and their constraints are given in the following.

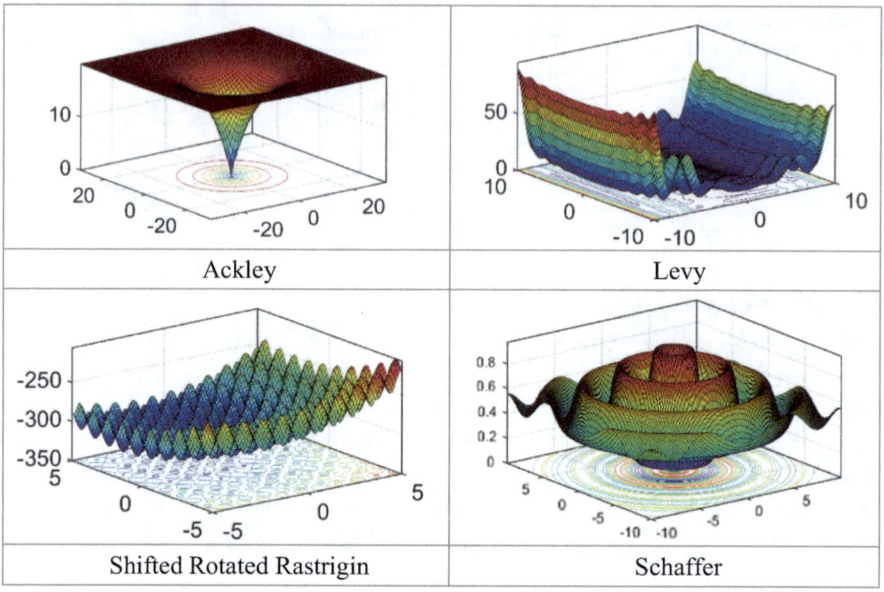

Fig. 6.1 2D surfaces for selected function

6.3.2.1 A Pressure Vessel Cost Optimization

In this section, a pressure cost minimization of pressure vessel production is considered. The system is shown in Fig. 6.3. The objective function and related constraints for this problem are given in Table 6.7. Best optimal solution and the mean values and standard deviation (Std.) of optimal solutions obtained are reported in Table 6.8. Based on the given data, ISA can find the lowest manufacturing cost. Also, based on the statistical reports, this method provides the most stable behavior during the optimization process (Table 6.9).

6.3.2.2 A Gear System Optimal Design

In this section, the optimal design of the gear system shown in Fig. 6.4 is considered. The objective function of this problem and its related constraints are given in Table 6.10. The variables of x_1, x_2, x_3 and x_4 indicate the number of teeth in the parts A, B, C and D parts, respectively. The obtained optimal solutions and statistical information for the optimization process are given in Table 6.11. Based on the given information, the DSO method could provide the most optimal design parameters. Also, based on the addressed statistical data, this method with the lowest standard deviation value (Std.) demonstrates the most stable behavior among all other selected methods (Fig. 6.5).

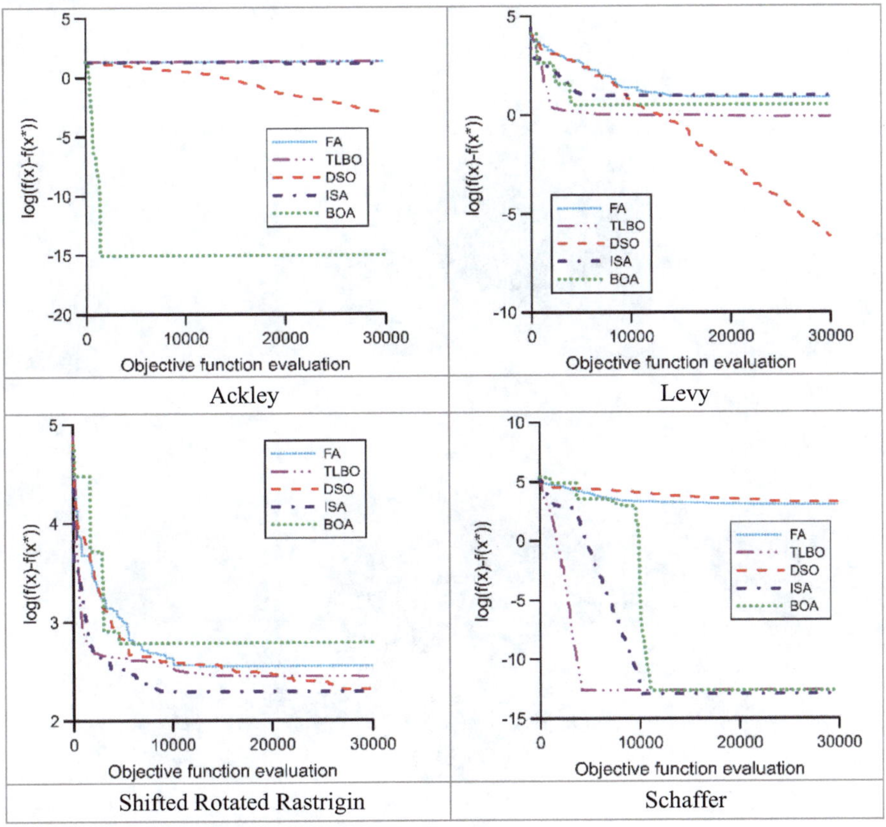

Fig. 6.2 Convergence diagrams for selected functions

6.3.3 Structural Design Problem

In this section, a structural optimal design problem is considered. During the minimization problem, the weight of the structural system is taken as objective function of the problem. The objective function for such a problem is defined as below:

Find $\mathbf{X}$

$$\min. W(\mathbf{X}) = \sum_{i=1}^{ns} \rho_i L_i A_i$$

where,

$$x_{i\_\min} \le x_i < x_{i\_\max}$$

Subjected to,

$$\Omega_j \ge \Omega_j^* \text{for some natural frequency } j$$

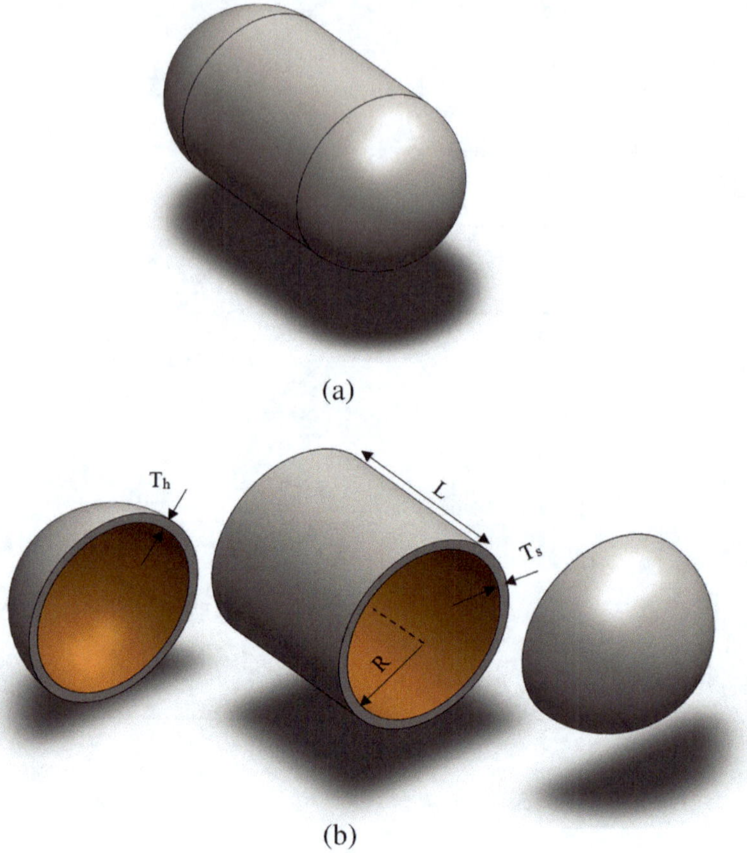

Fig. 6.3 Pressure vessel **a** assembled, **b** ripped apart

$$\Omega_k \leq \Omega_k^* \text{for some natural frequency} k \quad (6.6)$$

where $\boldsymbol{X}$ is the vector of design variables, and ρ_i, $L_i$andA_i, respectively, indicate the density the material, the length, and the cross-sectional area for ith member of the structural system. The natural frequency of the system is shown by Ω. Also, x_i shows the ith components of the design vector and $x_{i\_\min}$ and $x_{i\_\max}$, respectively, designates its upper and lower bounds, and $W(\mathbf{X})$ returns the total weight of the structure.

6.3.3.1 A 72-bar Spatial Truss Structure

In this section, the 72-bar truss system shown in is considered as the first structural example. For all nodes, the allowable nodal displacement in all principle directions is limited to ±0.25 in. For all structural members, the maximum allowable stress should

Table 6.7 Parameter setting for the applied algorithms

Title	Formulation
Ackley	$f(\mathbf{X}) = -20\exp\left(-0.2\sqrt{\sum_{i=1}^{D}\frac{x_i^2}{D}}\right) - \exp\left(\sum_{i=1}^{D}\cos\frac{(2\pi x_i)}{D}\right) + 20 + e$
Levy	$f(\mathbf{X}) = \sin^2(\pi w_1) + \sum_{i=1}^{D-1}(w_i - 1)^2\left[1 + 10\sin^2(\pi w_i + 1)\right] + (w_D - 1)^2\left[1 + \sin^2(2\pi w_d)\right]$ where, $w_i = 1 + \frac{x_i - 1}{4}$, for all $i = 1, \ldots, D$
Shifted rotated rastrigin	$f(\mathbf{X}) = \left(\sum_{i=1}^{D}\left(\sum_{j=1}^{i} z_j\right)^2\right) * (1 + 0.4\lvert N(0,1)\rvert) - 450$
Schaffer	$f(X) = 0.5 + \frac{\sin^2(x^2+y^2)^2 - 0.5}{(1+0.001(x^2+y^2))^2}$

Table 6.8 Objective function and constraints for pressure vessel problem

Property	Mathematical formulation
Objective function	$f(\mathbf{X}) = 0.6224x_1x_3x_4 + 1.7781x_2x_3^2 + 3.1661x_1^2x_4 + 19.84x_1^2x_3$
Constraints	$g_1(\mathbf{X}) = -x_1 + 0.0193x_3 \le 0$ $g_2(\mathbf{X}) = -x_2 + 0.00954x_3 \le 0$ $g_3(\mathbf{X}) = -\pi x_3^2x_4 - \frac{4}{3}\pi x_3^3 + 1,296,000 \le 0$ $g_4(\mathbf{X}) = x_4 - 240 \le 0$
Variables' bounds	$0 \le x_1 \le 100$ $0 \le x_2 \le 100$ $10 \le x_3 \le 200$ $10 \le x_4 \le 200$

Table 6.9 Obtained results for the pressure vessel problem

Variables	FA	TLBO	DSO	ISA	BOA
$x_1(T_s)$	0.794906	0.812500	0.778180	0.778221	0.854355
$x_2(T_h)$	0.407677	0.434500	0.384655	0.384674	0.434521
$x_3(R)$	40.795672	42.089181	40.320256	40.322124	43.358609
$x_4(L)$	195.705044	176.758731	199.991129	199.965183	161.73127
$f(\mathbf{X})$					
Best	6058.8022	6051.5639	5884.9064	5884.8400	6182.4520
Mean	6061.8187	6070.2585	5898.2598	5893.1934	8965.8712
Std.	24.12	35.12	45.36	8.13	1897.98
OFEs	12,850	15,875	25,365	1150	2345

Fig. 6.4 The gear system

Table 6.10 Objective function and related constraints the gear system design

Property	Mathematical formulation
Objective function	$f(\mathbf{X}) = \frac{1}{6.931} - \frac{x_1 x_2}{x_3 x_4}$
Variables' bounds	$12 \le x_1 \le 60$ $12 \le x_2 \le 60$ $12 \le x_3 \le 60$ $12 \le x_4 \le 60$

Table 6.11 Optimal result for the gear system problem

Variables	FA	TLBO	DSO	ISA	BOA
x_1	50.05234	55.62869	48.22301	51.32931	26.04295
x_2	24.38656	16.71727	18.77492	21.02347	12.00000
x_3	14.02798	21.27387	21.13716	14.79391	12.00000
x_4	46.3547	44.31047	57.03833	47.82131	38.33421
$f(\mathbf{X})$					
Best	6.5210E−13	3.0910E−14	1.4410E−14	3.0810E−14	1.5442E−09
Mean	1.1512E−13	3.3467E−12	9.8246E−12	7.6226E−11	8.1526E−02
Std.	4.3462E−08	4.0937E−09	2.9924E−09	5.6434E−08	8.2746
OFEs	2020	3285	5780	2050	1005

be equal or less than ±25 ksi. The node number 17 of the truss system is subjected to 5 kips constraint load in the *x*-, *y*- and gravitational directions. The modulus of elasticity is 10.0 Msi, and the density of the material is 0.1 lb/in^3. Attained the best, the mean optimal solutions and standard deviation (Std.) of the optimization process

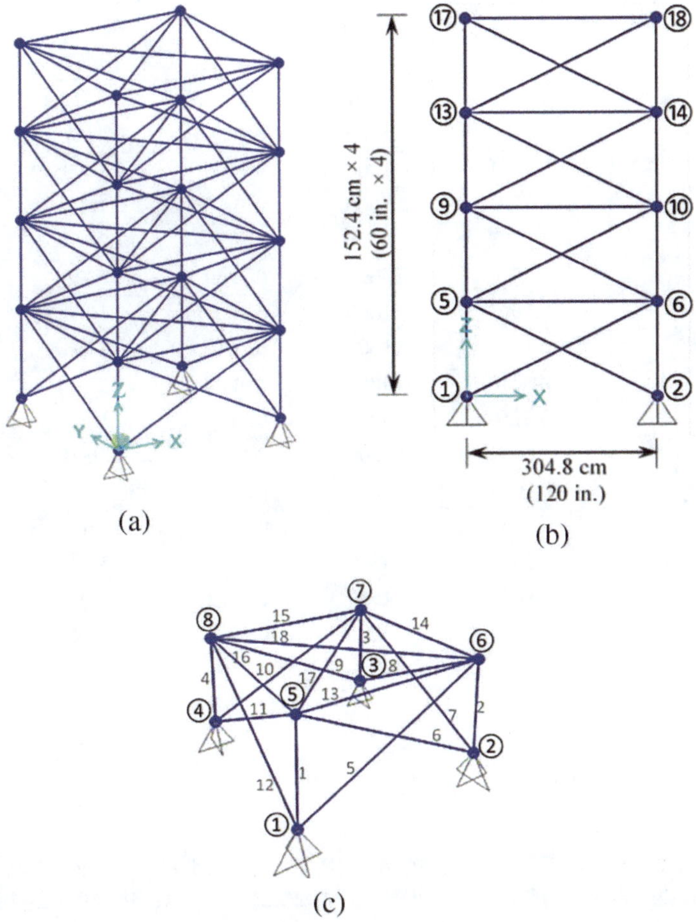

Fig. 6.5 72-bar truss system **a** 3D view, **b** side view and **c** modular view

are shown in Table 6.12. Based on the given information, the ISA method could find the lightest structure, and based on the reported standard deviation values, this method shows the most stable behavior among the selected method.

6.3.3.2 A 200-bar Planar Truss Structure

In the current section, the weight of the 200-bar planar truss structure shown in Fig. 6.6 is minimized. Maintaining symmetry, the structural members are put into 29 independent groups. The grouping information is given in Table 6.13. The material density is 0.283 lb/in^3, and the module of elasticity is 30 Msi. The minimum allowable cross-sectional area is set to 0.1 in^2. The allowable stress for both tensile and compressive members is limited up to ± 10.0 ksi. The structure is subjected to three

Table 6.12 Optimal result for 72-bar truss system

Variables (in^2)	FA	TLBO	DSO	ISA	BOA
A_1	1.9028	1.8601	1.8577	1.8329	1.7430
A_2	0.5180	0.5209	0.5059	0.5127	0.5181
A_3	0.1001	0.1000	0.1000	0.1000	0.1000
A_4	0.1003	0.1000	0.1000	0.1000	0.1000
A_5	1.2787	1.2710	1.2476	1.2635	1.3079
A_6	0.5074	0.5090	0.5269	0.5228	0.5190
A_7	0.1003	0.1000	0.1000	0.1000	0.1000
A_8	0.1003	0.1000	0.1012	0.1000	0.1000
A_9	0.5240	0.4850	0.5209	0.5273	0.5140
A_{10}	0.5150	0.5009	0.5172	0.5131	0.5460
A_{11}	0.1002	0.1000	0.1004	0.1000	0.1000
A_{12}	0.1015	0.1000	0.1005	0.1000	0.1090
A_{13}	0.1564	0.1680	0.1565	0.1568	0.1610
A_{14}	0.5494	0.5840	0.5507	0.5336	0.5089
A_{15}	0.4029	0.4330	0.3922	0.4110	0.4970
A_{16}	0.5504	0.5201	0.5922	0.5849	0.5620
Best (lb)	379.694	380.62	379.850	378.37	381.91
Mean (lb)	379.896	381.49	382.080	378.65	553.01
Std. (lb)	0.079	0.778	1.912	0.382	120.081
NSAs	15,600	13,000	19,600	10,480	13,850

load cases as (i): 1, 6, 15, 20, 29, 34, 43, 48, 57, 62 and 71 are loaded by (1, 0, 0) kips, (ii) nodes 1–6, 8, 10, 12, 14–20, 22, 24, 26, 28–34, 36, 38, 40, 42–48, 50, 52, 54, 56–62, 64, 66, 68, 70–75 are loaded by (0, -10, 0) kips, and (iii) both load cases (i) and (ii) act simultaneously. Achieved results are reported in Table 6.13. Based on the given information, the ISA could find the lightest structure system while the TLBO shows the most stable behavior during the optimization process.

6.3.4 Project Management Problem

The determination of the most appropriate time–cost alternatives for successful project completion within the shortest period and at the lowest cost is defined as time–cost trade-off problem (TCTP). TCTP is an optimization problem, which tries to minimize time and cost of a project at the same time. Due to this characteristic, TCTP is a multi-objective optimization problem (MOOP), and contrary, the optimization problem with single objective, in MOOP, there is no unique solution to be indicated as being the global optimal. Therefore, the solution of an TCTP offers a

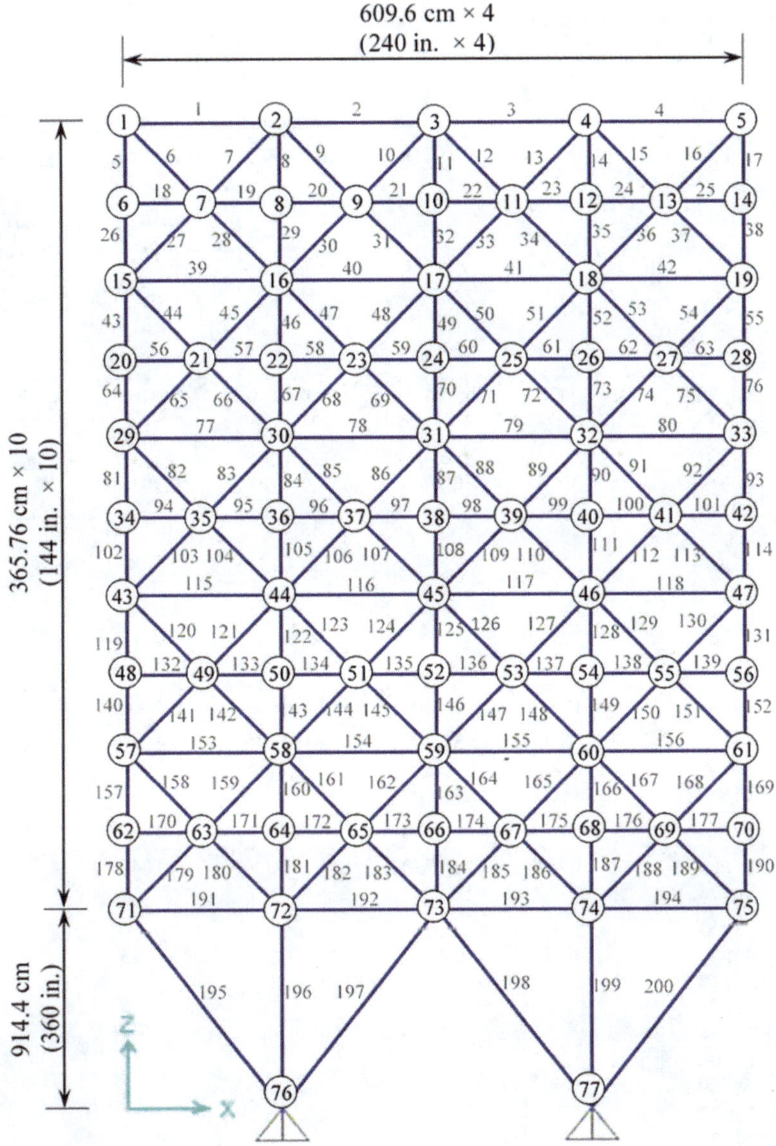

Fig. 6.6 The 200-bar truss system

Table 6.13 Optimal result for 200-bar truss structure

Group (in^2)	Member	FA	TLBO	DSO	ISA	BOA
A_1	1–4	0.1540	0.1460	0.1500	0.1469	0.7590
A_2	5, 8, 11, 14, 17	0.9410	0.9410	0.9460	0.9447	0.9032
A_3	19–24	0.10	0.10	0.1010	0.10	1.1
A_4	18, 25, 56, 63, 94, 101, 132, 139, 170, 177	0.10	0.1010	0.1	0.10	0.9952
A_5	26, 29, 32, 35, 38	1.9420	1.9410	1.9450	1.9405	2.1350
A_6	6, 7, 9, 10, 12, 13, 15, 16, 27, 28, 30, 31, 33, 34, 36, 37	0.3010	0.2960	0.2960	0.2958	0.4193
A_7	39–42	0.10	0.1000	0.1000	0.1000	1.0041
A_8	43, 46, 49, 52, 55	3.1080	3.1210	3.16101	3.1040	2.8052
A_9	57-62	0.10	0.1000	0.10200	0.10000	1.0344
A_{10}	64, 67, 70, 73, 76	4.1060	4.1730	4.1990	4.104	3.7842
A_{11}	44, 45, 47, 48, 50, 51, 53, 54, 65, 66, 68, 69, 71, 72, 74, 75	0.4090	0.4010	0.4010	0.4035	0.5269
A_{12}	77–80	0.1910	0.1810	0.1810	0.1916	0.4302
A_{13}	81, 84, 87, 90, 93	5.4280	5.4230	5.4310	5.4279	5.2683
A_{14}	95–100	0.10	0.10	0.1	0.10	0.9685

(continued)

Table 6.13 (continued)

Group (in^2)	Member	FA	TLBO	DSO	ISA	BOA
A_{15}	102, 105, 108, 111, 114	6.4270	6.4220	6.4280	6.4279	6.0473
A_{16}	82, 83, 85, 86, 88, 89, 91, 92, 103, 104, 106, 107, 109, 110, 112, 113	0.5810	0.5710	0.5710	0.5736	0.7825
A_{17}	115-118	0.1510	0.1560	0.1560	0.1338	0.5920
A_{18}	119, 122, 125, 128, 131	7.9730	7.9580	7.9610	7.97326	8.18580
A_{19}	133–138	0.10	0.10	0.1	0.10	1.03620
A_{20}	140, 143, 146, 149, 152	8.9740	8.9580	8.9590	8.9723	9.20620
A_{21}	120, 121, 123, 124, 126, 127, 129, 130, 141, 142, 144, 145, 147, 148, 150, 151	0.7190	0.7200	0.7220	0.7054	1.47740
A_{22}	153-156	0.4220	0.4780	0.4910	0.42076	1.83360
A_{23}	157, 160, 163, 166, 169	10.8920	10.8970	10.909	10.8669	10.6110
A_{24}	171–176	0.10	0.10	0.1010	0.10	0.98510
A_{25}	178, 181, 184,187, 190	11.8870	11.8970	11.9850	11.8673	12.5090
A_{26}	158, 159, 161, 162, 164, 165, 167, 168, 179, 180, 182, 183, 185, 186, 188, 189	1.0400	1.0800	1.0840	1.0349	1.97550

(continued)

Table 6.13 (continued)

Group (in^2)	Member	FA	TLBO	DSO	ISA	BOA
A_{27}	191–194	6.6460	6.4620	6.4640	6.68437	4.51490
A_{28}	195, 197, 198, 200	10.8040	10.7990	10.8020	10.8073	9.80000
A_{29}	196, 199	13.8700	13.9220	13.9360	13.8448	14.5310
Best (lb)		25,491.9	25491.9	25,542.5	25,450.86	28,537.8
Mean (lb)		25,508.3	25,497.32	25,889.2	25,475.02	29,102.3
Std. (lb)		34.56	10.54	183.07	14.66	1523.13
NSAs		15,450	28,059	28,080	10,550	25,980

set of solutions known as Pareto solutions. They cannot be preferred to each other in order to be declared as the best solution, and the decision maker can select anyone among them depending on his/her experiences [21–26].

In this section, the performances of the metaheuristic methods explained in this chapter are tested on TCTP of an 18-activity project. Corresponding problem is defined through the following mathematical expressions.

$$\begin{aligned} &\text{Find } \mathbf{X} \\ &\min. F_T(\mathbf{X}) \text{and} F_C(\mathbf{X}) \\ &x_{i\_\min} \le x_i \le x_{i\_\max} \end{aligned} \tag{6.7}$$

where **X** is the vector of design variables, which shows the number of activities for the related construction project, $F_T(\mathbf{X})$ and $F_C(\mathbf{X})$, respectively, are two conflicting objectives to be minimized simultaneously. The first objective represents the project duration (T), while the second one is the total project cost (C). x_i shows the ith components of the design vector, and $x_{i\_\min}$ and $x_{i\_\max}$ are its upper and lower bounds, respectively. Since an activity of project will be realized at least one execution option, which is a combination of resources such as labor, material, equipment, etc., $x_{i\_\min}$ is equal to 1. However, $x_{i\_\max}$ is the number of all different execution options for that activity. The two objectives can be calculated using the following equations.

$$F_T(\mathbf{X}) = \sum_{i=1}^{na} D_i^n \tag{6.8}$$

where D_i^n is the duration of activity (i) on the critical path using execution option (n), na is the number of activities.

$$F_C(\mathbf{X}) = \sum_{i=1}^{na} DC_i^n + T \times IDCR \tag{6.9}$$

where DC_i^n is the direct cost of activity (i) using execution option (n), T is the project duration, *IDCR* is the daily indirect cost rate in \$/day.

The fitness function presented in the following mathematical expression is employed to display the solutions' qualities.

$$F(\mathbf{X}) = S_T + S_C + (S_C/S_T) \tag{6.10}$$

where $S_T = 1 - \left(Z_T/Z_T^{\max}\right)$ and $S_C = 1 - \left(Z_C/Z_C^{\max}\right)$ are the normalized time and cost values for the corresponding solution, respectively. $Z_T^{\max}$ and $Z_C^{\max}$ represent maximum time and total cost values in the current population, Z_T and Z_C are the time and the total cost values of the **X** th solution in that population. The term, (S_C/S_T), is added to Eq. (6.10) for eliminating the small value effects.

The metaheuristic methods are applied to solve a TCTP of 18-activity project. The activities, precedence relations and time–cost alternatives for the test problem are given in Table 6.14, and the activity-on-node diagram for the project is presented in Fig. 6.7. For the problem, the indirect costs are adopted as \$1500/day. In the solution process, parameters are set as population size = 70 and maximum iteration = 40. Ten independent runs are carried out for each test instance due to stochastic nature of associated metaheuristic methods.

The performance of the metaheuristic methods utilized in the current chapter for the current problems are summarized in Tables 6.15 and 6.16. The obtained five patterns are given in Table 6.15. These patterns could be found by all methods. The Pareto frontier diagram for this example is plotted in Fig. 6.8. The CPU time for all tested techniques are reported in Table 6.16. Based on the given times, the ISA demands the least computational time to obtain the solutions.

6.4 Conclusions

In the current section, the search performance of the five metaheuristic algorithms as firefly algorithm (FA), teaching and learning-based optimization (TLBO), Drosophila food-search optimization (DSO), interactive search algorithm (ISA) and butterfly optimization Algorithm (BOA) are tested on solving four mathematical functions, two mechanical manufacturing problems, two structural weight minimization problems and a time–cost trade-off of a project as a multi-objective optimization problem. The mathematical functions have unconstrained search domains, while to reach feasible solutions, the search spaces of the engineering problems are restricted with different constraints. On the other hand, the solved project management problem not only contains dual contradictory objective functions, but also it is the integer optimization problem. It means, unlike the other selected problems, its decision variables should be selected from a discrete set of integers. The different features of selected problems challenge the solver methods from different aspects. The achieved results are given trough the illustrative tables and diagrams. Reported outcomes indicate

Table 6.14 Options for 18-activity project

Activity	Predecessors	Option	Duration (days)	Direct cost ($)
1	–	1	14	2400
		2	15	2150
		3	16	1900
		4	21	1500
		5	24	1200
2	–	1	15	3000
		2	18	2400
		3	20	1800
		4	23	1500
		5	25	1000
3	–	1	15	4500
		2	22	4000
		3	33	3200
4	–	1	12	45,000
		2	16	35,000
		3	20	30,000
5	1	1	22	20,000
		2	24	17,500
		3	28	15,000
		4	30	10,000
6	1	1	14	40,000
		2	18	32,000
		3	24	18,000
7	5	1	9	30,000
		2	15	24,000
		3	18	22,000
8	6	1	14	220
		2	15	215
		3	16	200
		4	21	208
		5	24	120
9	6	1	15	300
		2	18	240
		3	20	180
		4	23	150
		5	25	100

(continued)

Table 6.14 (continued)

Activity	Predecessors	Option	Duration (days)	Direct cost ($)
10	2, 6	1	15	450
		2	22	400
		3	33	320
11	7, 8	1	12	450
		2	16	350
		3	20	300
12	5, 9, 10	1	22	2000
		2	24	1750
		3	28	1500
		4	30	1000
13	3	1	14	4000
		2	18	3200
		3	24	1800
14	4, 10	1	9	3000
		2	15	2400
		3	18	2200
15	12	1	16	3500
16	13, 14	1	20	3000
		2	22	2000
		3	24	1750
		4	28	1500
		5	30	1000
17	11, 14, 15	1	14	4000
		2	18	3200
		3	24	1800
18	16, 17	1	9	3000
		2	15	2400
		3	18	2200

that for the mathematical functions with smooth search spaces (i.e., without shifting and/or rotating), BOA method shows a superior performance, while for the other functions, with higher complexity, the ISA and DSO give a better search capacity and could find more accurate optimal solutions. For designated mechanical problems, DSO and ISA comparatively provide more promising results. For the selected structural optimization problem, ISA could find the lightest truss system. For the multi-objective project management problem, although all methods can find the same optimal solutions, ISA demanding the lowest computational cost shows the fastest search process. Consequently, acquired results show that each method has its

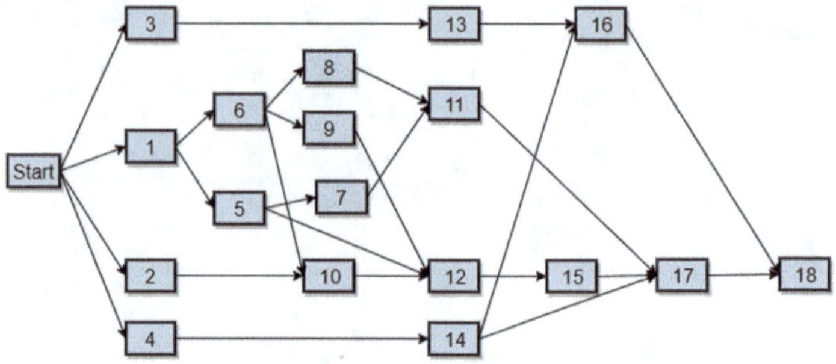

Fig. 6.7 Activity-on-node diagram for the 18-actvity project

Table 6.15 Acquired optimal options for 18-activity project

Duration	Cost	Activities																	
104	289,370	1	5	3	3	3	1	3	5	1	1	1	1	3	2	1	3	1	1
105	285,820	1	5	3	3	4	1	3	1	1	1	3	1	3	2	1	3	1	1
108	283,058	1	5	3	3	4	2	3	4	1	1	1	1	3	1	1	3	1	1
114	277,100	1	4	3	3	4	3	3	1	1	1	1	1	2	2	1	1	1	1

Table 6.16 CPU timing for 18-activity project

Method	CPU time (s)
BOA	159
ISA	153
DSO	542
TLBO	285
FA	161

Fig. 6.8 Pareto frontier for the 18-activity project

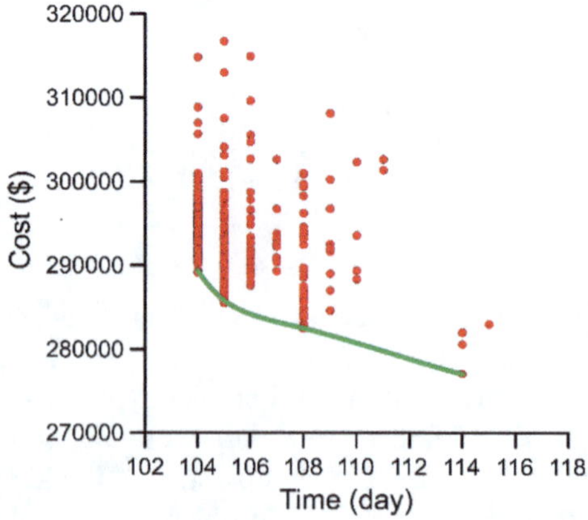

capabilities in solving different problems (or class of problems), and it is mostly due to the stochastic essence of the metaheuristic techniques.

References

1. Moloodpoor M, Mortazavi A, Ozbalta N (2019) Thermal analysis of parabolic trough collectors via a swarm intelligence optimizer. Sol Energy 181:264–275
2. Toğan V, Mortazavi A (2017) Sizing optimization of skeletal structures using teaching-learning based optimization. Optim Control Theories Appl 7:12
3. Mortazavi A (2020) A new fuzzy strategy for size and topology optimization of truss structures. Appl Soft Comput 93:106412
4. Mortazavi A, Toğan V, Daloğlu A, Nuhoğlu A (2018) Comparison of two metaheuristic algorithms on sizing and topology optimization of trusses and mathematical functions. Gazi Univ J Sci 31:416–435
5. Mortazavi A (2019) Interactive fuzzy search algorithm: a new self-adaptive hybrid optimization algorithm. Eng Appl Artif Intell 81:270–282
6. Pavithr RS, Gursaran (2016) Quantum inspired social evolution (QSE) algorithm for 0-1 knapsack problem. Swarm Evol Comput 29:33–46
7. Rao RV, Savsani VJ, Vakharia DP (2011) Teaching–learning-based optimization: a novel method for constrained mechanical design optimization problems. Comput Aided Des 43:303–315
8. Kennedy J, Eberhart R (1995) Particle swarm optimization. In: Proceedings IEEE international conference on neural networks, 1995, vol 1944, pp 1942–194
9. Das KN, Singh TK (2014) Drosophila Food-Search Optimization. Appl Math Comput 231:566–580
10. Mirjalili S, Mirjalili SM, Lewis A (2014) Grey Wolf Optimizer. Adv Eng Softw 69:46–61
11. Cheng M-Y, Prayogo D (2014) Symbiotic organisms search: a new metaheuristic optimization algorithm. Comput Struct 139:98–112
12. Rashedi E, Nezamabadi-pour H, Saryazdi S (2009) GSA: a gravitational search algorithm. Inf Sci 179:2232–2248
13. Baykasoğlu A, Akpinar Ş (2015) Weighted superposition attraction (WSA): a swarm intelligence algorithm for optimization problems – part 2: constrained optimization. Appl Soft Comput 37:396–415
14. Kirkpatrick S, Gelatt CD, Vecchi MP (1987) Optimization by simulated annealing. In: Fischler MA, Firschein O (eds) Readings in computer vision. Morgan Kaufmann, San Francisco (CA), pp 606–615
15. Holland JH (1984) Genetic algorithms and adaptation. In: Selfridge OG, Rissland EL, Arbib MA (eds) Adaptive control of ill-defined systems. Springer, US, Boston, MA, pp 317–333
16. Storn R, Price K (1997) Differential evolution—a simple and efficient heuristic for global optimization over continuous spaces. J Glob Optim 11:341–359
17. Gupta N, Khosravy M, Patel N, Sethi I (2018) Evolutionary optimization based on biological evolution in plants. Procedia Comput Sci 126:146–155
18. Mortazavi A, Toğan V, Nuhoğlu A (2018) Interactive search algorithm: A new hybrid metaheuristic optimization algorithm. Eng Appl Artif Intell 71:275–292
19. Arora S, Singh S (2019) Butterfly optimization algorithm: a novel approach for global optimization. Soft Comput 23:715–734
20. Yang X-S (2009) Firefly algorithms for multimodal optimization. In: Watanabe O, Zeugmann T (eds) Stochastic algorithms: foundations and applications. Springer, Berlin Heidelberg, Berlin, Heidelberg, pp 169–178
21. Eirgash Mohammad A, Togan V, Dede T (2019) A multi-objective decision making model based on TLBO for the time—cost trade-off problems. Struct Eng Mech 71:139–151

22. Toğan V, Eirgash MA (2019) Time-cost trade-off optimization of construction projects using teaching learning based optimization. KSCE J Civ Eng 23:10–20
23. Tran Duc H (2020) Optimizing time–cost in generalized construction projects using multiple-objective social group optimization and multi-criteria decision-making methods. Eng Constr Architectural Manage 27:2287–2313
24. Feng C-W, Liu L, Burns SA (1997) Using genetic algorithms to solve construction time-cost trade-off problems. J Comput Civ Eng 11:184–189
25. Hegazy T (1999) Optimization of construction time-cost trade-off analysis using genetic algorithms. Can J Civ Eng 26:685–697
26. Koo C, Hong T, Kim S (2015) An integrated multi-objective optimization model for solving the construction time-cost trade-off problem. J Civ Eng Manage 21:323–333

Chapter 7
The Design of Trapezoidal Corrugated Web Beams Using Firefly Method

Ferhat Erdal, Osman Tunca, Erkan Dogan, and Ramazan Ozcelik

7.1 Introduction

In steel I-section profile beams, the web element carries the compressive stress caused by the loads and transmits the shear forces within the I-section, while the upper part of flanges responds to the applied loads. Considering material saving and load-bearing capacities together, the principle of the delicate web against thick flanges is adopted for these beams. In this case, compressive stress on thin and flat bodies due to overloads causes loss of beam stability and large deformations.

The corrugated beams with trapezoidal and sinusoidal web forms can be designed to be joined with the upper and lower flanges which are made of the corrugated sheet with a thin cross section and which are steel plate with the same material properties as the web. The fabrication process has been started with rolling of thin web plates and then making corrugated steel webs ready for welding as shown in Fig. 7.1.

In the second process, welding of corrugated thin webs and upper and lower parts of plane flanges have been done (Fig. 7.2).

Finally, corrugated steel web beams have been subjected to sandblasting and antirust painting before using as displayed in Fig. 7.3.

F. Erdal (✉) · R. Ozcelik
Department of Civil Engineering, Akdeniz University, Antalya, Turkey
e-mail: eferhat@akdeniz.edu.tr

R. Ozcelik
e-mail: rozcelik@hotmail.com

O. Tunca
Department of Civil Engineering, Karamanoglu Mehmetbey University, Karaman, Turkey
e-mail: osmantunca@kmu.edu.tr

E. Dogan
Department of Civil Engineering, Celal Bayar University, Manisa, Turkey
e-mail: erkan.dogan@cbu.edu.tr

S. Carbas et al. (eds.), *Nature-Inspired Metaheuristic Algorithms for Engineering Optimization Applications*, Springer Tracts in Nature-Inspired Computing,
https://doi.org/10.1007/978-981-33-6773-9_7

Fig. 7.1 Rolling of corrugated web plates

Fig. 7.2 Rolling of corrugated web plates

Fig. 7.3 Sandblasting and antirust application for corrugated beams

The corrugated web beams prevent large deformations and loss of stability; they prevent the beam from losing stiffness before it reaches the plasc limit under load combinations applied; they prevent buckling and increase the load-bearing capacity

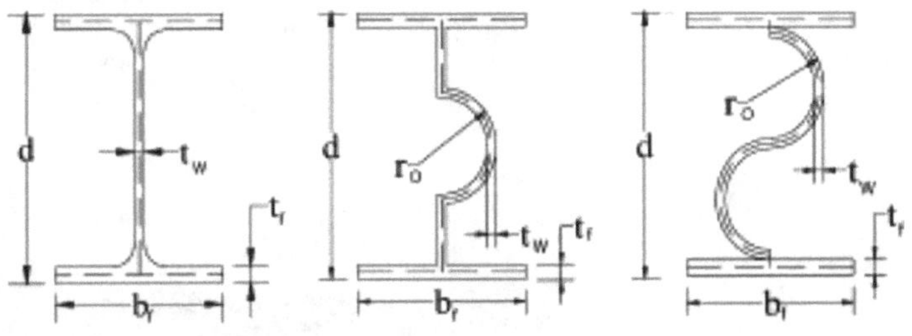

Fig. 7.4 Plane, one-arch and two-arch web beams

of the beam against local buckling by shear forces. These sinusoidal and trapezoidal structures of the beam web reduce the beam's own weight and increase its load-bearing capacity. These beams, which have been used effectively in advanced countries in recent times, may be used effectively in long spans with comparison to the steel frame systems or castella beams which are commonly preferred in floor beams or as columns under normal force effect.

Initial researches on these new-generation steel beams are intensive on trapezoidal troughs. Firstly, besides Sherman and Fisher, Elgaaly et al. examined the overall buckling analysis caused by shear and bending under different load conditions, respectively [1–3]. As a result of these studies, they suggested that the slender web can be ignored in bending analysis owing to the trivial effect on the bearing capacity of the beam. It was investigated by Johnson and Cafolla that collapse conditions due to shear and bending depend on the position of the loads and the trapezoidal structure of the corrugation [4]. In the experimental studies performed by Li et al., the trapezoidal web beams have been found to have at least 1.5 times more buckling strength compared to plane web beams [5]. The fact that buckling strength in thin web thicknesses of corrugated web beams is comparable to beams with flat and higher web thickness was investigated by Pasternak et al. [6]. Then, Khalid et al. have tested steel beams with longitudinal and vertical sinusoidal web as shown in Fig. 7.4 under single point loading [7].

In Dabon and Elamary's study, the effect of slenderness subject to buckling of trapezoidal web beams has been investigated with real tests as shown in Fig. 7.5 [8].

The shear stress effect of web openings on sinusoidal corrugated steel web beams is searched by Kıymaz et al. [9]. The have used Abaqus finite element modelling software for these analyses which is displayed in Fig. 7.6.

In 2009, Denan et al. have studied the moment of inertias on global axis of trapezoidal web beams [10]. The comparison of trapezoidal web beams and original I-section beams has been done (Fig. 7.7).

And then, Pasternak and Kubienec have studied (Fig. 7.8) the effect of welding on flanges for these beams under two-point loading [6].

Besides the bearing capacities of sinusoidal corrugated beams, the connection of the mentioned beams with adjacent elements has been searched by Pimenta et al. [11]. The detailed demonstration of connection and deformation values have been displayed in Fig. 7.9.

Fig. 7.5 Deformation of test beam under loading

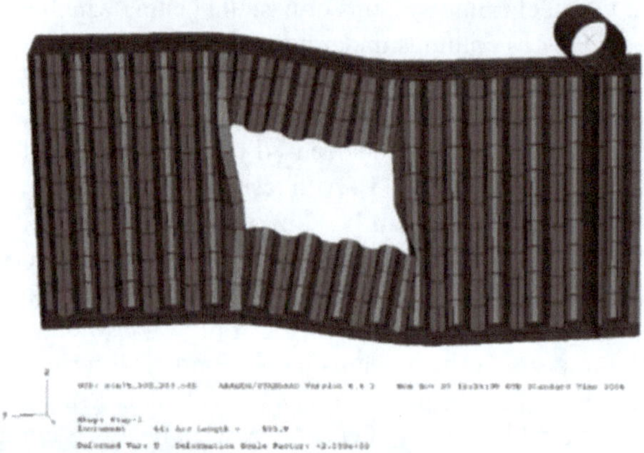

Fig. 7.6 Finite element model of sinusoidal beam with rectangular opening

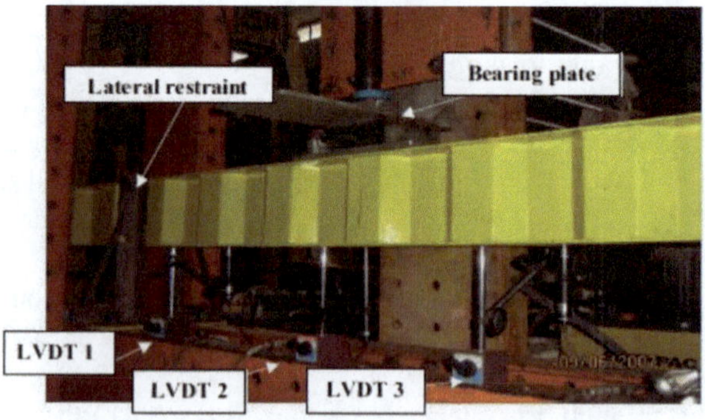

Fig. 7.7 Test set-up of trapezoidal web beams

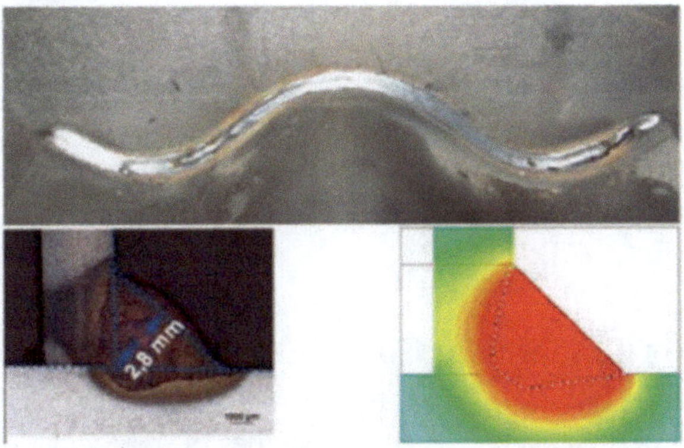

Fig. 7.8 Simulation of welding on sinusoidal web beam

Fig. 7.9 Deformation and appearance of connection

In the studies mentioned above, always the constant parameters related to the web structure of the beam have been used. During the design within this project, thickness and depth parameters of corrugated web structure and flange thicknesses will be evaluated in terms of cost and taken as design variables according to optimization principles.

In this chapter, the optimum dimensioning of the trapezoidal beams will be made by using new optimization methods such as to minimize its weight. Besides the height and thickness of corrugated web, the slope of corrugation, the distance of centres of trapezoidal webs, flange width and thickness in construction beams with corrugated web beams were also accepted as design variables. With these parameters, values that will minimize the beam weight under the behaviour and geometric limiters will be obtained. The solution of formulated optimization problems will be obtained by adaptation of firefly method, and the performance of the mentioned search algorithm in finding the minimum weight of the beam will be examined.

The trapezoidal corrugated web beams will be designed considering the design criteria specified in BS EN 1993-1-5, DIN 18-800 Part 1-2 and *DAST-Ri. 015* specifications [12–15]. With the results obtained, it is aimed to determine how to make the optimum design of corrugated web beams, for which collapse situations under load differentiate from standard I-section profiles and whose designs cannot be made by ordinary analysis methods.

7.2 Design of Trapezoidal Corrugated Web Beam

Figure 7.10 demonstrates the geometric properties of trapezoidal corrugated steel web beam, which is combined of a thin-walled corrugated web and flat plate for upper and lower flanges.

Design constraints include displacement limitations, buckling capacities of web and flanges, bearing and torsional–flexural buckling capacities of flanges.

7.2.1 *Yielding Capacity of Trapezoidal Web Beams*

On the basis of analytic studies, real tests and finite element models, the design processes have been recommended for yielding of trapezoidal webs. The shear stress reasoning an element of trapezoidal web to yield when it is subjected to pure shear stress can be described using von Mises strength of the steel beam:

$$\tau_Y = \frac{f_y}{\sqrt{3}} \tag{7.1}$$

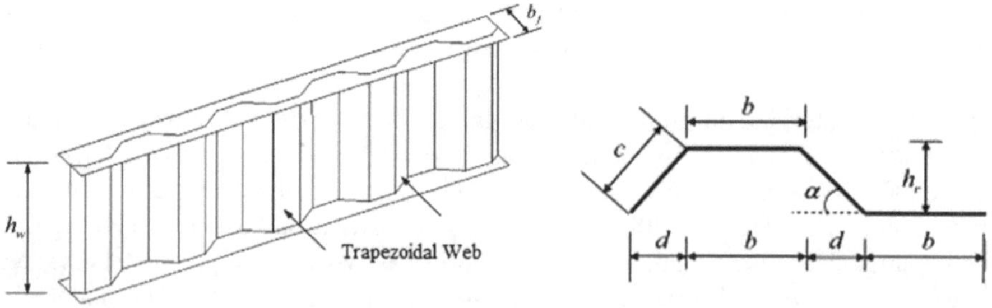

Fig. 7.10 Dimensions of trapezoidal web beam

7.2.2 Local Buckling Capacity of Flanges

As a result of plane flanges between vertical edges and a large width to thickness ratio, local buckling occurs as a failure mode. Considering the instability of a steel panel simply supported between two folds, corrugated web in this mode of failure acts as a series of flat plate sub-panels that are mutually supported by the flanges at their horizontal edges. These plane flanges are directly subjected to shear buckling risks.

$$\tau_{\mathrm{cr},L} = k_L \frac{\pi^2 E}{12(1-\nu^2)} \left(\frac{t_w}{w}\right)^2 \tag{7.2}$$

The shear stress coefficient is a function of the boundary restraints and the panel aspect ratio.

$$k_L = 5.34 + 2.31\left(\frac{w}{h_w}\right) - 3.44\left(\frac{w}{h_w}\right)^2 + 8.39\left(\frac{w}{h_w}\right)^3 \tag{7.3}$$

7.2.3 Global Buckling Capacity of Flanges

Global shear buckling is characterized by the formation of diagonal buckles through the entire web similarly to a flat plate web. Hence, the global elastic buckling stress can be calculated as

$$\tau_{\mathrm{cr},G} = 36\beta \frac{{D_y}^{0.25}{D_x}^{0.75}}{t_w {h_w}^2} \tag{7.4}$$

The flexural stiffness factors (D_x and D_y) per unit corrugation are defined for trapezoidal corrugation profile as follows:

$$D_x = \frac{E\, t_w^3\left[\left(\frac{h}{t_w}\right)^2 + 1\right]}{6\eta} \text{ and } D_y = \frac{E t_w^2}{12(1-\upsilon^2)}\eta \tag{7.5}$$

7.3 Firefly Optimization Method

Firefly optimization technique, firstly, searched the solution of design problems on the basis of idealized behaviour of flashing characteristics of mentioned insects [16–19]. There are three main rules for design algorithm of firefly method. The first rule is attracting one firefly to the other one. The second and third rules are flashing and brightness, respectively. Mathematical expressions of these three main instructions are given in [16] with more detail.

In the mathematical expressions, brightness is described as I, a particular location of a firefly is x, $I(\mathrm{x}) \propto f(\mathrm{x})$ and $f(\mathrm{x})$ is defined as the objective function. The attractiveness is designated as β for judging by other fireflies. Besides, r_{ij} is defined as distance between fireflies i and j. Finally, the attractiveness function is taken to be proportional to the light intensity by adjacent fireflies.

$$\beta(r) = \beta_0 \mathrm{e}^{(-\gamma r^m)}, \quad (m \geq 1) \tag{7.6}$$

where β_0 is the attractiveness at $r = 0$.

The distance between any two fireflies i and j at $x_i$and x_j is calculated as

$$r_{ij} = \|x_i - x_j\| = \sqrt{\sum_{k=1}^{d} \left(x_{i,k} - x_{j,k}\right)^2} \tag{7.7}$$

where $x_{i,k}$ is the kth component of spatial coordinate x_i of the ith firefly. The movement of a firefly i which is attracted to another more brighter firefly j is determined by

$$x_i = x_i + \beta_0 \mathrm{e}^{-\gamma r_{ij}^2}(x_j - x_i) + \alpha\left(rand - \frac{1}{2}\right) \tag{7.8}$$

where the second term is due to the attraction while the third term is randomization with α being the randomization parameter. "*rand*" is a random number generator uniformly distributed in [0, 1]. The values of parameters in the above equations are generally taken as $\beta_0 = 1$ and α ϵ 1 [0, 1]. Randomization term can be extended to a normal distribution $N(0,1)$ or other distributions. γ characterizes the variation of the attractiveness, and its value determines the speed of convergence and performance of the firefly algorithm. In applications, its value is taken between 0 and 100.

The pseudocode of FFO method is given in [16] which is also displayed in Fig. 7.11. On the basis of this pseudocode, benchmark design problems of FFO technique are also mentioned in [17]. In reference [18], firefly optimization technique is used to describe the design of six optimization problems. The performance of FFO method is compared with other stochastic search methods [18]. The objective function results obtained from FFO technique algorithm clearly demonstrate that FFO algorithm is more efficient than other four metaheuristic techniques. In [19], the permutation flow shop is formulated as a mixed integer programming problem which is classified as hard to solve nonlinear programming problem. FFO algorithm

```
begin;
  Objective function f(x), (x) = (x1, … , xd)
  Generate initial population of fireflies xi, (i =1, … , n)
  Light intensity Ii at xi is determined by f(xi)
  Define light absorption coefficient γ
while (until the termination criteria is satisfied)
      for i = 1 : n all n fireflies
      for j = 1 : i all n fireflies
if (Ij > Ii)
  Move firefly i towards j in d-dimension
end if
  Attractiveness varies with distance r via exp[−γr²]
Evaluate new solutions and update light intensity
end for j
end for i
  Rank the fireflies and find the current best
end while
  Post-process results and visualization
end procedure;
```

Fig. 7.11 Pseudocode of firefly algorithm

is applied to find the optimum result of this design problem. It is concluded that the preliminary results indicated that FFO performs better than the ant colony algorithm on benchmark problems taken from the literature.

7.4 Benchmark Minimization Design Example

Firstly, a well-known benchmark problem is selected for the application of firefly search algorithm to mathematical optimization problems. In this design problem, the objective function (see Eq. 7.9) has two design variables and six design and side constraints (Eqs. 7.9–7.13). The optimization process is related to reaching the best outcome of a given benchmark problem while satisfying design and side constraints.

$$\min f(x) = -\left(x_2 - 1.275{x_1}^2 + 5x_1 - 6\right)^2 - 10\left(1 - \frac{1}{8\pi}\right)\cos(\pi x_1) - 10 \quad (7.9)$$

$$g_1(x) = -\pi x_1 - x_2 \le 0 \quad (7.10)$$

$$g_2(x) = -\pi^2 {x_1}^2 + 4x_2 \le 0 \quad (7.11)$$

$$-1.5 \le x_1 \le 3.5 \quad (7.12)$$

$$0 \leq x_2 \leq 15 \tag{7.13}$$

The benchmark problem is also solved by six different algorithm methods including wolf colony algorithm [20], particle swarm optimizer [21], bat algorithm [22], ant colony optimizer [23], glow-worm algorithm [24] and artificial bee colony algorithm [25]. The optimum designs of the problem obtained by seven metaheuristic techniques are given in Table 7.1, and the design histories of each algorithm are displayed in Fig. 7.12. The optimum solution belongs to the firefly algorithm which is −195.32 value.

In an effort to compare the optimum value of FFO algorithm with those other search techniques, four different methods yield the same maximum value for the same

Table 7.1 The minimum objective function results for search techniques

	PSO	ACO	ABC	BAT	FFO	GLS	WLF
x_1	2.467	2.876	2.467	2.466	2.467	2.467	2.467
x_2	15.000	6.061	15.000	14.985	14.999	15.000	15.000
$g_1(x)$	−22.746	−22.749	−22.746	−22.727	−22.746	−22.746	−22.746
$g_2(x)$	0.000	−0.227	0.000	−0.007	−0.0006	0.000	0.000
$f(x)$	−195.33	−194.84	−195.33	−194.98	−195.32	−195.33	−195.33

PSO particle swarm optimization; ACO ant colony optimization; ABC artificial bee colony; BAT bat algorithm; GLS glow-worm optimization; WLF wolf colony optimization algorithms

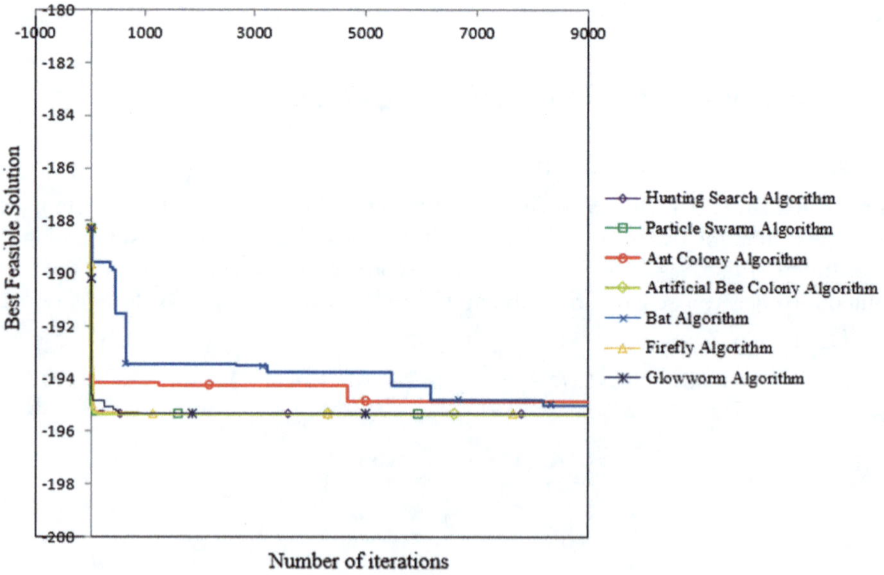

Fig. 7.12 Design history curve for seven search algorithms

benchmark problem: 195.33 [wolf colony algorithm (WLF), particle swarm optimizer (PSO), artificial bee colony algorithm (ABC), glow-worm algorithm (GLS)]. On the other hand, ant colony algorithm (ACO) finds −194.84, and bat optimization algorithm finds −194.98 value, respectively.

7.5 Benchmark Maximization Design Example

Secondly, a different benchmark problem is selected for the application of firefly search algorithm to mathematical optimization problems. In this maximization design problem, the objective function (see Eq. 7.14) has two design variables and six design and side constraints (Eqs. 7.15–7.18).

$$\max f(x) = \left[\sin^3(2\pi x_1)\sin(2\pi x_2)\right]/\left[{x_1}^3(x_1 + x_2)\right] \tag{7.14}$$

$$g_1(x) = {x_1}^2 - x_2 + 1 \leq 0 \tag{7.15}$$

$$g_2(x) = 1 - x_1 + (x_2 - 4)^2 \leq 0 \tag{7.16}$$

$$0 \leq x_1 \leq 10 \tag{7.17}$$

$$0 \leq x_2 \leq 10 \tag{7.18}$$

Besides firefly algorithm, the maximization benchmark problem is also solved by six different algorithm methods including wolf colony algorithm (WLF), particle swarm optimizer (PSO), bat algorithm (BAT), ant colony optimizer (ACO), glow-worm algorithm (GLS) and artificial bee colony algorithm (ABC). The optimum designs of the problem obtained by seven metaheuristic techniques are given in Table 7.2, and the design histories of each algorithm are displayed in Fig. 7.13. The optimum solution belongs to the firefly algorithm which is –0.0956 value.

Table 7.2 The maximum objective function results for seven search techniques

	PSO	ACO	ABC	BAT	FFO	GLS	WLF
x_1	1.2286	1.4701	1.2286	1.2287	1.2286	1.1919	1.2286
x_2	4.2475	4.3319	4.2475	4.2474	4.2474	4.3797	4.2475
$g_1(x)$	−1.7380	−1.739	−1.7381	−1.7376	−1.7379	−1.9591	−1.738
$g_2(x)$	−0.1673	−0.166	−0.1673	−0.1674	−0.0167	−0.0476	−0.1674
$f(x)$	0.0956	0.0955	0.0956	0.0956	0.0956	0.0598	0.0956

PSO particle swarm, ACO ant colony, ABC artificial bee, BAT bat algorithm, GLS glowworm, WLF wolf colony algorithms

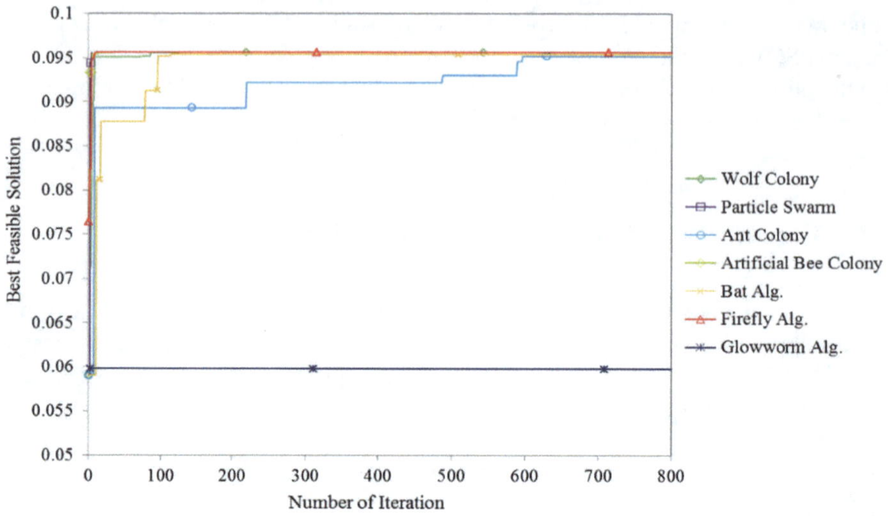

Fig. 7.13 Design history curve for maximization problem with search techniques

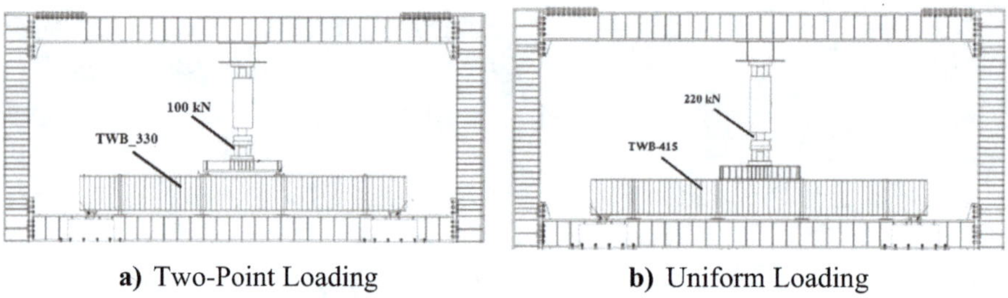

Fig. 7.14 Demonstration of loads on trapezoidal corrugated web beam

In an effort to compare the optimum value of FFO method with those other search techniques, five different methods yield the same maximum value for the same benchmark problem: 0.0956 [wolf colony algorithm (WLF), particle swarm optimizer (PSO), bat algorithm (BAT), ant colony optimizer (ACO) and artificial bee colony algorithm (ABC)] and glow-worm algorithm (GLS) finds 0.0598 value.

7.6 Design of Corrugated Beam

The design of a corrugated web beam (CWB) requires the selection of width and thickness of a plate from which the corrugated web is to be produced, distance between the peak points of each corrugate, the length of corrugated web, the selection of width and thickness of a plate for upper and lower flanges in the beam. For this purpose, a design pool is prepared. The optimum design problem formulated considering the design constraints explained in the previous sections yields the following mathematical model [6–9]. Find a integer design vector $\{I\} = \{I_1, I_2, I_3, I_{4,}, I_5\}^{\mathrm{T}}$ where I_1 is the sequence number for the width of upper and lower flanges, I_2 is the sequence number for the thickness values of upper and lower flanges, I_3 is the thickness of corrugated web, I_4 is distance between the peak points of each corrugated web and I_5 is the height of trapezoidal corrugate. Hence, the design problem turns out to minimize the weight of CWB.

$$W_{\mathrm{SOB}} = \rho_s\big((2 \times b_f \times t_f \times L) + (h \times t_w \times L_{d\ddot{u}z})\big) \tag{7.19}$$

where ρ_s is density of steel, b_f the width of flange, t_f thickness of flange, L span of beam, h height of corrugated web, t_w thickness of corrugated web and $L_{d\ddot{u}z}$ span of beam before corrugation process. Design of a corrugated beam requires the satisfaction of some geometrical restrictions that are formulated through Eqs. (7.20–7.27).

Web dimensions:

$$333\mathrm{mm} \le h \tag{7.20}$$

$$h \le 1500\mathrm{mm} \tag{7.21}$$

$$1.5\mathrm{mm} \le t_w \tag{7.22}$$

$$t_w \le 5\mathrm{mm} \tag{7.23}$$

Flange dimensions:

$$120\mathrm{mm} \le b_f \tag{7.24}$$

$$b_f \le 450\mathrm{mm} \tag{7.25}$$

$$t_f \le 30\mathrm{mm} \tag{7.26}$$

$$6.0\mathrm{mm} \le t_f \tag{7.27}$$

7.7 Optimum Design Problem of Trapezoidal Web Beam

In this problem, optimum design of laterally supported steel trapezoidal corrugated webs is realized. The span of trapezoidal web beam is 5 m as shown in Fig. 7.14. Trapezoidal steel web beam is subjected to a 100 kN two-point loading and 220 kN partially distributed loading, respectively. The upper flange of the beam is laterally supported by the floor system that it supports. The values of maximum displacement and modulus of elasticity are taken as 17 mm and 205 kN/mm^2, respectively.

The firefly optimization algorithm (FFO) is used for solving the mentioned design problem. The size of the initial population and the maximum number of iterations in FFO are taken as 40 and 10,000, respectively. Firefly algorithm parameters are taken as $\beta_0 = 0.25$, $\beta = 1$, $\alpha = 0.01$, $\gamma = 1$, respectively. The optimum design results of trapezoidal corrugated web beams for the FFO parameters are given in Table 7.3.

The optimum shape of the TWB-330 and TWB-415 corrugated web beams for the loading considered is shown in Figs. 7.15 and 7.16, respectively.

Firefly optimization algorithm finds the least weight for trapezoidal corrugated web beam which is equal to 136.35 kg under two-point loading. The maximum value of the strength ratio is 0.95 which is almost upper bound. This reveals the fact that the strength constraints are dominant in the problem. FFO method mentioned

Table 7.3 Cardiovascular parameters

Optimum section	t_w (mm)	h (mm)	t_f (mm)	H_c (mm)	Minimum weight (kg)
TCW-330	2.0	330	160	43	136.35
TCW-415	2.5	415	200	43	271.89

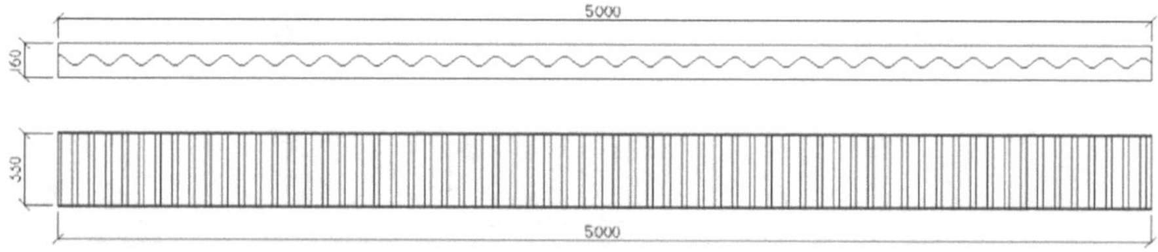

Fig. 7.15 Optimum geometry 5-m span TWB-330 corrugated web beam

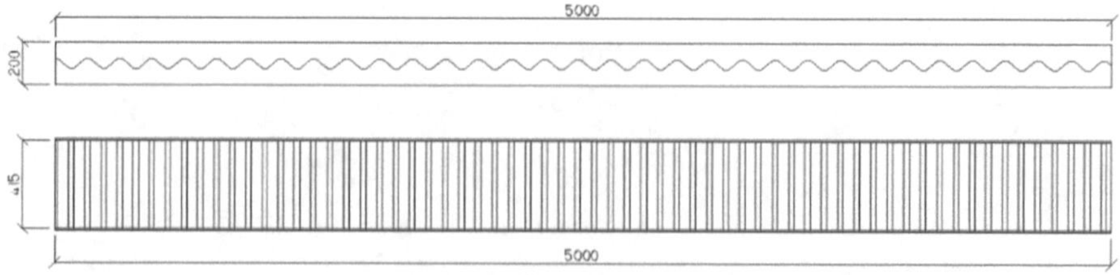

Fig. 7.16 Optimum geometry 5-m span TWB-415 corrugated web beam

chooses TWB-330 for the root beam. The optimum trapezoidal web beam should be produced 2 mm as thickness of corrugation, 330 mm as the depth of web, 8 mm as thickness of upper and lower flanges and 160 mm as the width of upper and lower flanges. FFO produces 271.89 kg weight for under partially distributed loading, too. It is apparent from Table 7.1 that FFO method performs 2.5 mm as thickness of corrugation, 415 mm as the depth of web, 10 mm as thickness of upper and lower flanges and 200 mm as the width of upper and lower flanges, in this design problem.

7.8 Conclusions

In this chapter, a firefly algorithm is successfully applied to optimum design of trapezoidal corrugated web beam problems to demonstrate the effect of the proposed algorithms. FFO method is also used to find minimizing and maximizing of two benchmark problems under design constraints. The benchmark problem results clearly show that the algorithm of FFO is mathematically simple but very effective in finding the solutions of benchmark problems. Fly-back mechanism is employed for handling the problem constraints, and feasible ones being candidate solutions to give the minimum value are determined. Trapezoidal corrugated web beam examples are also designed to illustrate the efficiency of the proposed algorithm. Optimum designs attained by method clearly show that FFO technique is an efficient and robust technique that can successfully be used in optimum design of steel trapezoidal corrugated web beams and design of other structures.

Acknowledgements This chapter is partially based on research supported by the Akdeniz University Research Funding (*FBA-2018-3668*), which is gratefully acknowledged.

References

1. Elgaaly M, Hamilton RW, Seshadri A (1996) Shear strength of beams with corrugated webs. J Struct Eng (ASCE) 122(4):390–398
2. Elgaaly M, Hamilton RW, Seshadri A (1997) Bending strength of steel beams with corrugated webs. J Struct Eng (ASCE) 123(6):772–782
3. Elgaaly M, Seshadri A (1998) Depicting the behavior of girders with corrugated web sup to failure using non-linear finite element analysis. Adv Eng Softw 29(3–6):195–208
4. Johnson RP, Cafolla J (1998) Local flange buckling in plate girders with corrugated webs. Struct Build (ICE) 123:148–156
5. Li Y, Zhang W, Zhou Q, Qi X, Widera GEO (2000) Development and research on H-beams with wholly corrugated webs. J Mater Process Technol 101(1–3):115–118
6. Pasternak H, Kubieniec G (2010) Plate girders with corrugated webs. J Civ Eng Manage 16(2):166–171
7. Khalid YA, Chan CL, Shari BB, Hamouda AMS (2004) Bending behavior of corrugated web beams. J Mater Process Technol 150(3):242–254
8. Dabon M, Elamary AS (2006) Flange compactness effect on the behavior of steel beams with corrugated webs. J Eng Sci 34(5):1507–1523

9. Kıymaz G, Coskun E, Cosgun C, Seckin E (2010) Transverse load carrying capacity of sinusoidally corrugated steel web beams with web openings. Steel Compos Struct 10(1):69–85
10. Denan F, Saad S, Osman MHB (2009) Determination of the second moment of area of trapezoid web profile steel section. In: The conference of Meniti Pembangunan Lestari dalam Kejuruteraan Awam
11. Pimenta RJ, Queiroz G, Diniz SMC (2015) Reliability-based design recommendations for sinusoidal-web beams subjected to lateral-torsional buckling. Eng Struct 84:195–206
12. British Standards, BS 5950 (2000) Structural use of steelworks in building, Part 1. Code of practice for design in simple and continuous construction, hot rolled sections, British Standard Institution, London, U.K.
13. DAST—Richtlinie 015 (1990) Träger mit schlanken Stegen (German recomendations for girders with slender web plates)
14. DIN V ENV 1993-1--, EUROCODE 3: Design of Steel Structures; Part 1—General Rules and rules for buildings
15. DIN 18 800 Teil1-3, Stahbauten; Bemessung und Konstruktion
16. Yang XS (2008) Nature-inspired metaheuristic algorithms. Luniver Press, UK
17. Yang XS (2009) Firefly algorithms for multimodal optimization. In: Watanabe O, Zeugmann T (eds) Stochastic algorithms: foundations and applications, SAGA 2009, Lecture notes in computer science, 5792, Springer, Berlin, pp 169–178
18. Yang XS (2010) Firefly algorithm, stochastic test functions and design optimization. Int J Bio-Inspired Comput 2(2):78–84
19. Yang XS, He X (2013) Firefly algorithm: recent advances and applications. Int J Swarm Intell Inderscience Publishers, Artificial Intelligence and Intelligent Systems, vol 1, pp 36–50. Science, Engineering and Technology
20. Liu CY, Yan XH, Wu H (2011) The wolf colony algorithm and its application. Chin J Electron 20(2):212–216
21. Perez RE, Behdinan K (2007) Particle swarm approach for structural design optimization. Comput Struct 85(19–20):1579–1588
22. Yang XS, Gandomi AH (2012) A novel approach for global engineering optimization. Eng Comput 29(5):464–483
23. Dorigo M, Christian B (2005) Ant colony optimization theory: a survey. Theoret Comput Sci 344:243–278
24. Marinaki M, Marinakis Y (2016) A glowworm optmization algorithm for the vehicle routing problem with stochastic demands. Expert Syst Appl 46:145–163
25. Karaboga D, Basturk B (2008) On the performance of artificial bee colony (ABC) algorithm. Appl Soft Comput 8:687–697

Chapter 8
Designing Fuzzy Controllers for Frame Structures Based on Ground Motion Prediction Using Grasshopper Optimization Algorithm: A Case Study of Tabriz, Iran

Mahdi Azizi

8.1 Introduction

In the last few years, there has been an increasing concern in seismic hazard analysis that results in some realistic models for describing seismic sources, seismic attenuation, earthquake recurrence frequency, and probability of earthquake occurrence at a site. In most cases, this analysis is conducted based on the seismological records of past earthquakes. Probabilistic seismic hazard analysis (PSHA) is an analytical approach that estimates the probability of exceeding various levels of ground motion at a site (or a map of sites) for all possible earthquakes. In this approach, based on the behavior of the earth, some random components such as source-to-site distance (R), magnitude (M), and the deviation measure of the strong ground motion from a predicted or median value (epsilon) are utilized. The results of the PSHA are presented as estimated probabilities and frequencies per unit time or as expected number of events per year. In order to make the results of a PSHA more useful, a deaggregation process should be conducted. In this process, the exceedance probability of earthquakes based on the contributions of different magnitude–distance pairs is represented as a guide for selection of response spectra or acceleration time histories for dynamic analysis.

The PSHA has developed over decades, and the empirical statistical approaches have been largely renewed by the analytical and numerical attempts. Cornell [1] conducted the first analytical method for PSHA, and many researchers improved it subsequently. The most comprehensive research in this field was presented by the Senior Seismic Hazard Analysis Committee (SSHAC) report [2], which covers many methodological issues. Many PSHA researches have been conducted around

M. Azizi (✉)
Department of Civil Engineering, University of Tabriz, Tabriz, Iran
e-mail: mehdi.azizi875@gmail.com

S. Carbas et al. (eds.), *Nature-Inspired Metaheuristic Algorithms for Engineering Optimization Applications*, Springer Tracts in Nature-Inspired Computing,
https://doi.org/10.1007/978-981-33-6773-9_8

the world especially for Iran [3–6], while some of them discussed the city of Tabriz [7–10] which is located in northwest (NW) of Iran. The NW of Iran is a region of intense seismicity and deformation situated between two thrust belts of the Caucasus and the Zagros Mountains from north to south. The north Tabriz fault, which is one of the well-known seismogenic faults in this region, is passing within striking distance of urban area and has a major history of intense seismic activity [11].

Ground motion prediction model (GMPM), which is also renown as ground motion prediction equation (GMPE), attenuation relation, ground motion model or ground motion relation, estimates the level of ground shaking and its related uncertainty regarding any given location or site, based on some parameters such as source-to-site distance, earthquake magnitude, fault mechanism, and soil conditions. The results of GMPM are utilized in a wide range such as seismic analysis and design of structures for a specific site, development of seismic hazard maps, and seismic loss estimation. As discussed by Douglas [12], some of the most common intensity measures (IMs) used in GMPMs are peak ground motion values such as peak ground acceleration (PGA), peak ground velocity (PGV), and peak ground displacement (PGD) [13] alongside the elastic response spectral ordinations for acceleration, velocity, and displacement [14]. In addition, some other challenges in considering IMs are conducted based on the Arias intensity (AI) [15], cumulative absolute velocity (CAV) [16], Fourier spectral amplitude (FSA) [17], maximum absolute unit elastic input energy (IE) [18], inelastic response spectral ordinate (ISO) [19], Japanese Meteorological Agency (JMA) seismic intensity [20], macro-seismic intensity (MI) [21], mean period (MP) [22], relative significant duration (RSD) [23], and vertical-to-horizontal (VH) response spectral ratio [24].

One of the possible options in designing engineering structures is to install a smart system to provide the stiffness or damping required to withstand severe dynamic loads. In recent years, using control systems to reduce the response of structures under different types of dynamic loads such as earthquake, wind, and shock loads is a worldwide concern. Active control systems are a kind of intelligent systems that use an external energy source to generate control power for structural response control. Determination of control force in active control system is dealt with a proper control algorithm. Fuzzy logic controller (FLC) is one of the well-known control algorithms which has been of great attention in recent years due to its capability in vibration control of complex systems. Most of the times, the FLCs are formulated based on human knowledge which is not capable of providing acceptable control strategy in dealing with complicated design problems so optimal tuning of these control systems is necessary [25].

Optimum design of FLCs has been of great importance in recent years. Fu et al. [26] utilized the well-known genetic algorithm (GA) for optimization of a nonlinear fuzzy controller to use in the magnetorheological elastomer isolators. Soltani et al. [27] utilized particle swarm optimization (PSO) to design a fuzzy sliding mode controller by means of parallel-distributed compensator to intelligently deal with the problem of inappropriate selection in parameters of sliding surface. Chen et al. [28] discussed parameter evaluation of a sliding mode fuzzy controller utilized in hydraulic systems of turbine regulators with imperialist competitive algorithm (ICA).

Olivas et al. [29] generalized the type-2 fuzzy logic control algorithm for dynamically parameter evaluation with ant colony optimization (ACO) algorithm in order to optimize the membership functions and the rule base. Amador-Angulo and Castillo [30] utilized bee colony optimization (BCO) method for optimal distribution of the membership functions for FLCs implemented in complex type of nonlinear plants. Chamorro et al. [31] proposed the application of differential evolution (DE) algorithm for optimal tuning of a fuzzy controller in order to improve the synthetic inertia control in power systems. Ray et al. [32] discussed the parameter tuning of fuzzy proportional–integral–derivative (PID)-based power system stabilizer under different operating conditions by firefly algorithm (FA). Chrouta et al. [33] presented optimal control of an irrigation station process based on the FLC with Takagi–Sugeno (TS) model using cuckoo search algorithm (COA). Giri and Bera [34] developed a fuzzy proportional–integral (PI) controller for damping the oscillation of system frequency of distributed energy generation in wind turbines and optimized the gains of this controller with gray wolf optimization (GWO) algorithm. Chao et al. [35] utilized the harmony search (HS) algorithm to find optimal network parameters for better performances of fuzzy cerebellar model articulation controller. Li et al. [36] designed a fuzzy controller of type PID for optimal design of hydraulic nonlinear turbines by means of the gravitational search algorithm (GSA). Among these researches, some comparative studies based on utilizing different metaheuristic algorithms have been widely studied [37–45].

Most of the times, the optimization of FLCs is considered by utilization of linear system problems including structural systems with linear behavior [46]. Besides, the utilized mathematical models for control and structural systems are not accurate models and the important aspects of optimization problem including the objective functions are not formulated properly [47]. The main contribution of this chapter is optimization of FLCs implemented in structural systems by delicate utilization of grasshopper optimization algorithm (GOA). The GOA, recently proposed by Saremi et al. [48], is a novel metaheuristic algorithm that simulates the idealized and fully specified swarming behaviors of grasshopper insects (species) in nature. This algorithm has very fast convergence rate with a special adaptive mechanism that strikes a balance between the exploration and exploitation phases. It could be noted that this algorithm is capable of finding the global solution and is simple in implementation. In this chapter, the GOA is utilized for optimum design of FLCs which are implemented in a steel building structure with 20 stories in which the mathematical model of the structural systems is a kind of precise model and the nonlinear behavior of the structure is also considered in this model. The seismic inputs are considered based on the PSHA information from the Moghaddam et al. [10] research for Tabriz city and the GMPM based on the Ambraseys et al. [49] research for Middle East and Europe. The earthquake records in optimization problem are selected based on the energy-based ground motion selection and modification (EBGMSM) method proposed by Marasco and Cimellaro [50]. The EBGMSM is an efficient method with lower variability in predefined parameters and has accuracy in preserving the demanded median requirements for the specific hazard scenarios. The selected earthquake records based on the mentioned method are utilized as seismic inputs for the

20-story structure, and the capability of the GOA is examined with respect to the destructive effects of these ground motions.

8.2 Ground Motion Prediction

The north Tabriz fault is an active fault with lateral strike-slip displacement and a vertical displacement on the north side. This fault has an average strike over a length of about 150 km similar to the vertical direction in the dip [51]. Some of the most destructive historical earthquakes generated by this fault are the 1042, 1721, 1780, and 1786 earthquakes with magnitudes of 7.6, 7.7, 7.7, and 6.3, respectively [52]. A summary of historical and instrumental earthquakes that have occurred in Tabriz is displayed in Fig. 8.1.

As discussed by Masson et al. [53], based on the seismic events during the 700 years, the recurrent time interval for Tabriz is about 250 years. According to the paleoseismological studies over the past 3600 years, Hessami et al. [11] suggested 821 ± 176 years for this parameter. Regarding the absence of ground motion seismic

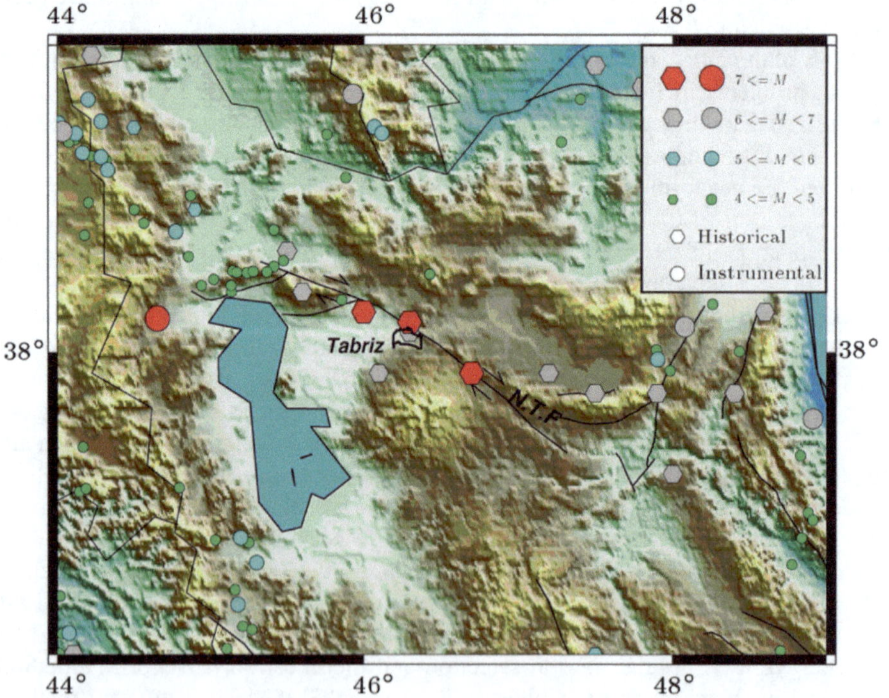

Fig. 8.1 Historical and instrumental earthquakes around north Tabriz fault [10]

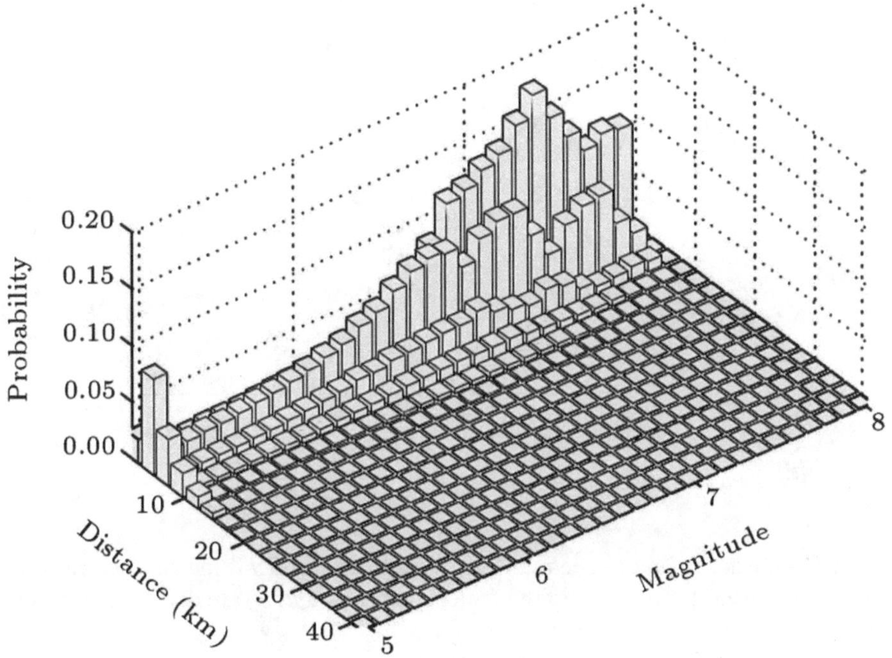

Fig. 8.2 Tabriz deaggregation plot for 10% exceedance probability in 50 years [10]

records in this area, seismic hazard analysis and ground motion prediction are necessary. Moghaddam et al. [10] studied the PSHA and the seismic hazard deaggregation for Tabriz. Based on a maximum likelihood method, they estimated the earthquake magnitudes and source-to-site distances for a 10% exceedance probability within 50 years (a returning period of 475 years). Deaggregation plot for this study is displayed in Fig. 8.2. It should be noted that the mean magnitude and distance for a response period of one second are calculated as 7.4 and 4.39 km, respectively.

As mentioned before, Ambraseys et al. [49] conducted a study by means of surface wave magnitudes (Ms), source-to-site distances, and site geology to find GMPEs for prediction of absolute spectral acceleration ordinates in Middle East and Europe. They used 422 triaxial records generated from 157 earthquakes with magnitude from 4.0 to 7.9, source-to-site distances of up to 200 km, and focal depth less than 30 km. The final GMPE for mentioned regions is as follows:

$$\log(y) = -1.48 + 0.266 \times M - 0.922 \times \log(r) + 0.117 \times S_A + 0.124 \times S_S \tag{8.1}$$

$$r = \sqrt{d^2 + 3.5^2} \tag{8.2}$$

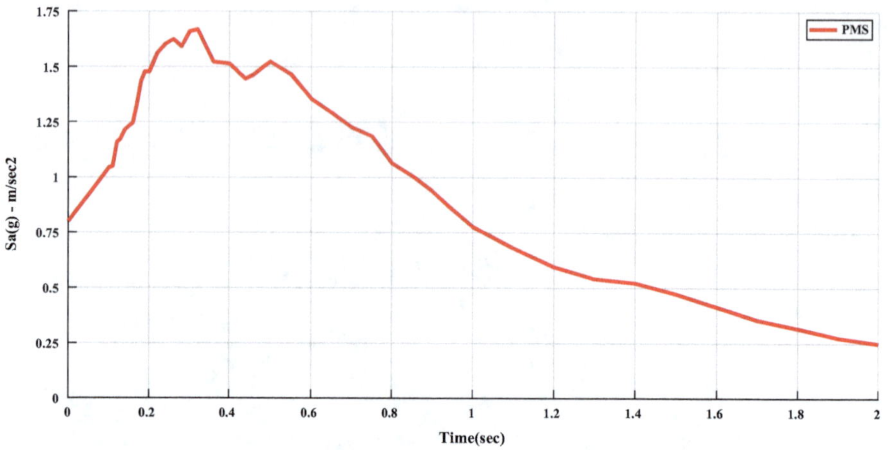

Fig. 8.3 Predicted mean spectrum for Tabriz

where y is the acceleration spectra, M is the magnitude and d is the distance between site and the surface projection of the fault rupture. S_A and S_S are defined in terms of the shear wave velocity (V_S) at the site.

Based on the PSHA for Tabriz [10], M, d, and V_S are calculated as 7.4, 4.39 km, and 3.2 km/s, respectively. The predicted mean spectrum (PMS) for these values based on the Ambraseys et al. [49] GMPE is depicted in Fig. 8.3.

In order to perform a dynamic analysis, selection and modification of earthquake records are concerned. In this study, the earthquake records in optimization problem are selected based on the EBGMSM method. This method comes out of comparing a predefined set of horizontal seismic ground motions at multiple (different) frequency ranges with a target spectrum. As mentioned by Marasco and Cimellaro [50], the predicted mean spectrum (PMS), conditional mean spectrum (CMS), design spectrum (DS), and uniform hazard spectrum (UHS) are some of the options for considering as target spectrum and the records with compelling contribution to the target spectrum are selected. In this method, a set of seismic ground motions with close seismic activity that matches the predetermined target spectrum regarding the specified period of interest (T_{ref}) are considered. Two of the most common ground motion databases such as the Pacific Earthquake Engineering Research (PEER) Center and the European Strong-Motion Data (ESMD) are considered for data selection. The PEER database includes 600 shallow crustal events with 21,336 three-component records that has a magnitude range between 3 and 7.9, range of 0.05–1533 km for rupture distance, and the shear wave velocity (V_S) at the site (in the top 30 m) range of 94–2100 m/s. The ESMD database consists of 462 triaxial ground motion seismic records regarding 110 earthquakes recorded at 261 individual stations in Middle East and Europe. The scaled mean spectrum of the ground motions is required to have an equivalent Housner intensity (HI) for the periods of $0.2T_{ref}$ to $2T_{ref}$. By utilizing a specific index in terms of the energy–frequency trend's shape and its scattering degree nearby the mean value, the horizontal components for each band of frequency are obtained which results in a spectrum-compatible record sets considering approximately low dispersion for the structural response parameters and also identical severity [50].

By performing the EBGMSM method with PMS as the target spectrum and utilizing the OPENSIGNAL as a computer-based software for signal processing [54, 55], the specifications and principles of the selected earthquake ground motion records are presented in Table 8.1.

The considered earthquake ground motion records have a magnitude (M) range from 5.8 to 7.62 with a source-to-site distance (R) range from 0.27 to 7.05 km to the fault. In Fig. 8.4, the acceleration and mean spectra for the selected earthquakes are compared with the target spectrum (PMS). It can be concluded that the mean spectrum and the PMS are compatible for a wide range of periods.

The north–south (N-S) components of the considered earthquake ground motion records are utilized for dynamic structural analysis. In Fig. 8.5, the acceleration time histories of the N-S components are presented.

In Fig. 8.6, the acceleration spectrum alongside the velocity and also the displacement spectra relating the N-S earthquake components based on the considered earthquake ground motion records are depicted for damping of 5% while the mean spectra are also represented.

8.3 Fuzzy Logic Controller

In recent years, qualitative and quantitative information has been of great importance in formulation of intelligent systems. The theory of fuzzy systems and sets represents the implementation of human-related approximate deductions and verbal expressions for specific purposes. The FLC is some kind of control system in which the requirement for a specific mathematical model of the system problem is not concerned. Most of the time, the FLC can be implemented to a closed-loop control system (Fig. 8.7).

The general aspects of an FLC are "fuzzification" which represents the modification of the crisp inputs to fuzzy inputs, "rule base" which concerns the required IF-THEN rules, "fuzzy inference mechanism" which calculates the fuzzy outputs based on the rules, and "defuzzification" which provides the crisp outputs.

8.4 Grasshopper Optimization Algorithm (GOA)

Grasshoppers are some kind of insects that make damage to agriculture and also crop productions. These creatures join in the large swarms with continental scale that is a nightmare for farmers. This swarming behavior is seen in both of the adulthood

Table 8.1 Characteristics of the considered earthquake ground motion records

Abbr.	Earthquake date	*M*	*R*	Fault mechanism	Station	Component	Scale factor	PGA (g)	PGV (cm/s)
Equation 1	San Salvador, 1986	5.8	6.3	Strike slip	Geot. Inves. Center	GIC090	1.10	0.9668	65.51
						GIC180	1.81	0.8609	87.61
Equation 2	Northridge (1), 1994	6.69	5.43	Reverse	Jensen Fil. Pl. Ad. Buil.	JEN022	1.39	0.7932	105.66
						JEN292	0.60	0.6158	40.17
Equation 3	Imperial Valley, 1979	6.53	7.05	Strike slip	El Centro Array #4	E04140	1.44	0.6976	53.73
						E04230	1.58	0.5685	120.84
Equation 4	Kobe, 1995	6.9	0.27	Strike slip	Takarazuka	TAZ000	0.87	0.6008	59.16
						TAZ090	0.95	0.6588	80.99
Equation 5	Chi-Chi, 1999	7.62	5.8	Reverse oblique	TCU071	TCU071E	1.05	0.5957	46.72
						TCU071N	1.02	0.6710	71.09
Equation 6	Northridge (2), 1994	6.69	5.3	Reverse	Sylmar—Ol. Vi. Med FF	SYL090	1.26	0.7641	98.69
						SYL360	0.90	0.7577	116.19
Equation 7	Tabas, 1978	7.35	2.05	Reverse	TABAS	TABL1	1.09	0.9122	106.65
						TABT1	1.17	0.9998	142.22

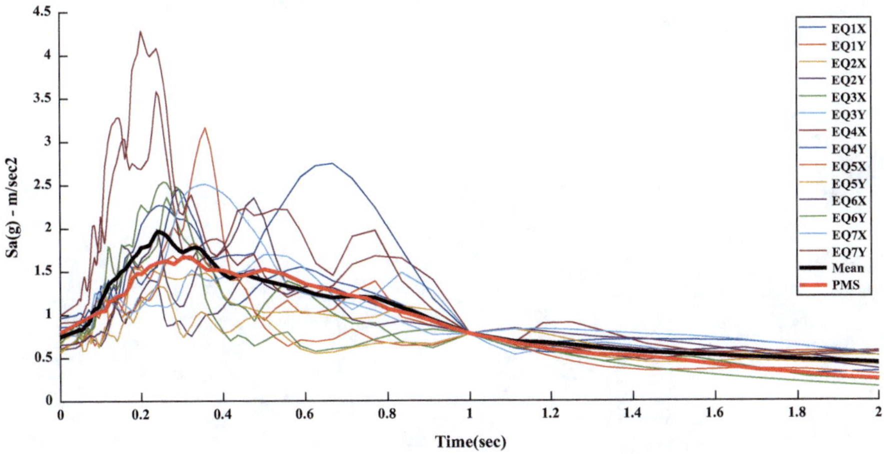

Fig. 8.4 Comparison of acceleration and mean spectra of selected earthquakes with the PMS

and nymph. The nymph grasshoppers move by rolling in the ground and eat all vegetation, while the adult ones form the swarm in the air and migrate over large distances. In the larva phase, the grasshoppers move slowly with small steps while in the adulthood, abrupt and long-range movements are seen [48]. In Fig. 8.8, a real grasshopper and the life cycle of grasshoppers are displayed.

Most of the nature-inspired optimization algorithms have two main phases such as exploitation and exploration. In this regard, search agents are encouraged to move or slide abruptly in the exploration phase, while in the exploitation phase these agents are motivated to move or slide locally. Grasshoppers perform the mentioned two functions alongside the target seeking, naturally, so the mathematical model of the optimization algorithm is relied on the characteristics of the grasshoppers. The position of the search agents or grasshoppers in the middle of the swarm demonstrates a feasible solution for the optimization problem. The ith grasshopper's position is presented as follows [48]:

$$X_i = S_i + G_i + A_i \tag{8.3}$$

where S_i represents the social interaction between grasshoppers, A_i demonstrates the wind advection, and G_i presents the gravity force related to the ith grasshopper. The mathematical presentation of the social interaction is provided as follows [48]:

$$S_i = \sum_{\substack{j = 1 \\ j \neq i}}^{N} s\left(d_{ij}\right)\widehat{d_{ij}} \tag{8.4}$$

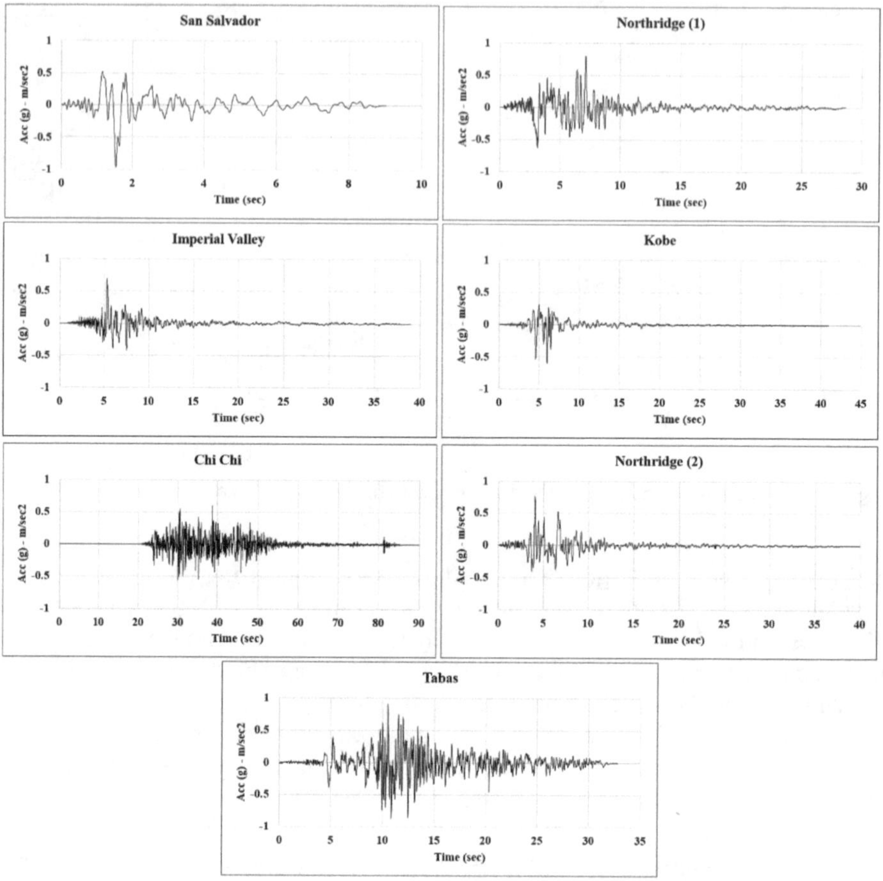

Fig. 8.5 Acceleration time histories of the selected earthquake records (N-S components)

where $d_{ij} = |x_j - x_i|$ is the distance between ith and jth grasshoppers, $\widehat{d_{ij}} = \frac{x_j - x_i}{d_{ij}}$ is a unit vector regarding the ith grasshopper toward the jth grasshopper, and N is the number of the grasshoppers. s is a function for determination of the strength of related social forces, while f represents attraction intensity and l defines the length scale for attraction as follows [48]:

$$s(r) = fe^{\frac{-r}{l}} - e^{-r} \tag{8.5}$$

The function s for $l = 1.5$ and $f = 0.5$ is illustrated in Fig. 8.9. Two kinds of forces are caused by this function between the grasshoppers such as repulsion and attraction. The grasshoppers repel each other to avoid collision as the distance between two grasshoppers is between [0, 2.079]. When the distance is in [2.079, 4] range, the attraction force increases to persevere the cohesion in the considered swarm. As the distance is equal to 2.079, no attraction and repulsion are considered which is

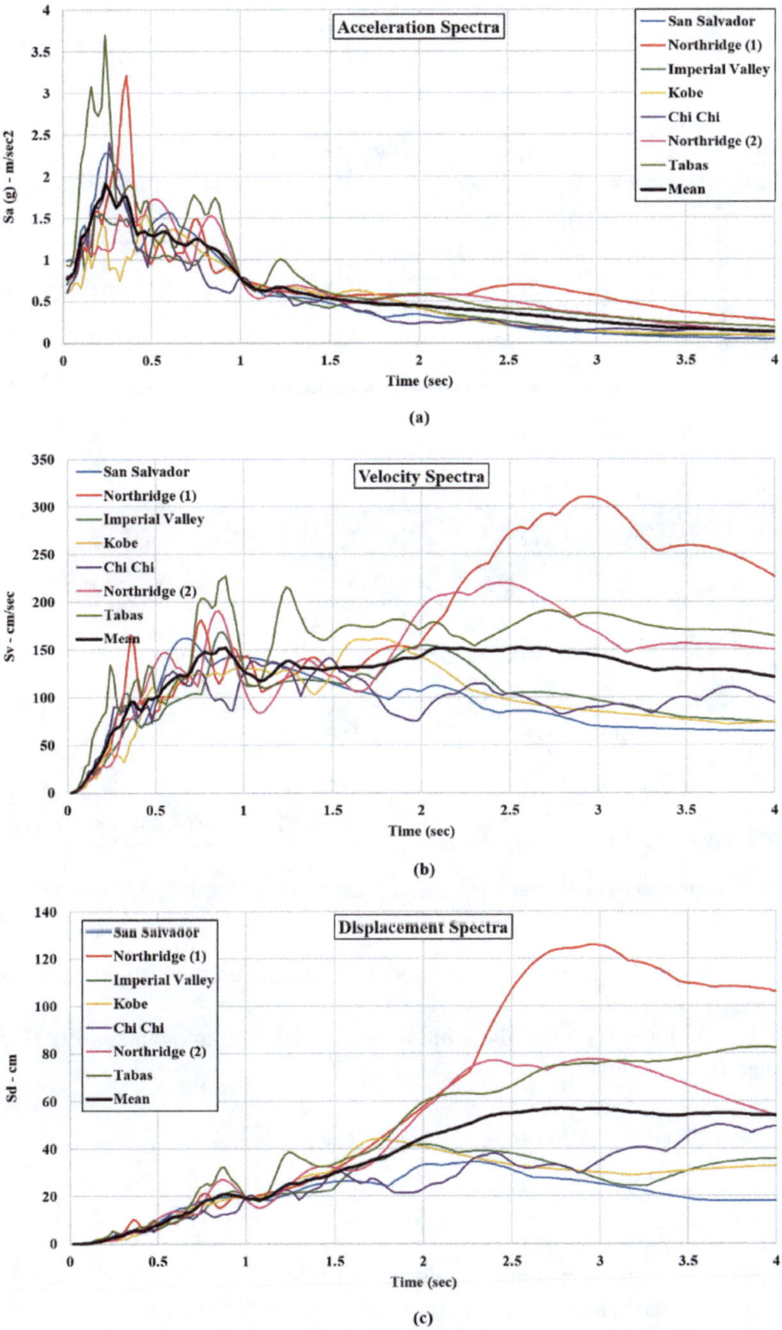

Fig. 8.6 Acceleration, velocity, and displacement spectra of the selected earthquake records (N-S components)

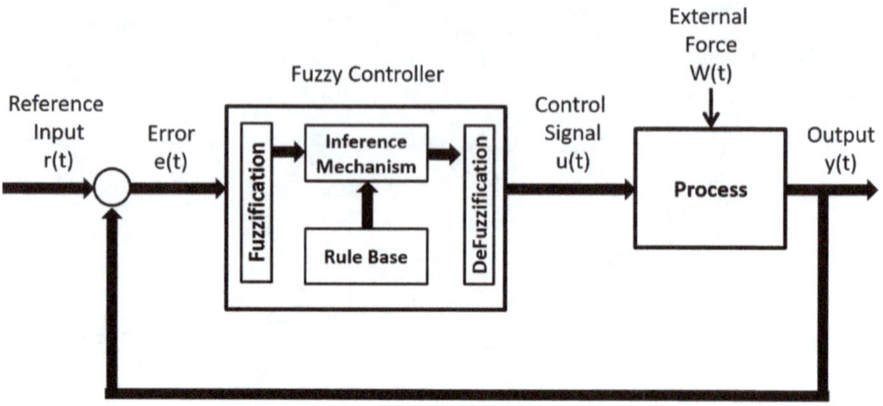

Fig. 8.7 Fuzzy logic control system

Fig. 8.8 Real grasshopper (left) and life cycle of grasshoppers (right) [48]

called as comfort zone. A conceptual model of interaction between grasshoppers is illustrated in Fig. 8.10.

The gravity force (G_i) and the wind advection (A_i) components in Eq. (8.3) are calculated in the following [48]:

$$G_i = -g\widehat{e_g} \tag{8.6}$$

$$A_i = u\widehat{e_w} \tag{8.7}$$

where g is the gravity constant, u is a drift constant, $\widehat{e_g}$ is a vector of unity directly to the center o f earth, and $\widehat{e_w}$ is a vector of unity toward wind direction.

By substituting these equations in Eq. (8.3), X_i is presented as follows [48]:

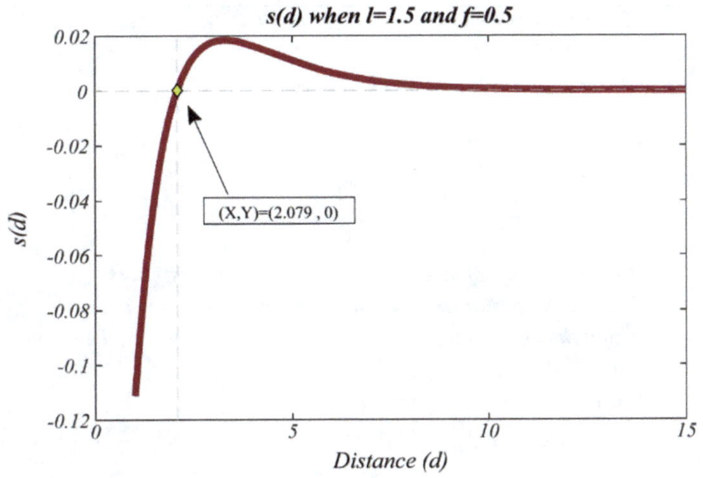

Fig. 8.9 Function s for specific values of l and f [56]

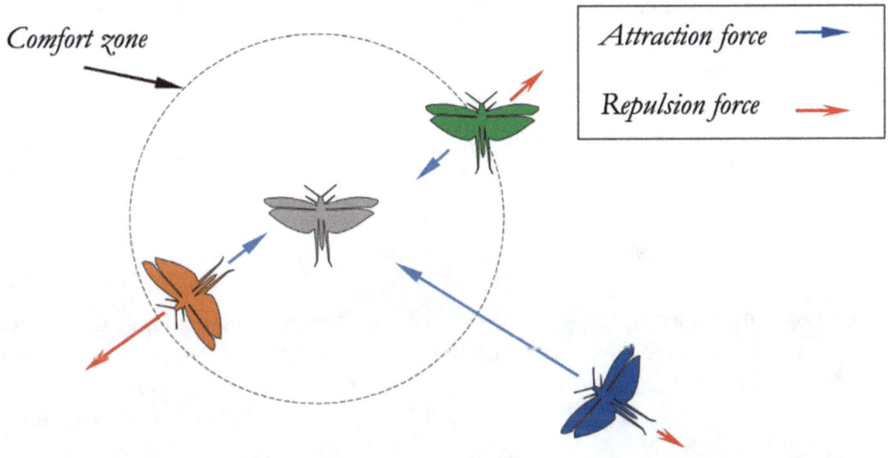

Fig. 8.10 Attraction, repulsion, and comfort zone in GOA [48]

$$X_i = \sum_{\substack{j=1 \\ j \neq i}}^{N} s\left(\left|x_j - x_i\right|\right)\frac{x_j - x_i}{d_{ij}} - g\widehat{e_g} + u\widehat{e_w} \tag{8.8}$$

This equation is capable of simulating the interactions between grasshoppers, but a mathematical model based on the exploration and exploitation is needed to formulate the optimization algorithm. Saremi et al. [48] proposed the following as the mathematical model:

```
procedure Grasshopper Optimization Algorithm (GOA)
Initialize the swarm X_i (i=1, 2, ..., N)
Initialize cmax, cmin and maximum number of iterations
Calculate the fitness of each search agent
T=the best search agent
While (l<maximum number of iterations)
    Update c using Eq. (10)
    for each search agent
        Normalize the distances between grasshoppers in [1,4]
        Update the position of the current search agent by the Eq. (9)
        Bring the current search agent back if it goes outside the boundaries
    end for
    Update T if there is a better solution
    l=l+1
end while
Return T
end procedure
```

Fig. 8.11 Pseudocode of the GOA

$$X_i^d = c\left(\sum_{\substack{j=1\\ j\neq i}}^{N} c\frac{ub_d - lb_d}{2} s\left(\left|x_j^d - x_i^d\right|\right)\frac{x_j - x_i}{d_{ij}}\right) + \hat{T}_d \tag{8.9}$$

where ub_d is the upper and lb_d is the lower bound in the dth dimension, $\hat{T}_d$ represents the amount of dth dimension regarding the best so far found solution as target, and c represents a coefficient for decreasing the comfort, repulsion, and attraction areas. Pseudocode of the GOA is described in Fig. 8.11, while the flowchart of this algorithm is depicted in Fig. 8.12.

8.5 Design Example

In this chapter, a building structure with 20 stories is considered as design example which has a plan of 30.48 m by 36.58 m and height of 80.77 m. The 3D and one of the N-S frames of this building are presented in Fig. 8.13.

For lateral resistance of this building, steel moment-resisting perimeter frames are utilized in which the beam and column elements are fabricated by 248 and 345 MPa

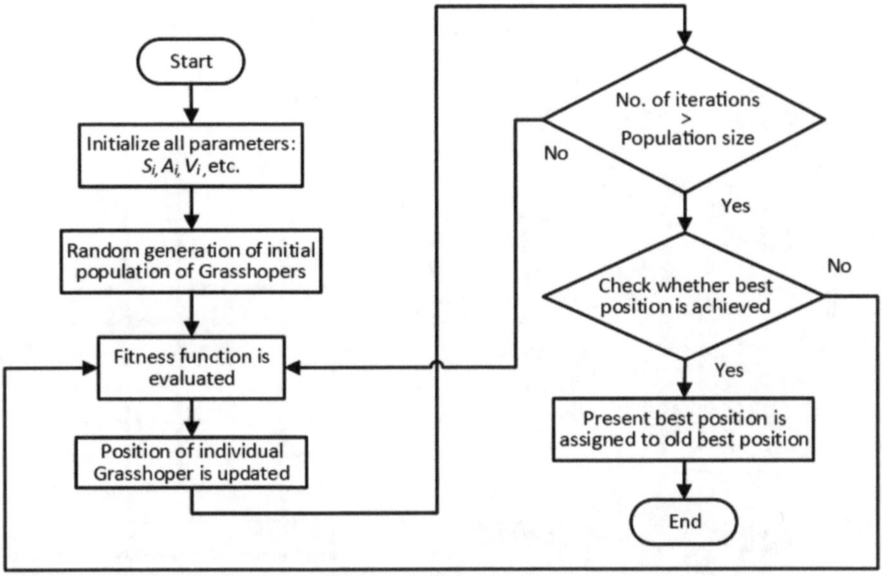

Fig. 8.12 Flowchart of the GOA [57]

steel wide-flange (W) sections while the corner columns are fabricated by box (B)-shaped sections. The seismic mass of the structure and the design sections of the structural members are described in Table 8.2.

By considering the previously presented severe earthquake ground motions by means of the EBGMSM method, structural elements are capable of yielding based on the nonlinear response behavior of the structure. In this regard, a bilinear hysteresis model (Fig. 8.14) is considered for structural members while a nonlinear dynamic procedure is also formulated for analysis purposes by means of the Newmark-β method [58–60].

For vibration control purposes, an active control strategy is also implemented in the 20-story structure by means of the FLC in which five individual accelerometers are placed in the 4th, 8th, 12th, 16th, and 20th stories while five control actuators are located in each of the structural levels of the building with 1000 kN as maximum control forces. The schematic configuration of this control system is presented in Fig. 8.15. In this figure, four earthquake records are selected as inputs of the system while a fuzzy logic-based feedback controller is also implemented for vibration serviceability issues. The ideal actuators are modeled through a single individual block which inserts the control force to the structural system, while the nonlinear structural analysis is conducted by a well-established block.

For performance evaluation of the control strategy in vibration control of the considered budging structure, nine performance criteria (PC) are determined which are presented in Table 8.3.

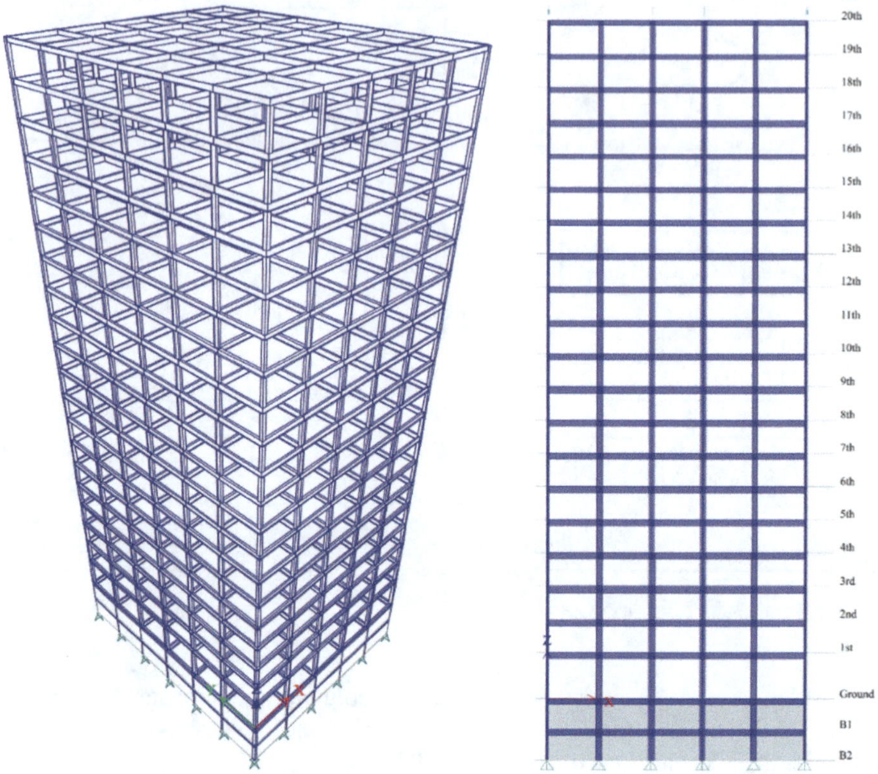

Fig. 8.13 Twenty-story steel building (left) and the N-S perimeter frame (right)

8.6 Statement of the Optimization Problem

In order to optimize the FLCs implemented in the 20-story building structure, optimization problem is formulated. In this regard, 11 fuzzy variables are utilized for formulation of FLC which are represented in Table 8.4.

The decision variables of the GOA are determined based on the characteristics of the FLCs in which 11 fuzzy input variables ($a_1, a_2, \ldots, a_{11}$) and 15 output variables ($b_1, b_2, \ldots, b_{15}$) are determined in this purpose and are presented in Fig. 8.16. Besides, 64 variables ($c_1, c_2, \ldots, c_{64}$) are also determined based on the fuzzy rule base which are provided in Table 8.5.

In this chapter, the objective function is determined by utilization of the previously mentioned PC in which the nonlinear responses of the considered structure are in perspective. The objective function for this study is formulated as follows:

$$\text{Obj} = \frac{\sum_{i=1}^{n} P_i \frac{\text{CR}_i}{\text{UR}_i}}{\sum_{i=1}^{n} P_i} \tag{8.10}$$

Table 8.2 Seismic mass and design sections of the building

Summary	Level	Properties
Column section (interior)	B-2 to 4th	W24 × 335
	4–10th	W24 × 229
	10–13th	W24 × 192
	13–16th	W24 × 131
	16–18th	W24 × 117
	18–20th	W24 × 84
Column section (corner)	B-2 to 1st	B38 × 5.08 cm
	1st to 4th	B38 × 3.18 cm
	4–13th	B38 × 2.54 cm
	13–18th	B38 × 1.91 cm
	18–20th	B38 × 1.27 cm
Beam section	B-2 to 4th	W30 × 99
	5–10th	W30 × 108
	11–16th	W30 × 99
	17–18th	W27 × 84
	19th	W24 × 62
	20th	W21 × 50
Seismic mass	Ground level	5.32×10^5 kg
	1st level	5.63×10^5 kg
	2nd to 19th level	5.52×10^5 kg
	20th level	5.84×10^5 kg
	Entire structure	1.11×10^7 kg

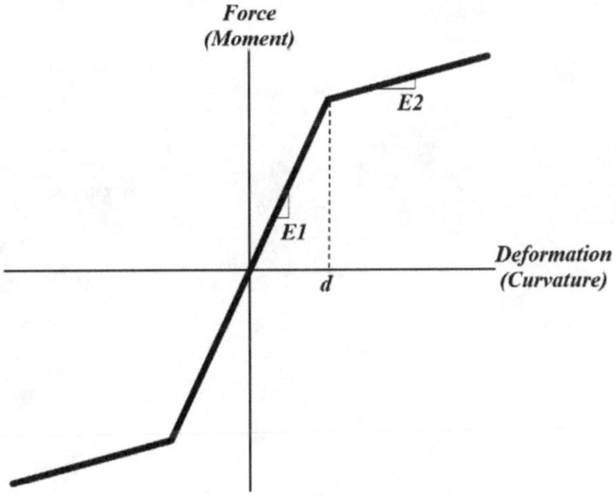

Fig. 8.14 Bilinear model for structural members

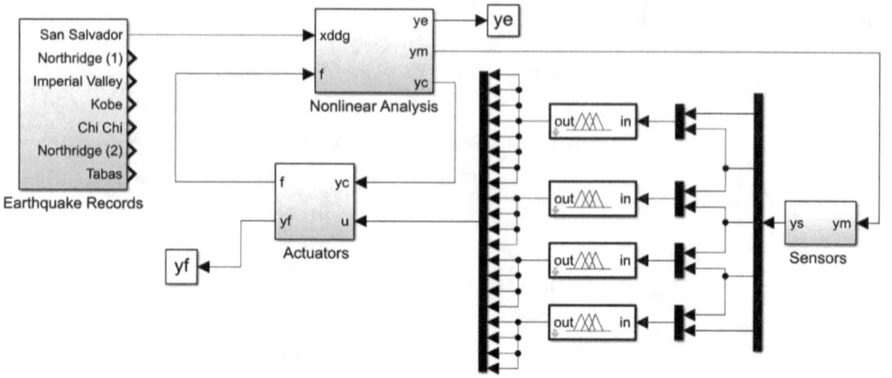

Fig. 8.15 Simulink [61] block diagram for the vibration control of 20-story building

Table 8.3 Summary of the performance criteria

Drift ratio	Story acceleration	Base shear
$PC_1 = \max_{7EQs}\left\{\frac{\max_{t,i}\frac{\lvert d_i(t)\rvert}{h_i}}{\delta^{\max}}\right\}$	$PC_2 = \max_{7EQs}\left\{\frac{\max_{t,i}\lvert\ddot{x}_{ai}(t)\rvert}{\ddot{x}_a^{\max}}\right\}$	$PC_3 = \max_{7EQs}\left\{\frac{\max_{t,i}\left\lvert\sum_i m_i\ddot{x}_{ai}(t)\right\rvert}{F_b^{\max}}\right\}$
Ductility	Dissipated energy	Plastic hinges
$PC_4 = \max_{7EQs}\left\{\frac{\max_{t,j}\frac{\lvert\varphi_j(t)\rvert}{\varphi_{yj}}}{\varphi^{\max}}\right\}$	$PC_5 = \max_{7EQs}\left\{\frac{\max_{t,j}\frac{\int dE_j}{F_{yj}\cdot\varphi_{yj}}}{E^{\max}}\right\}$	$PC_6 = \max_{7EQs}\left\{\frac{N_d^c}{N_d}\right\}$
Control force	Control device stroke	Control power
$PC_7 = \max_{7EQs}\left\{\frac{\max_{t,l}\lvert f_l(t)\rvert}{W}\right\}$	$PC_8 = \max_{7EQs}\left\{\frac{\max_{t,l}\lvert y_l^a(t)\rvert}{x^{\max}}\right\}$	$PC_9 = \max_{7EQs}\left\{\frac{\max_t\lvert\sum P_l(t)\rvert}{\dot{x}^{\max}W}\right\}$

Table 8.4 Fuzzy variables

Abbreviation	Variables
PVL	Positive and very large
PL	Positive and large
PM	Positive and medium
PS	Positive and small
PVS	Positive and very small
ZR	Zero
NVS	Negative and very small
NS	Negative and small
NM	Negative and medium
NL	Negative and large
NVL	Negative and very large

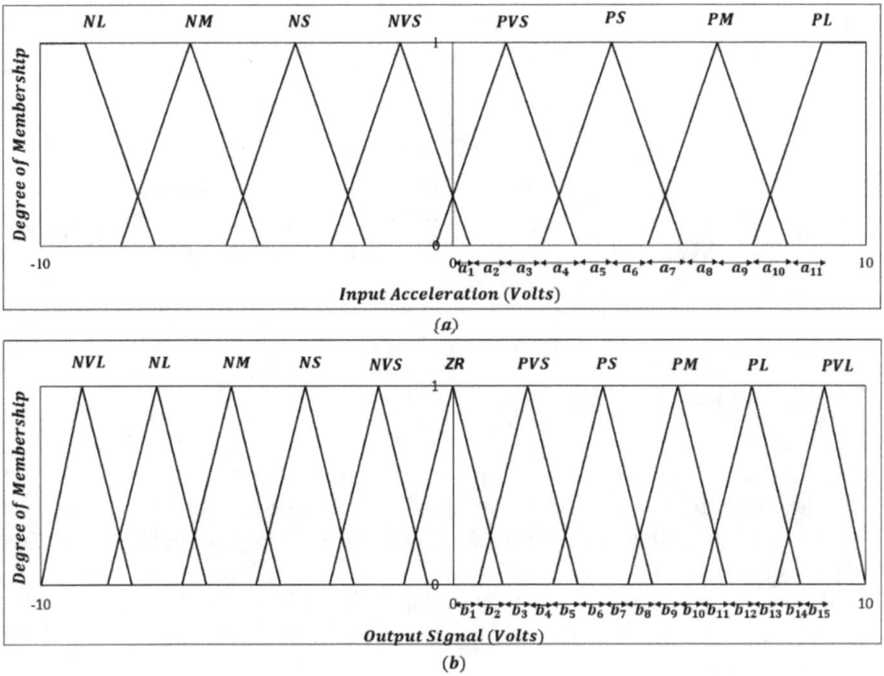

Fig. 8.16 Possible membership functions for the fuzzy inputs (**a**) and output (**b**)

Table 8.5 Variables for the fuzzy rule base

Control force								
Second input	First input							
	NL	NM	NS	NVS	PVS	PS	PM	PL
NL	PVL/c1	PL/c9	PM/c17	PS/c25	PVS/c33	ZR/c41	NVS/c49	NS/c57
NM	PL/c2	PM/c10	PS/c18	PS/c26	PVS/c34	ZR/c42	NVS/c50	NS/c58
NS	PM/c3	PS/c11	PS/c19	PVS/c27	PVS/c35	ZR/c43	NVS/c51	NS/c59
NVS	PM/c4	PS/c12	PVS/c20	PVS/c28	ZR/c36	NVS/c44	NS/c52	NM/c60
PVS	PM/c5	PS/c13	PVS/c21	ZR/c29	NVS/c37	NVS/c45	NS/c53	NM/c61
PS	PS/c6	PVS/c14	ZR/c22	NVS/c30	NVS/c38	NS/c46	NS/c54	NM/c62
PM	PS/c7	PVS/c15	ZR/c23	NVS/c31	NS/c39	NS/c47	NM/c55	NL/c63
PL	PS/c8	PVS/c16	ZR/c24	NVS/c32	NS/c40	NM/c48	NL/c56	NVL/c64

where P_i is a weighing coefficient which is determined based on the PGAs of the previously selected earthquake ground motions, the summation is concerned for considering multiple records, and CR_i is the controlled structural responses alongside the UR_i as the uncontrolled ones.

Based on the fact that the ductility of the high-rise building structures is of great importance, the objective function is determined by utilization of the ductility factor (PC_4) as follows:

$$\mathrm{Obj} = \frac{\begin{array}{l} 0.96 \times (PC_4)_{\text{San Salvador}} + 0.79 \times (PC_4)_{\text{Northridge(1)}} + 0.69 \\ \times (PC_4)_{\text{Imperial Valley}} + 0.60 \times (PC_4)_{\text{Kobe}} + 0.59 \times (PC_4)_{\text{ChiChi}} \\ +0.76 \times (PC_4)_{\text{Northridge(2)}} + 0.91 \times (PC_4)_{\text{Tabas}} \end{array}}{(0.96 + 0.79 + 0.69 + 0.60 + 0.59 + 0.76 + 0.91)} \quad (8.11)$$

8.7 Numerical Results

In this part, the numerical results of the fuzzy optimization process are presented in which the convergence history of the best comparative results of the GOA is provided while the optimized and non-optimized PCs for the selected earthquake records are also presented for comparative purposes.

The convergence history for the best results of the GOA for Obj is illustrated in Fig. 8.17. It should be noted that the GOA is capable of decreasing the considered objective function by up to *0.9365*.

In Table 8.6, the PC of the 20-story building structure based on the optimum design procedure of the GOA is presented. It is obvious that the PC_1 as the inter-story drift ratio of the considered design example for all of the considered earthquake ground motion records is reduced in which the maximum reduction rate is for San Salvador (up to 14%). The structural response relating the PC_2 is decreased by up to 11% for one of the earthquake records (Chi-Chi), while for rest of them it is increased by

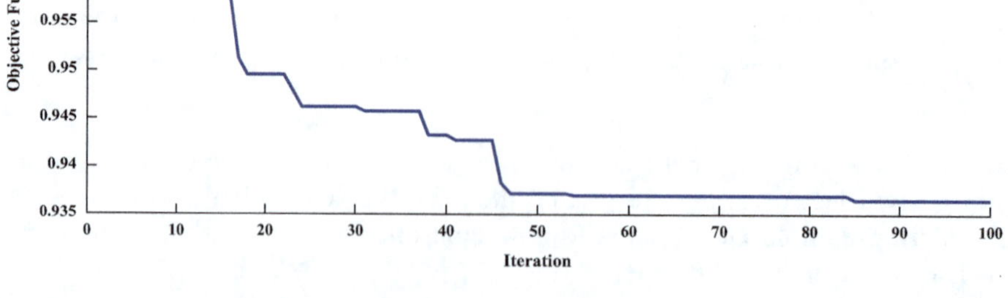

Fig. 8.17 Convergence history of Obj for the best results of the GOA considering 20-story building

Table 8.6 Optimized performance criteria for 20-story building

Performance criteria		Earthquakes						
		San Salvador	North. (1)	Imperial Valley	Kobe	Chi-Chi	North. (2)	Tabas
PC_1	Drift ratio	0.8649	0.9870	0.9655	0.9744	0.9759	0.9819	0.9758
PC_2	Story acceleration	1.0142	1.0301	1.0039	1.0386	0.8961	1.0315	1.0386
PC_3	Base shear	0.9864	0.9683	1.0236	1.0026	1.0309	1.0271	0.9605
PC_4	*Ductility*	*0.7919*	*0.9847*	*0.9348*	*0.9243*	*1.0613*	*0.9562*	*0.9593*
PC_5	Dissipated energy	0.8997	0.9647	1.0868	1.1873	0.9559	0.9972	1.1037
PC_6	Plastic hinges	0.9512	1.0287	0.9841	0.9649	1.0149	1.0119	1.0098
PC_7	Control force	0.0066	0.0070	0.0066	0.0048	0.0071	0.0048	0.0065
PC_8	Device stroke	0.1540	0.0726	0.1098	0.0969	0.0930	0.1077	0.1044
PC_9	Control power	0.0043	0.0039	0.0033	0.0044	0.0043	0.0032	0.0057

less than 4%. The optimal design process based on the GOA is capable of reducing PC_3 due to three of the considered earthquake ground motion records (San Salvador, Northridge (1), and Tabas) by up to 4%, while for rest of them it is increased by less than 3%. The maximum reduction rate for PC_4 is up to 21% (San Salvador), and the greatest reduction rate is for Chi-Chi earthquake (6%). PC_5 as the energy dissipated rate is reduced for four of the considered seismic records by up to 11%, while for other three (Imperial Valley, Kobe, and Tabas), it is increased by up to 18%. PC_6 as the total number of structural plastic hinges is reduced by up to 5% for three of the earthquake records, while for four of them, this ratio is increased slightly (by less than 3%).

Table 8.7 presents the non-optimized PC of the considered design example (20-story building) in which the knowledge-based FLC is utilized as control algorithm. It is obvious that the non-optimized PC has larger values than optimized ones which are near unity in most of the cases. In addition, the structural response of the building with non-optimized controllers is higher than the optimized ones with larger values for nonlinear parameters of the structure such as PC_4–PC_6.

In Fig. 8.18, the performance criteria for building responses and damage (PC_1–PC_6) with optimized and non-optimized fuzzy logic-based controller are compared individually. The results of the optimized and non-optimized FLC are selected from Tables 8.6 and 8.7, respectively. In most cases, the non-optimized PC is larger than the optimized ones and it leads to the fact that utilization of the optimization algorithm for optimal tuning of the fuzzy logic-based controllers and the related parameters

Table 8.7 Non-optimized performance criteria for 20-story building

Performance criteria		Earthquakes						
		San Salvador	North. (1)	Imperial Valley	Kobe	Chi-Chi	North. (2)	Tabas
PC_1	Drift ratio	0.9471	0.9901	0.9989	1.0625	1.0076	1.0102	1.0035
PC_2	Story acceleration	0.9767	1.0209	1.0178	0.9958	0.9199	1.0124	0.9825
PC_3	Base shear	0.9268	0.9678	0.9900	0.9288	1.0583	1.0087	0.9992
PC_4	Ductility	0.8813	0.9975	0.9963	1.0588	1.0775	0.9803	1.0059
PC_5	Dissipated energy	0.9664	0.9602	1.1988	1.2673	0.9636	0.9212	1.1404
PC_6	Plastic hinges	0.9512	1.0449	1.0000	1.0175	0.9851	1.0357	1.0000
PC_7	Control force	0.0046	0.0036	0.0028	0.0028	0.0045	0.0037	0.0046
PC_8	Device stroke	0.1686	0.0728	0.1136	0.1016	0.0930	0.1063	0.1073
PC_9	Control power	0.0043	0.0041	0.0032	0.0033	0.0039	0.0037	0.0046

produces smaller structural responses. The maximum difference between the optimized and non-optimized PCs for the first three criteria (PC_1–PC_3) is for PC_1 under San Salvador earthquake that is about 8%. In addition, the maximum difference in PC_4–PC_6 is for PC_4 under Kobe earthquake that is about 13%.

Regarding PC_7–PC_9 as critical requirements for control devices, the optimized and non-optimized FLCs are compared in Fig. 8.19. For PC_7, the optimized values are higher than the non-optimized ones but for PC_8 and PC_9, the optimized values are less than the non-optimized ones and the maximum difference is for PC_9 under Northridge (2) earthquake that is about 12%.

The ratio of the maximum curvature for the end pints of the structural elements to the yield curvature (ϕ_j/ϕ_{yj}) is summarized for uncontrolled, non-optimized and optimized structures in Fig. 8.20. For three earthquakes (Kobe, Chi-Chi, and Tabas), this ratio for the building structure with non-optimized fuzzy controller is higher than the related values for the uncontrolled budding structure which is definitely not acceptable for a structural system with implemented active vibration control system. In addition, this ratio for the building structures with metaheuristic-based optimized fuzzy controller is lower than the uncontrolled and the non-optimized ones for all of the selected earthquake ground motion records. The maximum difference between the optimized structure and the structure without control strategy and system is for San Salvador earthquake record that is about 20%, while the maximum difference between the non-optimized and the uncontrolled structure is about 11% for the same earthquake. The maximum response difference of the structures with optimized and non-optimized fuzzy controller is for Kobe earthquake that is about 12%.

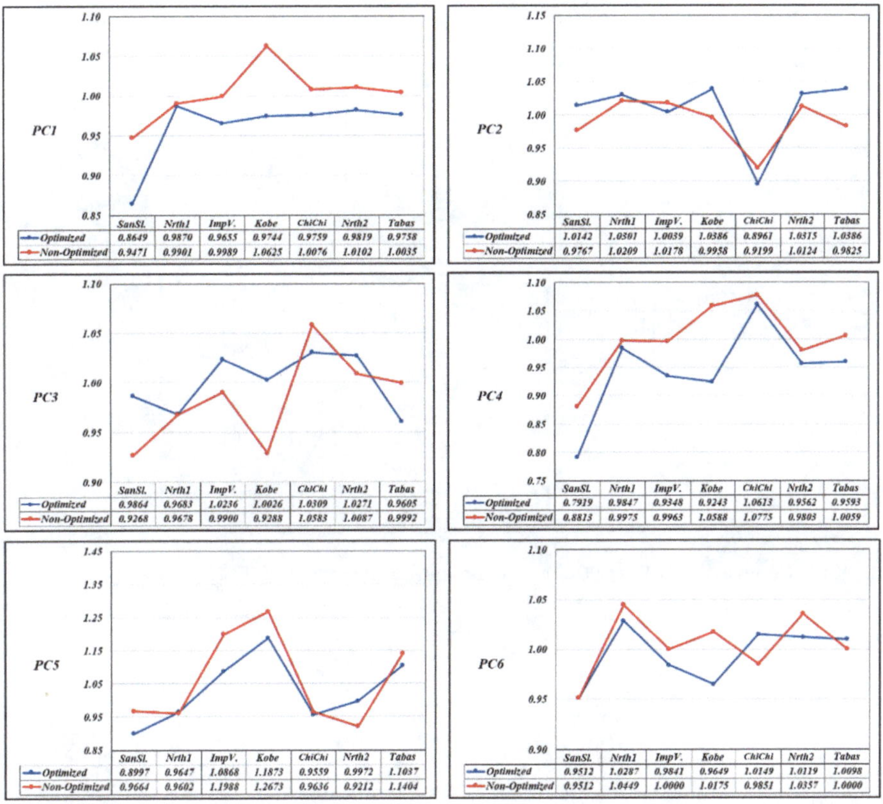

Fig. 8.18 Comparison of PC_1–PC_6 for non-optimized and optimized FLC

Figure 8.21 represents the variation of considered design variables for the FLCs during the optimization process. It is obvious that GOA is capable of converging to global and local optimum values properly and the searching process is properly handled by this algorithm.

8.8 Conclusions

This chapter represents the optimization of fuzzy logic controllers implemented in a 20-story steel structure with nonlinear behavior using grasshopper optimization algorithm. Regarding the absence of earthquake records for Tabriz, ground motion prediction is necessary for structural seismic analysis. The seismic inputs were considered based on the PSHA for city of Tabriz in Iran, while the GMPM for Middle East was selected to calculate the PMS. The earthquake records were selected based on the EBGMSM method. The numerical results show that the optimization

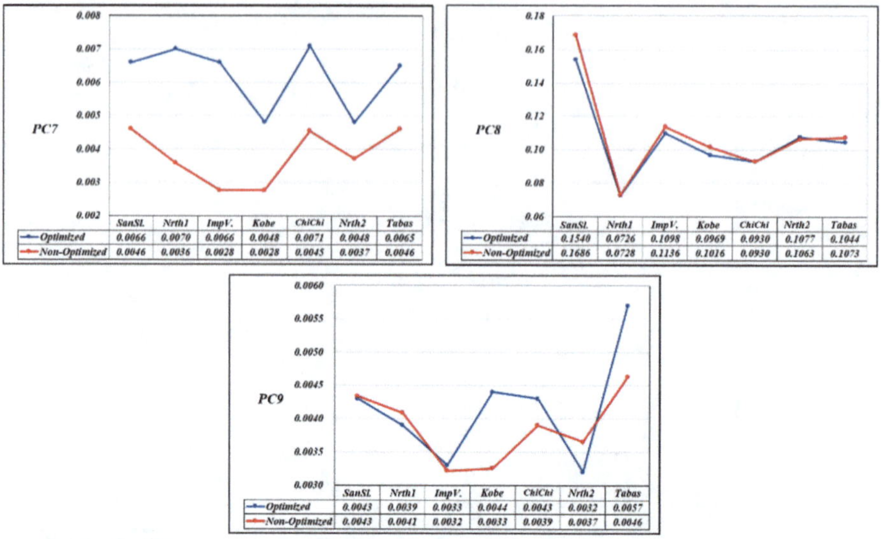

Fig. 8.19 Comparison of PC_7–PC_9 for non-optimized and optimized FLC

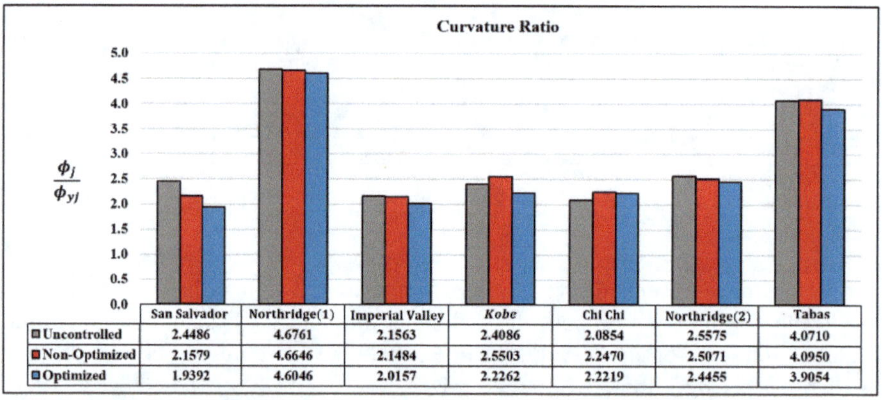

Fig. 8.20 Comparison of the ratio of maximum curvature to the yield curvature

method was able to reduce the objective function by up to 0.9365. For the optimized fuzzy controller, PC_1–PC_3 are reduced by up to 14%, 11%, and 4%, respectively, while for the non-optimized one, these PCs are almost unity. In this regard, it has to be mentioned that the knowledge-based fuzzy controller is not effective in vibration control of the budding structures. For PC_4–PC_6, the optimization process is capable of reducing the criteria by up to 21%, 11%, and 5%, respectively. Comparison of these criteria with corresponding values in non-optimized FLC demonstrates that the optimization technique is very vigorous for damage protection of structural elements. Regarding PC_7–PC_9 as control device criteria, it must be declared

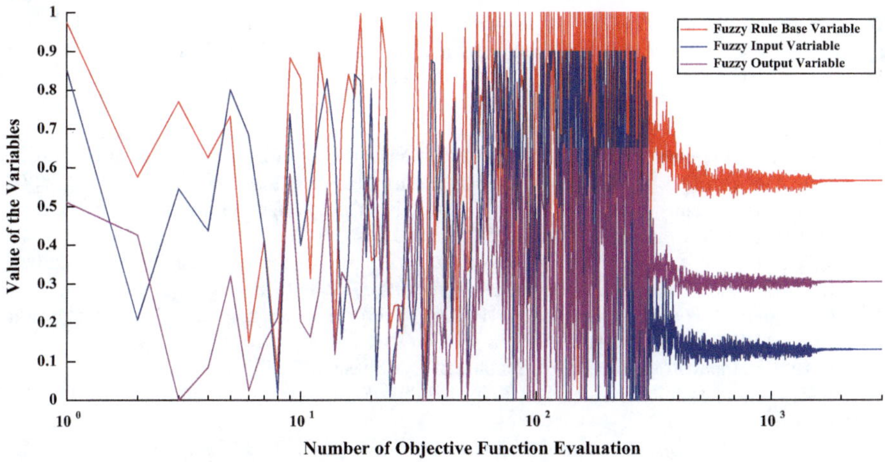

Fig. 8.21 Convergence of design variables including fuzzy input, fuzzy output, and fuzzy rule base

that the optimization technique is able to increase the maximum required control force so the performance of the controlled building structure is extremely better than the non-optimized one. Comparing the responses of the structure with optimized and non-optimized controller displays that when the selection of ground motions is considered based on the PSHA and GMPM for a specific site with high seismic activity (such as Tabriz), an optimization process should be implemented to maintain an efficient control strategy in order to prevent damages in structural members. It should be noted that utilizing metaheuristic algorithms such as GOA in dealing with complex optimization problems such as FLC tuning has important effects on achieving better results. By comparing the result of the GOA and the non-optimized knowledge-based FLCs, it can be concluded that GOA plays an important role in reducing the key structural responses of the building alongside the responses of the control system so utilizing metaheuristics in this purpose is necessary for future challenges.

References

1. Cornell CA (1968) Engineering seismic risk analysis. Bull Seismol Soc Am 58(5):1583–1606
2. Budnitz RJ, Apostolakis G, Boore DM (1997) Recommendations for probabilistic seismic hazard analysis: guidance on uncertainty and use of experts. Nuclear Regulatory Commission, Washington, DC, USA
3. Mahsuli M, Rahimi H, Bakhshi A (2018) Probabilistic seismic hazard analysis of Iran using reliability methods. Bull Earthq Eng 17:1117–1143
4. Keshavarz A, Moghaddam BM (2018) Probabilistic seismic hazard analysis and determination of uniform hazard spectrum of Bushehr province assuming linear source model. J Struct Constr Eng 5(1):127–142

5. Yaghmaei-Sabegh S, Ebrahimi-Aghabagher M (2017) Near-field probabilistic seismic hazard analysis with characteristic earthquake effects. Nat Hazards 87(3):1607–1633
6. Yazdani A, Nicknam A, Dadras EY, Eftekhari SN (2017) Near-field probabilistic seismic hazard analysis of metropolitan Tehran using region-specific directivity models. Pure APPL Geophys 174(1):117–132
7. Vafaie J, Taghikhany T, Tehranizade M (2008) Near field uniform seismic hazard spectra of Tabriz zone. In: The 14th world conference on earthquake engineering, Beijing, China, 2008
8. Vafaie J, Taghikhany T, Tehranizadeh M (2011) Near field effect on horizontal equal-hazard spectrum of Tabriz city in north-west of Iran. Int J Civ Eng 9(1):49–56
9. Razeghi HR, Amiri GG, Amrei SR, Rahimi MA (2016) Probabilistic seismic hazard analyses and obtaining uniform hazard spectra of Tabriz, Iran. KSCE J Civ Eng 20(5):1813–1823
10. Moghaddam H, Fanaie N, Hamzehloo H (2009) Uniform hazard response spectra and ground motions for Tabriz. J Sci Iran 16(3):238–248
11. Hessami K, Pantosti D, Tabassi H, Shabanian E, Abbassi MR, Feghhi K, Solaymani S (2003) Paleoearthquakes and slip rates of the North Tabriz Fault, NW Iran: preliminary results. Ann Geophys 46
12. Douglas J (2018) Ground-motion prediction equations 1964–2018. Pacific Earthquake Engineering Research Center, Berkeley, CA
13. Javan-emrooz H, Eskandari-Ghadi M, Mirzaei N (2018) Prediction equations for horizontal and vertical PGA, PGV, and PGD in Northern Iran using prefix gene expression programming. Bull Seismol Soc Am 108
14. Sedaghati F, Pezeshk S (2017) Partially nonergodic empirical ground-motion models for predicting horizontal and vertical PGV, PGA, and 5% damped linear acceleration response spectra using data from the Iranian Plateau. Bull Seismol Soc Am 107(2):934–948
15. Rajabi AM, Khamehchiyan M, Mahdavifar MR, Del Gaudio V (2010) Attenuation relation of Arias intensity for Zagros Mountains region (Iran). Soil Dyn Earthq Eng 30(3):110–118
16. Sandıkkaya MA, Akkar S (2017) Cumulative absolute velocity, Arias intensity and significant duration predictive models from a pan-European strong-motion dataset. Bull Earthq Eng 15(5):1881–1898
17. Gupta ID, Trifunac MD (2017) Scaling of Fourier spectra of strong earthquake ground motion in western Himalaya and northeastern India. Soil Dyn Earthq Eng 102:137–159
18. Alıcı FS, Sucuoğlu H (2016) Prediction of input energy spectrum: attenuation models and velocity spectrum scaling. Earthq Eng Struct Dyn 45(13):2137–2161
19. Hassani N, Amiri GG, Bararnia M, Sinaeian F (2017) Ground motion prediction equation for inelastic spectral displacement in Iran. Sci Iran Trans B Mech Eng 24(1):164
20. Oth A, Miyake H, Bindi D (2017) On the relation of earthquake stress drop and ground motion variability. J Geophys Res: Solid Earth 122(7):5474–5492
21. Yaghmaei-Sabegh S (2018) Macroseismic intensity attenuation in Iran. Earthq Eng Eng Vibr 17(1):139–148
22. Yaghmaei-Sabegh S (2015) New models for frequency content prediction of earthquake records based on Iranian ground-motion data. J Seismol 19(4):831–848
23. Afshari K, Stewart JP (2016) Physically parameterized prediction equations for significant duration in active crustal regions. Earthq Spectra 32(4):2057–2081
24. Haji-Soltani A, Pezeshk S, Malekmohammadi M, Zandieh A (2017) A study of vertical-to-horizontal ratio of earthquake components in the Gulf coast region. Bull Seismol Soc Am 107(5):2055–2066
25. Al-Dawod M, Samali B, Kwok K, Naghdy F (2004) Fuzzy controller for seismically excited nonlinear buildings. J Eng Mech 130(4):407–415
26. Fu J, Lai J, Liao G, Yu M, Bai J (2018) Genetic algorithm based nonlinear self-tuning fuzzy control for time-varying sinusoidal vibration of a magnetorheological elastomer vibration isolation system. Smart Mater Struct
27. Soltani M, Chaouech L, Chaari A (2018) Fuzzy sliding mode controller design based on Euclidean particle swarm optimization. In: Real-time modelling and processing for communication systems. Springer, Cham, pp 95–122

28. Chen Z, Yuan X, Yuan Y, Lei X, Zhang B (2018) Parameter estimation of fuzzy sliding mode controller for hydraulic turbine regulating system based on HICA algorithm. Renew Energy
29. Olivas F, Valdez F, Castillo O, Gonzalez CI, Martinez G, Melin P (2017) Ant colony optimization with dynamic parameter adaptation based on interval type-2 fuzzy logic systems. Appl Soft Comput 53:74–87
30. Amador-Angulo L, Castillo O (2018) A new fuzzy bee colony optimization with dynamic adaptation of parameters using interval type-2 fuzzy logic for tuning fuzzy controllers. Soft Comput 22(2):571–594
31. Chamorro HR, Riaño I, Gerndt R, Zelinka I, Gonzalez-Longatt F, Sood VK (2019) Synthetic inertia control based on fuzzy adaptive differential evolution. Int J Electr Power Energy Syst 105:803–813
32. Ray PK, Paital SR, Mohanty A, Eddy FY, Krishnan A, Gooi HB, Amaratunga GA (2018) Firefly algorithm scaled fractional order fuzzy PID based PSS for transient stability improvement. In 2018 19th international Carpathian control conference (ICCC) 28 May 2018. IEEE, pp 428–433
33. Chrouta J, Chakchouk W, Zaafouri A, Jemli M (2018) Modelling and control of an irrigation station process using heterogeneous cuckoo search algorithm and fuzzy logic controller. IEEE Trans Ind Appl
34. Giri S, Bera P (2018) Design of fuzzy-PI controller for hybrid distributed generation system using grey wolf optimization algorithm. In: Methodologies and application issues of contemporary computing framework. Springer, Singapore, pp 109–122
35. Chao F, Zhou D, Lin CM, Zhou C, Shi M, Lin D (2018) Fuzzy cerebellar model articulation controller network optimization via self-adaptive global best harmony search algorithm. Soft Comput 22(10):3141–3153
36. Li C, Mao Y, Zhou J, Zhang N, An X (2017) Design of a fuzzy-PID controller for a nonlinear hydraulic turbine governing system by using a novel gravitational search algorithm based on Cauchy mutation and mass weighting. Appl Soft Comput 52:290–305
37. Talatahari S, Azizi M (2020) Chaos Game Optimization: a novel metaheuristic algorithm. Artif Intell Rev 22:1–88. https://doi.org/10.1007/s10462-020-09867-w
38. Talatahari S, Azizi M (2020) Optimization of constrained mathematical and engineering design problems using chaos game optimization. Comput Ind Eng 20:106560. https://doi.org/10.1016/j.cie.2020.106560
39. Talatahari S, Azizi M (2020) Optimal design of real-size building structures using quantum-behaved developed swarm optimizer. Struct Des Tall Spec Build 15:e1747. https://doi.org/10.1002/tal.1747
40. Talatahari S, Motamedi P, Farahmand Azar B, Azizi M (2019) Tribe-charged system search for parameter configuration of non-linear systems with large search domains. Eng Optim. https://doi.org/10.1080/0305215X.2019.1696786
41. Azizi M, Ejlali RG, Ghasemi SA, Talatahari S (2019) Upgraded Whale Optimization Algorithm for fuzzy logic based vibration control of nonlinear steel structure. Eng Struct 1(192):53–70. https://doi.org/10.1016/j.engstruct.2019.05.007
42. Azizi M, Ghasemi SA, Ejlali RG, Talatahari S (2020) Optimum design of fuzzy controller using hybrid ant lion optimizer and Jaya algorithm. Artif Intell Rev 53(3):1553–1584. https://doi.org/10.1007/s10462-019-09713-8
43. Azizi M, Ghasemi SA, Ejlali RG, Talatahari S (2019) Optimal tuning of fuzzy parameters for structural motion control using multiverse optimizer. Struct Des Tall Spec Build 28(13). https://doi.org/10.1002/tal.1652
44. Talatahari S, Azizi M (2020) Optimum design of building structures using tribe-interior search algorithm. Structures
45. Azizi M, Ghasemi SA, Ejlali RG, Talatahari S (2021) Optimization of fuzzy controller for nonlinear buildings with improved charged system search. Struct Eng Mech
46. Kaveh A, Pirgholizadeh S, Khadem HO (2015) Semi-active tuned mass damper performance with optimized fuzzy controller using CSS algorithm. Asian J Civ Eng (BHRC) 16(5):587–606
47. Abubaker S, Nagan S, Nasar T (2016) Particle swarm optimized fuzzy control of structure with tuned liquid column damper. Global J Pure Appl Math 12(1):875–886

48. Saremi S, Mirjalili S, Lewis A (2017) Grasshopper optimisation algorithm: theory and application. Adv Eng Softw 105:30–47
49. Ambraseys NN, Simpson KU, Bommer JJ (1996) Prediction of horizontal response spectra in Europe. Earthq Eng Struct Dynam 25(4):371–400
50. Marasco S, Cimellaro GP (2018) A new energy-based ground motion selection and modification method limiting the dynamic response dispersion and preserving the median demand. Bull Earthq Eng 16(2):561–581
51. Karakhanian AS, Trifonov VG, Philip H, Avagyan A, Hessami K, Jamali F, Bayraktutan MS, Bagdassarian H, Arakelian S, Davtian V, Adilkhanyan A (2004) Active faulting and natural hazards in Armenia, eastern Turkey and northwestern Iran. Tectonophysics 380(3–4):189–219
52. Berberian M, Yeats RS (1999) Patterns of historical earthquake rupture in the Iranian Plateau. Bull Seismol Soc Am 89(1):120–139
53. Masson F, Djamour Y, Van Gorp S, Chéry J, Tatar M, Tavakoli F, Nankali H, Vernant P (2006) Extension in NW Iran driven by the motion of the South Caspian Basin. Earth Planet Sci Lett 252(1–2):180–188
54. Cimellaro GP, Marasco S (2014) OPENSIGNAL: a software framework for earthquake record processing and selection. In: Second European conference on earthquake engineering and seismology (2ECSEES), Istanbul, Turkey Google Scholar
55. Cimellaro GP, Marasco S (2015) A computer-based environment for processing and selection of seismic ground motion records: OPENSIGNAL. Front Built Environ 1:17
56. Mirjalili SZ, Mirjalili S, Saremi S, Faris H, Aljarah I (2018) Grasshopper optimization algorithm for multi-objective optimization problems. Appl Intell 48(4):805–820
57. Ullah I, Khitab Z, Khan MN, Hussain S (2019) An efficient energy management in office using bio-inspired energy optimization algorithms. Processes 7(3):142
58. Newmark NM (1959) A method of computation for structural dynamics. J Eng Mech Div 85(3):67–94
59. Subbaraj K, Dokainish MA (1989) A survey of direct time-integration methods in computational structural dynamics—II. Implicit methods. Comput Struct 32(6):1387–1401
60. Ohtori Y, Spencer Jr BF (1999) A MATLAB-based tool for nonlinear structural analysis. In: Proceedings of the 13th engineering mechanics conference, pp 13–16
61. MATLAB, the MathWorks Inc., MA, United States

Chapter 9
Optimization and Artificial Neural Network Models for Reinforced Concrete Members

Melda Yücel, Sinan Melih Nigdeli, Aylin Ece Kayabekir, and Gebrail Bekdaş

9.1 Introduction

In recent years, especially nowadays, artificial intelligence (AI) and machine learning technology that AI's applications are reflected to it are benefited very often in many different fields of study. As the leading cause of this, it can be demonstrated to be able to provide any unknown parameter with higher accuracy, quick as namely effective in terms of consuming time, besides of these, with the usage of less effort and manpower because of the mentioned technologies.

In the medical field, diagnosing of disease speedily by be defined of foreign cells; in chemistry or pharmacology, prediction of measures of ingredients within a medicine; or in civil engineering and structural area, detection what properties of materials such as concrete, steel, fly ash, lean concrete that they can ensure of sufficient resistance and safety in any structural member, can be given as examples to this case. Not only mentioned samples, it is possible to find many examples like these in lots of other study fields such as electronics and telecommunication, space researches, language science, and genetics.

This technology has an "*intelligence,*" which is artificial, as understood from its name, and this intelligence model was actually put forward by inspiring from humans. On the other side, machine learning is concerned with the usage and progress of

M. Yücel (✉) · S. M. Nigdeli · A. E. Kayabekir · G. Bekdaş
Department of Civil Engineering, Istanbul University-Cerrahpaşa, 34320 Avcılar, Istanbul, Turkey
e-mail: melda.yucel@yahoo.com.tr

S. M. Nigdeli
e-mail: melihnig@istanbul.edu.tr

A. E. Kayabekir
e-mail: ecekayabekir@gmail.com

G. Bekdaş
e-mail: bekdas@istanbul.edu.tr

S. Carbas et al. (eds.), *Nature-Inspired Metaheuristic Algorithms for Engineering Optimization Applications*, Springer Tracts in Nature-Inspired Computing,
https://doi.org/10.1007/978-981-33-6773-9_9

functions of this artificial intelligence. Just as if the intelligence of any human being was also prone to learn by obtaining information, to improve oneself by training, and at the same time to get experience from all of these, it is possible that all kinds of machine, computer, device, or software, which can use any type of intelligence, can learn and enhance themselves, too. Hence, this case also creates resource for a technology called machine learning.

In this chapter, it gives place to a review where explain of applications, which are performed via artificial intelligence and machine learning methods intended for the mentioned area. Due to that, the major area is civil engineering and structural optimization that we will handle in this chapter.

9.2 Review of AI and Machine Learning Applications for Structural Optimization

Structural engineering is one of the sub-disciplines of civil engineering, and in this area, there are various applications such as structural dynamics and earthquake analyses, static applications, structural mechanics, and construction management. Also, in any civil engineering structure like building, bridge, dam, etc., the process of making real of the mentioned applications can be performed with different techniques. On the other side, in analysis stages and designs, to achieve the most appropriate result can take plenty of time because these structures perform not a linear behavior, and this also causes unnecessary consumption of cost and effort losses. In this respect, various optimization procedures are benefited to preclude the mentioned cases. Especially, very different algorithms accepted as metaheuristic methods, which are inspired by nature and use the heuristic, are often preferred nowadays. However, the referred methods also can take really so long time in the analysis of these structures, which display nonlinear behavior, even it sometimes lasts until weeks. In this case, it comes in sight clearly that it is needed to propose methods, which can prevent redundant losses of time, cost, and effort, namely effectual for many respects, adapt to different circumstances, and advances. Within this context, AI, which is a technology that matches to the said conditions, and machine learning technique manifest themselves among the best choices, which can be benefited.

When we keep in view o f hese conditions, realizing these effective resolutions, which are needed by structural engineering and structural optimization, will provide great conveniences in these fields, without doubt. With this direction, below, a literature summary can be seen belonging studies conducted in these areas.

The protection of structures across dynamic/vibration factors such as wind or earthquake forces, various seismic excitations, or vehicle crashing is really important to provide safety and strength.

In this regard, Ramezani et al. [1] performed an application intended for tuning of mass dampers on high-rise buildings in the direction of providing protection to these structures against the wind forces. Here, as a sample, ANNs were preferred to determine by predicting these properties of an office building with 76 storeys, after reaching optimum properties of tuned mass dampers for structures, which have different freedoms between 10 and 80.

Yucel et al. [2] performed a decision-making model and prediction process for different structures, which possess tuned mass dampers (TMD). In this respect, this modeling was made possible by ensuring the optimization data via flower pollination algorithm (FPA), and later, the mentioned process was carried out with multilayer feedforward ANNs to observe optimum mechanical parameters of TMD comprised from damping ratio and period.

Moreover, Yang et al. [3] generated a model for the determination of dynamic increase values of reinforced concrete (RC) models with steel fiber, by combining a metaheuristic handled as firefly algorithm (FA) and random forest, which is one of the machine learning techniques. In this study, some parameters such as strain rate, matrix strength, fiber dosage, shape, or tensile strength were selected as inputs intended for usage in the prediction of targeted outputs besides that FA was benefited to adjust parameters that belong to the random forest technique.

On the other hand, various studies exist intended for applications performed to predict different parameter values or achieve an optimal design of members in any structures. As an example [4], a hybrid prediction model consisted of a kind of metaheuristic algorithm with ANNs was generated to determine the ultimate shear capacity of RC deep beams. In this respect, ANNs were assembled with harmony search (HS) to tune network parameters in training process of it.

An ANNs model [5] was developed intended for decision-making process with regard to train design parameters optimized via FPA and detect new design values for tubular column models. Furthermore, Chen et al. [6] generated a decision structure by hybridizing particle swarm optimization (PSO) with ANNs to make prediction for values of shear strength-intended RC walls with squat form. To realize this, they handled 139 different samples of wall data. In the study, it was carried out that selection of one of the models among 31 different ANNs structures, which are optimized with an algorithm known as multi-objective grey wolf optimization (GWO), to be used for the determination of compression strength for silica fume concrete [7].

Yucel et al. [8, 9] conducted a design process, which was applied as prediction with ANNs following optimization (by using FPA), with the purpose of ensuring minimum vertical deflection of beams with I-section under different conditions. In these studies, optimum design variables were tried to find related to the usage of lengths and loads applied to beam diversely from (in terms of value) each other. On the other side, in a study carried out by Nguyen et al. [10], a feedforward neural network's weights and bias values were adjusted by optimizing the usage of invasive weed optimization (IWO) to estimate axial capacity belonging rectangular steel tubes filled with concrete, which is under the compression force. In that study, this process was also realized with single-layer neural network, and they both were compared to each other.

As to the scope and purpose of this study, it is related to ensuring the decision-making and prediction models via machine learning techniques by using different data information provided via various optimization processes. In this regard, two different prediction models were generated for optimum structural designs by using artificial neural networks (ANNs). With this direction, design variables as section properties and the main objective functions can be predicted for both designs, directly and in a quick time by benefiting from this created decision-making model.

9.3 Artificial Neural Networks (ANNs)

Artificial neural networks (ANNs) are a type of algorithm, which is smart, learner, and can enhance of self. Reason of its smart can be seen as that this method's cooperation and working style with brain, besides resembling is to central neural system of humans and variety features of this system. Because, the mentioned system's functioning structure can process information acquired by humans, learn this information by stages, and also increase/enhance experiences by formatting reactions to different and not faced cases by this means these teachings. Hence, in time, just as a human being, who has mind and intellect, this formation makes the pointed decisions; besides that, it can exhibit behaviors, with high accuracy, and so, the mentioned method can be thought of as a sort of artificial intelligence and machine learning technique.

In addition to this, as aforesaid in the previous paragraph, the structure of ANNs is composed of constituents of central neural system of humans. Neural cells in this biologic structure and axons, which provide the transferring of information between cells within neural system, represent the connection nodes (input, hidden, and output) and node connections in the algorithm, respectively.

There are so many kinds of ANNs, and also some of these are perceptrons, single-layer ANNs, multilayer ANNs, feedforward back-propagation ANNs, cellular ANNs, convolutional ANNs, Hopfield ANNs, etc. In this study, multilayer feedforward back-propagation ANNs are used for training of the datasets, and prediction process regards design parameters that belong to structural beam models.

The used feedforward model for ANNs is related to the usage, and training of input data as transmission to onwards. On the other hand, this ANNs kind was handled by coordinating with back-propagation algorithm, which was developed by Rumelhart, Hinton, and Williams in the year 1986 and is one of the learning methods of data benefited for ANNs [11].

9.4 Metaheuristic Algorithms and Optimization

Optimization is to gain access to the most proper solution, and best parameters, which can provide this solution, intended for any problem within the direction of a specific goal (e.g., minimization of weight or cost of a structural model, maximization of income, which will be supplied from a project, minimizing of harmful agents in a chemical formulation, etc.). In this scope, from the past to the present, many optimization methods were benefited.

Metaheuristic algorithms are also one of the mentioned methods, and nowadays, they are pointed out as methods, which are used most frequently and widely, with the purpose of optimization. Nevertheless, the origin point of them is natural life, some natural processes together with genetic properties or abilities of livings, namely in fact nature. Hence, there exists not only a kind of metaheuristic algorithm; there are also many types of metaheuristic methods available that they are proposed by inspired from all vary events belonging to nature and features of numerous livings survived of own lives within this environment.

For instance, particle swarm optimization (PSO) [12], flower pollination algorithm (FPA) [13], grey wolf optimization (GWO) [14], genetic algorithm (GA) [15], method of artificial immune system (AIS) [16], harmony search (HS) [17], bat algorithm (BA) [18], teaching–learning-based optimization (TLBO) [19], electromagnetism-like algorithm (ELA) [20], big-bang big-crunch (BB-BC) [21], and gravitational search algorithm (GSA) [22] are developed based upon the natural behaviors of creatures lived as gregarious/coordinately, genetic features or capabilities of both humans and other creatures, some physical and chemical events realized specific to universe or material. Also, inspiring sources of some algorithms are not exactly clear. As an example, Jaya algorithm (JA) [23] works based on swarm intelligence, and so, it is one of the algorithms which is population-based one.

In this study, the optimum data of reinforced concrete applications were generated by using parameter-free algorithms. These algorithms are TLBO and JA. The general processes of these algorithms are shortly explained in this section.

9.4.1 Teaching–Learning-Based Optimization (TLBO)

Rao et al. [19] developed TLBO in the year 2011 by using the basic idea of the principles of learning via a teacher or self-learning by themselves in a class. Teachers have the most knowledge level in a class. Teachers affect students in the improvement of knowledge of students. Then, students improve their knowledge of communication and sharing their knowledge with each other.

In the optimization process of TLBO, after an initial solution matrix containing sets of design variables is generated, two phases of optimization are applied.

In teacher phase as the first one, the best solution named as teacher ($X_{i,\text{best}}$) was used with a randomly chosen teaching factor (TF) value to improve an existing solution (X_i) to find the improved solution ($X_{i,\text{new}}$). Also, a random number between 0 and 1 (rand ()) and mean of all existing values ($X_{i,\text{mean}}$) are used. This process and TF are formulated as Eqs. (9.1) and (9.2), respectively.

$$X_{i,\text{new}} = X_i + \text{rand}()\big(X_{i,\text{best}} - (\text{TF})X_{i,\text{mean}}\big) \tag{9.1}$$

$$\mathrm{TF} = \mathrm{round}\ (1 + \mathrm{rand}()) \tag{9.2}$$

In learner phase that is consequently applied after teacher phase, two random solutions ($X_{i,\,a}$ and $X_{i,\,b}$) are selected. Then, the updating of results is done via Eq. (9.3) according to the objective function values of these solutions (OF_a and OF_b).

$$X_{i,\mathrm{new}} = \begin{cases} \mathrm{OF}_a < \mathrm{OF}_b, & \text{and} \quad X_i + \mathrm{rand}()\left(X_{i,a} - X_{i,b}\right) \\ \mathrm{OF}_a > \mathrm{OF}_b, & \text{and} \quad X_i + \mathrm{rand}()\left(X_{i,b} - X_{i,a}\right) \end{cases} \tag{9.3}$$

9.4.2 Jaya Algorithm (JA)

In 2016, Rao [21] generated a single-phase and parameter-free algorithm called Jaya algorithm (JA) that has a name coming from Sanskrit word Jaya meaning victory. As shown in Eq. (9.4), it uses the best solution and the worst existing solution ($X_{i,\,\mathrm{worst}}$) in a single equation. It requires only common parameters such as population size and maximum number of iterations for the optimization process.

$$X_{i,\mathrm{new}} = X_i + \mathrm{rand}()\left(X_{i,\mathrm{best}} - |X_i|\right) - \mathrm{rand}()\left(X_{i,\mathrm{worst}} - |X_i|\right) \tag{9.4}$$

9.5 Machine Learning Applications via ANNs for Reinforced Concrete (RC) Structures

In the present chapter, an application with three stages including optimization, and prediction process via machine learning, which are realized as consecutively, are represented for reinforced concrete (RC) structures consisted from two different beam models, which were handled with the aim of optimum design.

Here, firstly, optimization is carried out for both problems with the usage of TLBO for solving T-shaped beam model and JA for carbon fiber reinforced polymer (CFRP) implemented beam model. Also, population and iteration numbers are respectively considered as 10 and 100000 for T-shaped beam, and 20 and 10000 for beam with CFRP structural optimization problems. The optimization is done to generate a dataset of optimum results for different design constants.

As the second stage of this application, decision-making models via ANNs were generated to predict design parameters for the different, unknown/not encountered, or not used alternatives of these structural models directly (without any optimization process). In this process, design variables are predicted as output while the differentiated values of design constants of optimization cases are taken as input. In other words, the optimum values that are found in different values of the output are used in machine learning to generate an ANNs model that predicts the results of design variables for any other input values without making an optimization process.

Finally, the third stage is the evaluation of accuracy and precision performance of basic prediction models by using test models comprised of different structural options of beam models. These models generally include cases which are not used in the machine learning process to verify the prediction model.

A general flowchart of the process of the study is shown in Fig. 9.1.

9.5.1 *T-Shaped RC Beam*

- **Optimum structural design of RC T-beam**

The first application is related to minimizing of total section for a T-shaped RC beam model (Fig. 9.2) in the direction of the rules of Eurocode 2 [24–26].

For this purpose, firstly, an optimization process was applied with metaheuristic algorithm. After making this, in the following section, optimum design results were obtained from the determined method that they were ensured to utilize in machine learning applications.

Furthermore, constants together with parameter limitations of design are shown in Table 9.1 [25]. Here, some from all constants are handled within a value range to generate a training dataset in the direction of using as a main model for prediction operation via ANNs.

As to the objective function (minimization of the cost) (Cost), it can be formulized with Eq. (9.5).

$$\text{Cost} = db_w + (b - b_w)h_f + \left(\frac{C_s}{C_c}\right)A_s + \left(\frac{C_f}{C_c}\right)(b + 2(d_s + d)) \quad (9.5)$$

Moreover, several design limitations must be applied for proper structural design in the way of providing safe, strength, and economic case. Thus, with this purpose, Eqs. (9.6)–(9.9) should be considered and calculated:

$$g_1(x) : w(1 - 0.5w) \leq 0.392 \quad (9.6)$$

$$g_3(x) : \rho \leq \rho\text{max} \quad (9.7)$$

$$g_4(x) : M_{\text{existing}} \leq M_{\text{available}} \quad (9.8)$$

$$g_4(x) : V_{\text{existing}} \leq V_{\text{available}} \quad (9.9)$$

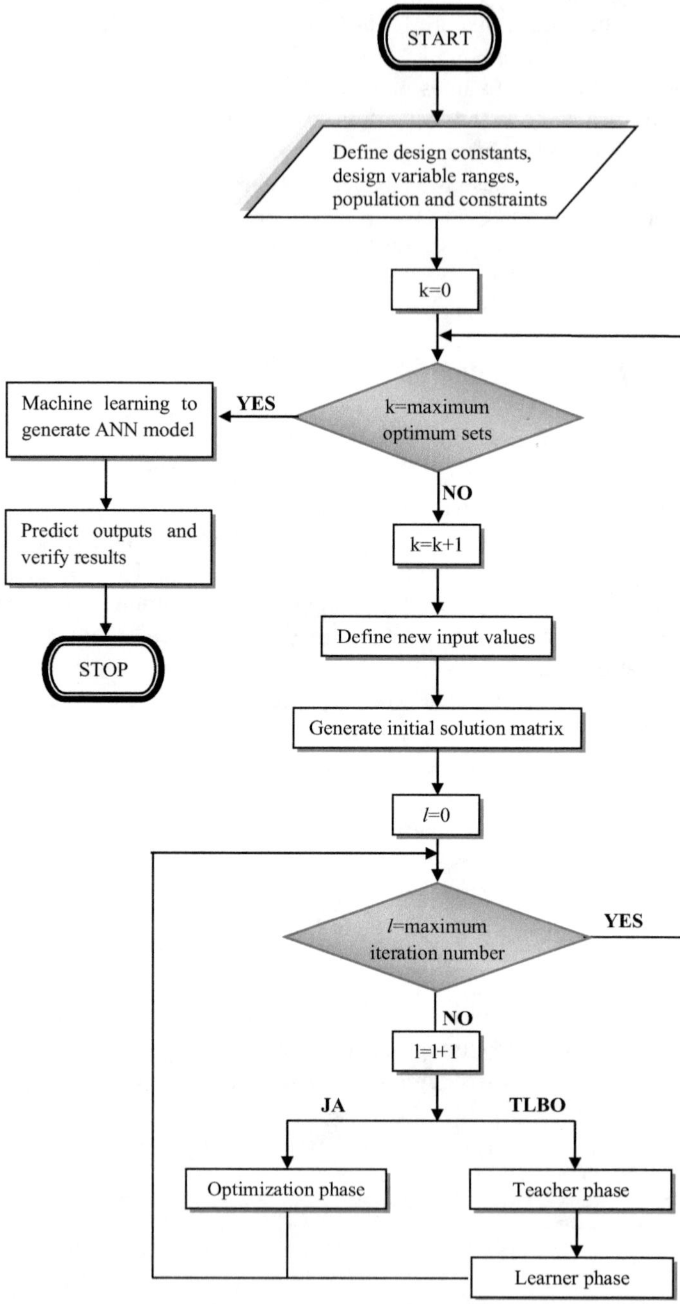

Fig. 9.1 Flowchart of the methodology

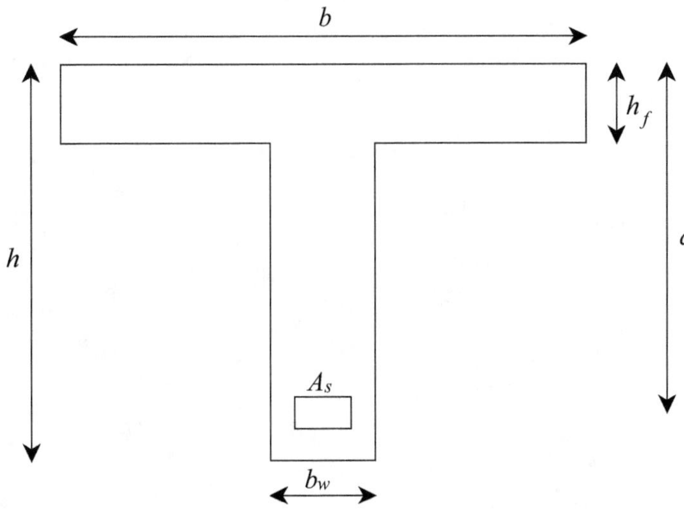

Fig. 9.2 T-shaped beam model and design details

The value used in the calculations (w) and reinforcement ratio (ρ) is formulized with Eqs. (9.10) and (9.11); besides, $M_{\text{available}}$ and $V_{\text{available}}$ express the capacities as bending moment and shear force belonging to beam, respectively.

$$w = \left(\frac{f_{yd}}{f_{cd}}\right)\left(\frac{A_s}{b_w d}\right) - \frac{(b - b_w)h_f}{b_w d} \tag{9.10}$$

$$\rho = \frac{A_s}{b_w d} \tag{9.11}$$

- **The main ANNs training and prediction model**

With the aim of usage during the prediction process by machine learning algorithm such as ANNs, outputs are considered from the design parameters such as h, b_w, b, h_f, A_s and also d, w with $\frac{\text{Cost}}{C_c}$ ratio. On the other hand, the inputs were selected in the same way consisting of M_{existing} and V_{existing} from all design constants. For this respect, a main dataset was generated by benefiting from optimum results of all constants and parameters, and a prediction process was carried out by training of ANNs with this dataset. To perform this operation, neural net fitting application from MATLAB R2018a software was utilized [27].

In this respect, for all design parameters, namely outputs, the results of ANN predictions, besides various metrics of errors corresponding to the structure of main ANNs (generated by considering basic dataset), are given in Table 9.2.

Here, mean absolute error (MAE) is an indicator, which expresses the average value of absolute errors calculated between actual and predicted values for all data. Also, mean squared error (MSE) is the mean square of errors occurred among whole

Table 9.1 Design information for T-shaped beam

	Explanation	Notation	Unit	Value/ranges/limits
Design constant	Beam effective depth	d	m	0.9 h
	Thickness of concrete cover	d_s		0.1 h
	Characteristic compressive strength of concrete	f_{ck}	MPa	20
	Design compressive strength of concrete	f_{cd}		$\frac{f_{ck}}{1.5}$
	Characteristic yield strength of steel	f_{yk}		400
	Design yield strength of steel	f_{yd}		$\frac{f_{yk}}{1.15}$
	Elastic modulus of steel	E_s		200000
	Minimum reinforcement ratio	ρ_{min}	–	
	Maximum reinforcement ratio	ρ_{max}		
	Unit cost of steel	C_s		36
	Unit cost of framework	C_f		0
	Unit cost of concrete	C_c		1
	Existing bending moment force	M_{existing}	Nm	0–25
	Existing shear force	V_{existing}	N	0–2.5
	Length of beam	L	m	20
Design parameter	Beam height	h	m	$\frac{L}{16} - 2.0$
	Breadth of beam web	b_w		0.2d−0.4d
	Breadth of slab	b		$b_w - min\ (0.2L + b_w, 8h_f)$
	Thickness of slab	h_f		0.15−d
	Area of tensile reinforcing bar	A_s	m^2	0–0.1

actual data and their predictions. As to root mean squared error (RMSE), metric is the root of MSE.

- **Evaluation of Test Models**

In this section, test models were analyzed with prediction results provided by ANNs model for 50 different test data. The generated test models and their optimization values are given in Table 9.3.

Also, the values of ANN predictions for test models with calculation of errors corresponding to optimization results are shown in Table 9.4. As it can be understood, all error values are really small with very sensitive decimal for all design variables.

Table 9.2 ANN predictions and error evaluations for outputs (for basic training dataset)

	Error performance metrics			Total data samples
	MAE	*MSE*	*RMSE*	
h	1.4789	2.2290	1.4930	450
b_w	1.3670	1.8989	1.3780	
b	0.0928	0.0159	0.1261	
h_f	0.2745	0.0886	0.2976	
d	1.7783	3.2072	1.7909	
A_s	1.8759	3.7135	1.9271	
w	1.7942	3.5660	1.8884	
$\frac{\text{Cost}}{C_c}$	1.2785	2.1740	1.4744	

MAE: mean absolute error; *MSE:* mean squared error; *RMSE:* root mean squared error

Table 9.3 20 different test model combinations for beam structure and optimum values of design parameters

Inputs		Outputs							
$M_{\text{available}}$ (Nm)	$V_{\text{available}}$ (N)	h (m)	b_w (m)	h_f (m)	b (m)	d (m)	w	A_s (m^2)	$\frac{\text{Area}}{C_c}$
20.368	2.264	2.000	0.516	0.337	2.597	1.833	0.477	0.038	2.999
23.929	0.976	2.000	0.632	0.361	2.892	1.812	0.502	0.045	3.593
17.064	0.687	2.000	0.418	0.311	2.489	1.842	0.377	0.030	2.511
9.374	0.738	2.000	0.387	0.163	1.300	1.934	0.514	0.017	1.521
10.492	1.854	2.000	0.372	0.208	1.660	1.862	0.461	0.019	1.649
2.683	2.266	2.000	0.401	0.150	0.401	2.007	0.160	0.004	0.957
5.937	1.292	1.808	0.325	0.150	1.200	1.627	0.494	0.013	1.121
12.041	2.028	2.000	0.428	0.257	1.546	1.963	0.416	0.021	1.874
8.243	1.618	2.000	0.361	0.169	1.351	1.803	0.501	0.016	1.395
11.011	1.375	2.000	0.382	0.263	1.334	1.909	0.511	0.020	1.709
0.932	0.657	1.250	0.216	0.267	0.216	1.081	0.410	0.003	0.346
23.542	0.586	2.000	0.615	0.359	2.874	1.821	0.484	0.044	3.519
23.277	1.666	2.000	0.600	0.419	2.425	1.850	0.532	0.044	3.464
9.999	1.121	2.000	0.373	0.201	1.584	1.863	0.457	0.018	1.594
23.055	2.088	2.000	0.574	0.373	2.958	1.821	0.377	0.042	3.439
14.789	1.146	2.000	0.389	0.262	2.094	1.869	0.531	0.027	2.149
22.603	0.871	2.000	0.580	0.361	2.890	1.837	0.398	0.041	3.377
14.656	2.406	2.000	0.470	0.277	1.969	1.846	0.495	0.028	2.273
15.710	2.046	2.000	0.393	0.313	2.012	1.966	0.395	0.026	2.231
0.304	0.812	1.335	0.240	0.150	0.240	1.201	0.081	0.001	0.316

Table 9.4 ANN predictions of outputs for structural test combinations of T-beam

Error metrics	Outputs							
	h (m)	b_w (m)	h_f (m)	b (m)	d (m)	w	A_s (m^2)	$\frac{\text{Area}}{C_c}$
	1.991	0.508	0.350	2.540	1.867	0.428	0.038	2.981
	2.004	0.620	0.360	2.910	1.821	0.449	0.046	3.575
	1.998	0.420	0.331	2.264	1.886	0.443	0.028	2.480
	1.965	0.362	0.186	1.537	1.827	0.493	0.018	1.523
	1.996	0.374	0.200	1.499	1.874	0.525	0.016	1.649
	2.045	0.417	0.532	0.357	1.979	0.211	0.008	0.993
	1.835	0.330	0.104	1.128	1.659	0.529	0.013	1.140
	1.987	0.393	0.287	1.598	1.896	0.535	0.018	1.849
	1.958	0.354	0.113	1.335	1.796	0.523	0.015	1.387
	2.012	0.369	0.190	1.665	1.890	0.464	0.022	1.675
	1.242	0.215	0.245	0.288	1.055	0.325	0.006	0.329
	1.998	0.609	0.387	2.861	1.833	0.457	0.044	3.528
	2.009	0.613	0.347	2.813	1.828	0.469	0.042	3.490
	1.972	0.362	0.186	1.583	1.843	0.481	0.021	1.566
	2.019	0.605	0.353	2.777	1.857	0.452	0.042	3.443
	2.005	0.390	0.287	2.039	1.901	0.439	0.026	2.142
	2.000	0.579	0.377	2.792	1.842	0.449	0.042	3.385
	2.015	0.440	0.427	1.860	1.942	0.440	0.032	2.183
	2.001	0.415	0.303	2.050	1.903	0.459	0.026	2.277
	1.277	0.217	0.265	0.124	1.092	0.241	0.002	0.254
MAE	0.0263	0.0142	0.0550	0.1033	0.0407	0.0614	0.0017	0.0243
MSE	0.0012	0.0003	0.0090	0.0225	0.0025	0.0060	0.0000	0.0012
RMSE	0.0346	0.0178	0.0946	0.1500	0.0498	0.0773	0.0020	0.0346

MAE: mean absolute error; MSE: mean squared error; RMSE: root mean squared error

9.5.2 Beam with Carbon Fiber Reinforced Polymer (CFRP)

- **Optimum structural design for carbon fiber reinforced polymer (CFRP) beam**

In the second application, an optimized model for T-shaped RC beam with carbon fiber reinforced polymer (CFRP) is provided with the aim of increasing the shear capacity of members of beam structure by using the rules of ACI 318 Building Code Requirements for structural concrete [28–30]. After optimization process, ANNs learning decision-making model is created to determine design parameters for different T-shaped beam models, in quick time with effective way without more

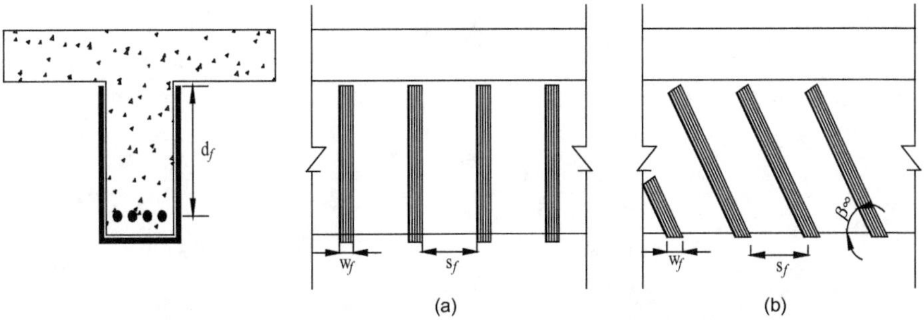

Fig. 9.3 Structural model and design parameters of the T-beam with CFRP

effort, time, and cost consumption. The mentioned structural design of beam and design parameters are shown in Fig. 9.3.

As shown in Fig. 9.3, the objective function of optimization problem for CFRP beam model is related to minimizing the required total CFRP area in unit meter length (A) (Eq. (9.12)), by making it possible that design variables will provide the necessary structural safety. Here, the mentioned design parameters consist of optimum spacing (s_f), width (w_f), and angle (β) for CFRP strips. Also, b_w is beam breadth, besides that d_f symbolizes the depth of beam where wrapping of CFRP is applied, and it can be calculated as Eq. (9.13), where h_f is slab thickness and d is effective depth of the beam (distance from tensile reinforcements to top point of beam flange).

$$A = \frac{w_f\left(\frac{2d_f}{\sin\beta} + b_w\right)}{s_f} \times 1000 \tag{9.12}$$

$$d_f = d - h_f \tag{9.13}$$

On the other hand, the constraint functions for design are applied by considering ACI 318 Regulation. All constraints are expressed in Eqs. (9.14)–(9.16):

$$g_1(x) : s_f \leq d/4 \tag{9.14}$$

$$g_2(x) : 0.7R\frac{(2t_f w_f f_{fe})(\sin\beta + \cos\beta)d_f}{s_f + w_f} > V_{\text{add}} \tag{9.15}$$

$$g_3(x) : \frac{(2t_f w_f f_{fe})(\sin\beta + \cos\beta)}{s_f + w_f} \leq \frac{2\sqrt{f_c'}b_w d}{3} - V_s \tag{9.16}$$

V_{add} and V_s are the required additional and total values of shear force capacity of reinforcing bar, respectively. Also, R is reduction factor and t_f is CFRP strip thickness;

Table 9.5 Information for design constants and parameters

	Explanation	Notation	Unit	Value/ranges/limits
Design constant	Breadth of beam	b_w	mm	200–500
	Height of beam	h		300–800
	Effective depth of beaö	d		0.9 h
	Thickness of CFRP strips	t_f		0.165
	Slab thickness	h_f		80–120
	Reduction factor	R	–	0.5
	Concrete compressive strength	f_c'	MPa	20
	Effective tensile strength of CFRP	f_{fe}		3790
	Additional shear force	V_{add}	N	50000–200000
	Shear capacity of reinforcing bar	V_s		50000
Design parameter	Width of CFRP strips	w_f	mm	10–1000
	Spacing among CFRP strips	s_f		0–d/4
	Wrapping angle of CFRP strips	β	°	0–90

besides that, f_{fe} and f_c' represent the effective tensile strength of CFRP and concrete compressive strength, respectively.

Additionally, in Table 9.5, values/ranges of design constants and upper and lower limits of design parameters are explained. As shown in this table, the values of some design constants have a specific range, without an exact number. The cause of this is to use these values in training dataset as input variables during following the application operated via ANNs.

- **The main ANNs training and prediction model**

Values of design parameters (s_f, w_f and β) provided in optimization process together with several design constants were handled to generate a training dataset. Design constants and parameters were defined as inputs and outputs in prediction applications by ANNs, respectively. Also, inputs were determined as h, b_w, h_f, V_{add} within all constants. But, there is something which needs to be remarked, a parameter as combination of s_f and w_f was proposed substituted for the usage of both. For this reason, the third output, which is called as CFRP rate and meant to how much there is width of CFRP until to starting of other consecutive CFRP strip, and symbolized as r, formulized as Eq. (9.17).

$$r = \frac{w_f}{w_f + s_f} \tag{9.17}$$

Table 9.6 Predictions ensured via main ANNs model and results of error evaluations for β and r

	Error performance metrics					Total data samples
	R	*MAE*	*RMSE*	*RAE*	*RRSE*	
$\beta(°)$	99.69%	0.0908	0.1155	8.0829%	7.9472%	231
CFRP rate (r)	99.70%	0.0121	0.0175	6.4224%	7.6681%	

R: correlation coefficient; *MAE:* mean absolute error; *RMSE:* root mean squared error; *RAE:* relative absolute error; *RRSE:* root relative squared error

After network parameters are determined, to realize the prediction process intended for target variables via ANNs, multilayer perceptrons (MLPs) were used with the same way of the first application. In this regard, MLPs algorithm was also preferred because it was operated in a machine learning software called Weka with 3.8.4 version [31].

In Table 9.6, for β and r, we can see the prediction results and error evaluations of main training model (generated by considering basic dataset), which explain the convergence performance of ANNs model to real optimum values.

Here, *R* is related to measure the similarity/closeness between actual and predictions belonging to data samples. Metric of relative absolute error (RAE) expresses how much prediction errors of all samples differ from actual value of deviation [32]. And finally, root relative squared error (RRSE) indicates the value of the root of relative squared error (RSE), which is meant to differ from the total of squared errors between actual-predicted values according to the total deviation of actual data [33].

As shown in Table 9.6, correlation coefficient, which shows the similarity of actual and predicted values, is very high (almost 100%) intended for both outputs. Also, error ratios are very successful with respect to deviation from the real value of data samples. In this regard, the main ANNs model can be benefited to observe the target values for any differences and unused structural combinations.

- **Evaluation of Test Models**

In this section, test models were evaluated with the predictions of ANNs main model corresponding to different test combinations that their values of inputs were generated between predefined ranges. The mentioned combinations of test models including optimum parameters (outputs) and constants (inputs) are shown in Table 9.7.

In the same way, ANNs predictions for test combinations and its measures of errors according to optimum data are shown in Table 9.8. Here, all error metrics are pretty small, especially for r parameter. For this reason, it can be said that the main prediction model makes convergence to real output values of data samples effectively and determines these values sensitively.

Table 9.7 Test model combinations for beam structure and optimum values of design parameters provided via JA

Inputs				Outputs			
h (mm)	b_w (mm)	h_f (mm)	V_{add} (N)	β (°)	w_f (mm)	s_f (mm)	CFRP rate (r)
350	200	120	70000	65.016	35.170	61.271	0.365
530	420	85	55000	63.324	18.639	54.053	0.256
750	370	100	120000	63.593	83.198	59.184	0.584
680	500	95	110000	62.608	58.911	53.272	0.525
420	450	120	70000	62.753	61.816	110.000	0.360
300	300	100	100000	64.259	99.541	104.068	0.489
770	220	90	80000	64.974	58.013	93.878	0.382
400	510	85	90000	62.683	76.496	106.484	0.418
550	200	80	50000	64.839	31.058	102.775	0.232
800	300	100	120000	64.217	97.372	68.646	0.587

Table 9.8 ANNs predictions for $\beta°$ and CFRP rate (r) intended for structural combinations of test models

CFRP rate (r)	Errors compared JA		β(°)	Errors compared JA	
	Absolute error	Squared error		Absolute error	Squared error
63.657	0.239	0.057	0.604	1.359	1.847
63.644	0.016	0.000	0.240	0.320	0.102
64.902	0.207	0.043	0.377	1.309	1.714
63.930	0.153	0.023	0.372	1.322	1.747
61.858	0.096	0.009	0.456	0.895	0.801
61.944	0.421	0.177	0.910	2.315	5.358
65.969	0.162	0.026	0.220	0.995	0.991
61.508	0.127	0.016	0.545	1.175	1.382
65.498	0.020	0.000	0.212	0.659	0.435
65.531	0.234	0.055	0.353	1.314	1.726
	MAE	*MSE*		*MAE*	*MSE*
Average	0.168	0.041		1.166	1.610
RMSE		0.202			1.269

MAE: mean absolute error; MSE: mean squared error; RMSE: root mean squared error

9.6 Conclusions

In the present study, two applications were conducted with the aim of the creation of optimum structural engineering designs related to different beam models. The first is the optimization of section area for T-shaped beam, and the second one is providing optimum design for a beam wrapped with CFRP strips as minimum wrapping area.

For these models, firstly, an optimization by using metaheuristics and secondly, the prediction process via ANNs are performed. Besides, a test model was also generated for both designs and their predicted values were evaluated by comparing with real optimum values. In all these durations, different error calculations and measurements were applied and considered in respect of making evaluations for predictions in correct and effective ways.

In the first design, for the main model, 450 different data samples were utilized in order to understand the success of prediction models by considering error metrics as MAE, MSE, and RMSE. As shown in Table 9.2, all error metrics are pretty good for the eight outputs; especially, they are very small for b and h_f. On the other side, for 50 combinations of test models, the calculated error measurements are excessive little according to the main models. Even, for almost all outputs, these error values are quite low. For this reason, the created ANNs prediction model can be accepted as a very successful tool with regard to reaching the target values.

As to the second design, for the main prediction model, all error calculations were ensured by using 231 data samples, and four different metrics (MAE, RMSE, RAE, and RRSE) together with correlation coefficient (R) were evaluated.

For both outputs as β and CFRP rate (r), correlation coefficients are extremely high and above 95%. Also, MAE and RMSE are 0.0908 and 0.1155 for β; 0.0121 and 0.0175 for r, respectively.

In error calculations for test combinations, it can be said that the decision-making model is a good correctly-predictor due to that values of three separate error metrics as MAE, MSE, and RMSE were occurred as really low level for test outputs.

In the direction of all these deductions, we can accept and think that ANNs models from machine learning techniques make a good job about numerical prediction and determination of target values of parameters for any structural engineering and optimization problem, too. Moreover, with applications in the current study, it can be also understood that machine learning methods can study and collaborate with optimization processes or data from optimization.

References

1. Ramezani M, Bathaei A, Ghorbani-Tanha AK (2018) Application of artificial neural networks in optimal tuning of tuned mass dampers implemented in high-rise buildings subjected to wind load. Earthq Eng Eng Vib 17(4):903–915
2. Yucel M, Bekdaş G, Nigdeli SM, Sevgen S (2019) Estimation of optimum tuned mass damper parameters via machine learning. J Build Eng 26:100847
3. Yang L, Qi C, Lin X, Li J, Dong X (2019) Prediction of dynamic increase factor for steel fibre reinforced concrete using a hybrid artificial intelligence model. Eng Struct 189:309–318
4. Khalilzade Vahidi E, Rahimi F (2016) Investigation of ultimate shear capacity of RC deep beams with opening using artificial neural networks. Adv Comput Sci Int J 5(4):57–65
5. Yucel M, Bekdas G, Nigdeli SM, Sevgen S (2018) Artificial neural network model for optimum design of tubular columns. Int J Theor Appl Mech 3:82–86

6. Chen XL, Fu JP, Yao JL, Gan JF (2018) Prediction of shear strength for squat RC walls using a hybrid ANN–PSO model. Eng Comput 34(2):367–383
7. Behnood A, Golafshani EM (2018) Predicting the compressive strength of silica fume concrete using hybrid artificial neural network with multi-objective grey wolves. J Clean Prod 202:54–64
8. Yücel M, Nigdeli SM, Bekdaş G (2019) Estimation model for generation optimization of design variables for I-beam vertical deflection minimization. In: 4th Eurasian conference on civil and environmental engineering (ECOCEE), İstanbul, Turkey
9. Yucel M, Nigdeli SM, Bekdaş G (2019) Generation of an artificial neural network model for optimum design of I-beam with minimum vertical deflection. In: 12th HSTAM international congress on mechanics. Thessaloniki, Greece
10. Nguyen HQ, Ly HB, Tran VQ, Nguyen TA, Le TT, Pham BT (2020) Optimization of artificial intelligence system by evolutionary algorithm for prediction of axial capacity of rectangular concrete filled steel tubes under compression. Materials 13(5):1205
11. Graupe, D (2013) Backpropagation. In: Principles of artificial neural networks, vol 7, 3rd ed. World Scientific, Singapore, p 59
12. Kennedy J, Eberhart RC (1995) Particle swarm optimization. In: Proceedings of IEEE international conference on neural networks No. IV, Perth Australia, November 27–December 1, pp 1942–1948
13. Yang XS (2012) Flower pollination algorithm for global optimization. In: International conference on unconventional computing and natural computation. Heidelberg-Berlin, Springer, pp 240–249
14. Mirjalili S, Mirjalili SM, Lewis A (2014) Grey wolf optimizer. Adv Eng Softw 69:46–61
15. Holland JH (1975) Adaptation in natural and artificial systems. University of Michigan Press
16. Farmer JD, Packard NH, Perelson AS (1986) The immune system, adaptation, and machine learning. Physica D 22(1–3):187–204
17. Geem ZW, Kim JH, Loganathan GV (2001) A new heuristic optimization algorithm: harmony search. Simulation 76(2):60–68
18. Yang XS (2010) A new metaheuristic bat-inspired algorithm. In: González JR, Pelta DA, Cruz C, Terrazas G, Krasnogor N (eds) Nature inspired cooperative strategies for optimization (NICSO 2010). Springer, Berlin, Heidelberg
19. Rao RV, Savsani VJ, Vakharia DP (2011) Teaching–learning-based optimization: a novel method for constrained mechanical design optimization problems. Comput Aided Des 43(3):303–315
20. Birbil Şİ, Fang SC (2003) An electromagnetism-like mechanism for global optimization. J Global Optim 25(3):263–282
21. Erol OK, Eksin I (2006) A new optimization method: big bang–big crunch. Adv Eng Softw 37(2):106–111
22. Rashedi E, Nezamabadi-Pour H, Saryazdi S (2009) GSA: a gravitational search algorithm. Inf Sci 179(13):2232–2248
23. Rao R (2016) Jaya: a simple and new optimization algorithm for solving constrained and unconstrained optimization problems. Int J Ind Eng Comput 7(1):19–34
24. EN (Veranst.): EN 1992-1-1 Eurocode 2 (2005) Design of concrete structures. CEN, Brussels
25. Fedghouche F, Tiliouine B (2012) Minimum cost design of reinforced concrete T-beams at ultimate loads using Eurocode2. Eng Struct 42:43–50
26. Kayabekir AE, Bekdaş G, Nigdeli SM (2019) Optimum design of T-beams using Jaya algorithm. In: 3rd international conference on engineering technology and innovation (ICETI), Belgrad, Serbia
27. The MathWorks, Matlab R2018a (2018) Natick, MA
28. ACI-318 (2005) Building code requirements for structural concrete and commentary, metric version. American Concrete Institute
29. Yucel M, Kayabekir AE, Nigdeli SM, Bekdaş G (2020) Optimum design of carbon fiber-reinforced polymer (CFRP) beams for shear capacity via machine learning methods: optimum prediction methods on advance ensemble algorithms–bagging combinations. In: Artificial intelligence and machine learning applications in civil, mechanical, and industrial engineering. IGI Global, pp 84–102

30. Khalifa A, Nanni A (2000) Improving shear capacity of existing RC T-section beams using CFRP composites. Cem Concr Compos 22(3):165–174
31. Witten IH, Frank E, Hall MA, Pal CJ (2016) Data mining: practical machine learning tools and techniques, 4th edn. Morgan Kaufmann, Cambridge, Massachusetts, MA
32. Cichosz P (2015) Regression model evaluation. In: Data mining algorithms: explained using R. Wiley, Chichester, United Kingdom, pp 299–300
33. Witten IH, Frank E (2005) Credibility: evaluating what' s been learned. Data mining: practical machine learning tools and techniques, 2nd edn. Morgan Kaufmann, San Francisco, California, pp 177–178

Chapter 10
Statistical Investigation of the Robustness for the Optimization Algorithms

Esra Uray, Huseyin Hakli, and Serdar Carbas

10.1 Introduction

The basis of engineering is based on generating solutions with analysis to current problems which have become complicated with the developing technology and increasing population. For this purpose, optimization methods have been employed commonly since from time immemorial coming from times of Newton, Lagrange, and Cauchy to nowadays [1]. Optimization utilized to lessen the gain or to increase yield in many fields has applications in a widespread manner for the solution of many complicated engineering problems today. The process of obtaining the minimum or the maximum global point of a design problem by utilizing methods based on mathematical or heuristic has been designated as the optimization.

Until recently, traditional optimization methods were widely used in various engineering optimization applications. However, the mathematical solution algorithms based on the derivative, the principle of continuously accepting the design variables, and being inadequate for practicing of infinite design space have made it challenging to implement the methods in large-scale systems. Also, utilizing continues optimization methods could not create ideal results for the realistic designs of engineering structures that require sizing according to pre-determined ready sections.

E. Uray (✉)
Department of Civil Engineering, KTO Karatay University, Konya, Turkey
e-mail: esra.uray@karatay.edu.tr

H. Hakli
Department of Computer Engineering, Necmettin Erbakan University, Konya, Turkey
e-mail: hhakli@erbakan.edu.tr

S. Carbas
Department of Civil Engineering, Karamanoglu Mehmetbey University, Karaman, Turkey
e-mail: scarbas@kmu.edu.tr

S. Carbas et al. (eds.), *Nature-Inspired Metaheuristic Algorithms for Engineering Optimization Applications*, Springer Tracts in Nature-Inspired Computing,
https://doi.org/10.1007/978-981-33-6773-9_10

Genetic algorithms (GAs) developed by Goldberg [2] emerged as a method that can overcome these cases encountered in engineering optimization problems. This heuristic method working on the principle of simulating in the computer for concepts of nature has been effectively and prosperously conducted in several engineering disciplines like in hydraulics [3], vehicle routing [4], structural optimization [5, 6], geotechnical [7], power system [8], data clustering [9], and more of the same due to its ability to apply the discrete problems and to converge about to the global solution without considering the initial point of the design variables. This withstanding feedback, achieved by GAs, has been encouraged researchers to found new heuristic optimization algorithms like simulated annealing (SA) [10], tabu search (TS) [11, 12], ant colony optimization (ACO) [13], harmony search algorithm (HSA) [14], particle swarm optimization (PSO) [15], and so forth. Also, recently qualified heuristic optimization methods have been developed by Karaboğa as the artificial bee colony algorithm (ABC) [16], by Yang as the firefly algorithm (FA) [17], and by Rajabioun as the cuckoo search algorithm (CS) [18].

Each heuristic optimization method has the parameters effected on robustness and success of the algorithm for the process of finding the global optimum solution while satisfying the design constraints. For example, ABC algorithms' parameters like the number of food sources, the population size (the number of employed, onlooker, and scout bees), the limit, the maximum iteration number, etc., are responsible for checking of deterred of the local solution, stored of the global solution, dropped of the food source, the velocity of convergence, and the stopping criteria. Significant numbers of studies [19, 20] have been performed with employing test functions or benchmark problems to investigate the used current heuristic optimizer whose algorithm parameters give the expected result of the optimum case. Nevertheless, this exploratory requires a huge number of trials to ensure whether the parameter sets of the current algorithm are reasonable or not for the desired optimum design. Also, the value of the algorithm parameters which is given the best result in a benchmark problem [21] or a test function [22, 23] may not provide the same achievement in the optimum design of a geotechnical structural optimization problem due to the characteristic of that problem or behavior of the algorithm in seeking of the optimum. This situation may arise from having the continuous or discrete design variables, being constraint or unconstraint design, finding the optimum solution for single or multi-objective function, the number of the design variables, etc. Since achieving the best values of the algorithm parameters are possible with many trials and errors, it is taken much time. Besides, one cannot be sure that the best combine is achieved, considering the situation involving all combinations of algorithm parameters. Therefore, examining the effect of the parameters statistically on the optimum result has been essential to boost the robustness and effectiveness of the optimization algorithm without performing many trials for available values of the algorithm parameters.

Overcoming this phenomenon is possible by finding the best or the worst combine statistically via Taguchi method [24]. Taguchi method proposed by Genichi Taguchi as an experimental design is widely utilized in quality management and development. In the same manner, it may be utilized for investigation of the parameter effect in experiments or trials, which are the most indispensable elements of scientific research, and to obtain the best or the worst design case. Since the orthogonal array suggested by Taguchi allows a small number of experiments rather than all the experiments involving the whole parameters combination, it plays a significant role in the investigation of parameters effect. The effect of the design parameters on the stability of the cantilever retaining wall has acquired by performing just 16 design instead of 1024 design combinations in the study conducted by Uray et al. [25]. Although there are some other studies of the genetic algorithms [26] and the blind mole-rat algorithm [27] for investigating the algorithm parameters effect, this kind of study has not been encountered for the scatter search to the best of our knowledge in the literature, so far. This chapter focuses on the prediction of the algorithm parameters for the scatter search (SS), which is one of the novel heuristic methods.

Due to the SS has advantages such as flexible structure, ease of implementation, and systematic design, it was applied to many real-world engineering design problems. Russell and Chiang [28] used SS to solve the vehicle routing problem, and their study aimed to obtain an effective solution. They proved that the solution quality of the SS algorithm was robust, and it is comparable with other metaheuristic algorithms. The SS algorithm was implemented to bi-criteria coal production planning problem by Pendharkar [29]. A permutation flow shop problem was solved by SS, and restarts and local search features were employed in their approach [30]. Experimental results showed that SS was a state-of-the-art technique, and many best solutions were yielded by SS for the flow shop problem. Hakli and Ortacay [31] performed a new approach of the SS algorithm (ISS) with four different crossover techniques using ensemble search strategy, and this technique was applied to the incapacitated facility location problem. ISS algorithm was an efficient and robust tool for this problem considering experimental results. Furthermore, the SS algorithm is used to solve many design problems such as detecting credit card fraud [32], land partitioning problem [33], mixed network design [34], locating capacitated transshipment points [35], and aircraft landing problem [36].

In order to investigate the algorithm parameters of the SS algorithm effect on the optimum result and find the most suitable combination of them, the cantilever retaining wall design was taken into consideration as a design example (Fig. 10.1).

In geotechnical engineering, the cantilever retaining wall is used to resist lateral soil load between two soil levels. Obtaining optimum wall dimensions like the base length, the toe extension, the base thickness, and the front face angle of the wall, which has the minimum cost, is a kind of the discrete optimization problem. At the same time, the optimum wall design should satisfy stability conditions like sliding, overturning, slope stability, and bearing capacity for various soil conditions. The optimum design of the cantilever retaining wall commonly has been investigated by utilizing the genetic algorithms (GAs) [38], the simulated annealing (SA) algorithm [39], the particle swarm optimizer (PSO) [40], the big bang-big crunch (BB-BC) algorithm [41], the firefly algorithm (FA) [42], the charged system search algorithm (CSSA) [43], the teaching learning-based optimization (TLBO) [44], and so on.

In this chapter, reasonable values of the SS algorithms' parameters according to the best case are estimated by the orthogonal arrays suggested by Taguchi, and the effect of the parameters on results has been determined by variance analysis.

Fig. 10.1 Cantilever retaining wall [37]

Then, a real case and a predicted case are compared in terms of the relative error. Obtained available values of the SS algorithm parameters via Taguchi method have been presented for the optimum design of the cantilever retaining wall.

10.2 Optimization Analysis via Scatter Search

10.2.1 Scatter Search

The scatter search (SS) algorithm, one of the evolutionary-based optimization techniques, was proposed by Glover [45]. The principles of the SS algorithm were raised in the 1960s, so it has a long history. Generally, evolutionary-based optimization algorithms use a selection mechanism based on randomness and do not consider benefit from the best results enough. In the SS algorithm, systematic designs and methods perform the selection of population and determine a *reference set* (*RefSet*) to generate new solutions [33, 46]. *RefSet* contains not only the best solutions for the current population but also the most different solutions from these best solutions. Therefore, this systematic design provides a good balance between intensification and diversification for the SS algorithm.

The SS has five phases of its implementation [45, 46]:

- *Diversification Generation*: Generate the initial population by the arbitrary trial solution.
- *Improvement*: Try to improve these initial solutions using local search methods. This step is optional.
- *Reference Set Update*: First, select the best b_1 solutions from the population. Add the most different solutions from these best b_2 solutions to *RefSet*. The $b_1+ b_2$ represents the size of the *RefSet* and named as *RFC*. Update the reference set according to their quality or their diversity of new solutions.
- *Subset Generation*: Determine the subsets (parents) to generate new solutions.
- *Solution Combination*: Obtain the new solutions and prepare the solution pool by one or more search strategy using the subsets.

Figure 10.2 shows the flowchart of the SS algorithm [31] that is an example of a statistical-based optimization algorithm.

10.2.2 The Optimum Design of the Cantilever Retaining Wall

In this section, the design example taken into consideration for obtaining parameters effect of the SS algorithm via Taguchi method is presented in detail. The study of the optimum design of the cantilever retaining wall (CRW) [25] has been utilized partially as a design example.

In the optimal design problem, the base length (X_1), the toe extension (X_2), the base thickness (X_3), and the front face angle (X_4) are treated as discrete design variables. The design variables and acting loads on the CRW are demonstrated in Fig. 10.3.

The discrete design variables with their lower and upper limits to form the design space has been presented in Table 10.1.

The minimum wall weight to obtain the optimum cost of the CRW has been considered as the objective function. Equation 10.1 presents the mathematical expression of the objective function.

$$f_{\text{min}} = (X_1 X_3 + bH + 0.50 H^2 X_4)\gamma_c \tag{10.1}$$

Here, b is the top thickness of CRW; γ_c is the unit volume weight of concrete for CRW. The values of them are taken, respectively, as 0.25 m and 25 kN/m^3 in the analysis.

In the optimum design of the CRW, stability conditions of the wall should be satisfied with considering soil conditions. The safety factors of sliding, overturning, slope stability, and bearing capacity of soil have been taken as stability conditions.

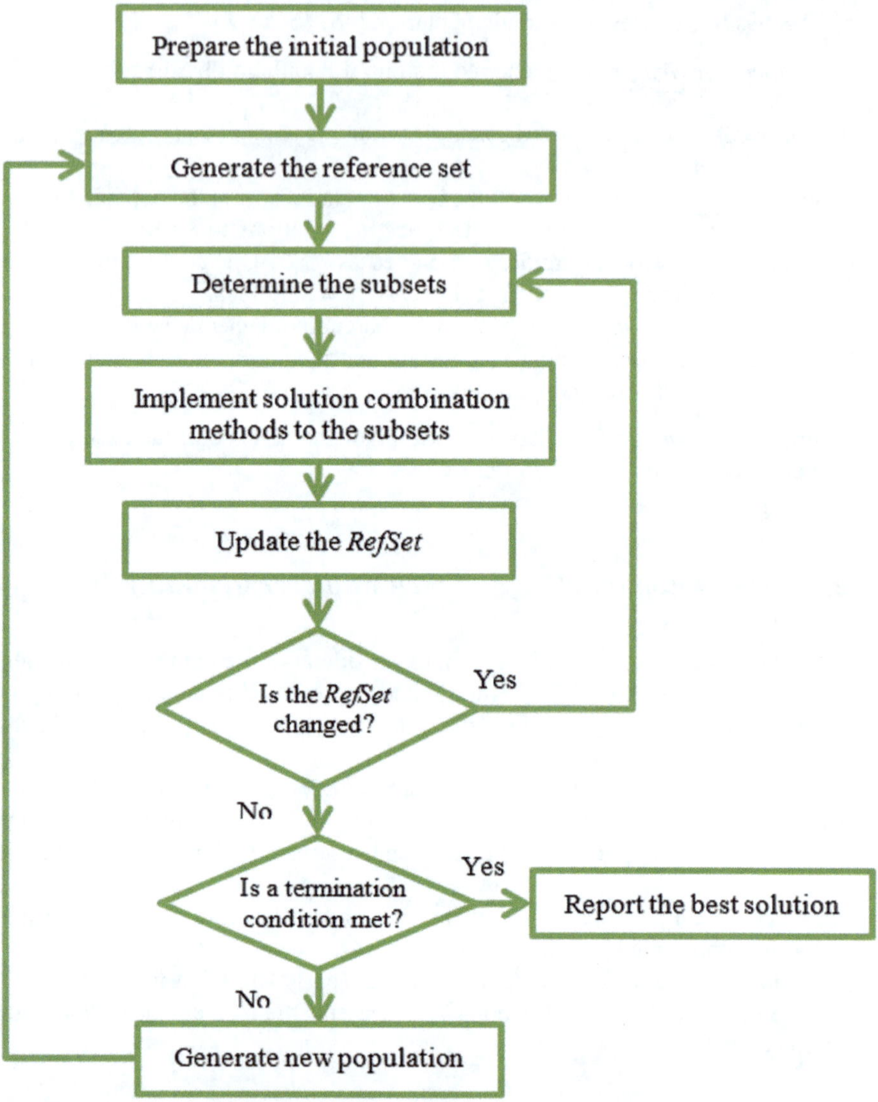

Fig. 10.2 Flowchart of SS algorithm

Constraint of sliding safety factor $g_1(x)$, overturning safety factor, $g_2(x)$, and slope stability safety factor, $g_3(x)$, must be equal or greater than 1.5 [48] for the condition that satisfying stability of the CRW. Equations 10.2–10.5, respectively, have demonstrated the normalized mathematical expressions of $g_1(x)$, $g_2(x)$, and $g_3(x)$.

$$g_1(x) = 1 - F_s/1.50 \tag{10.2}$$

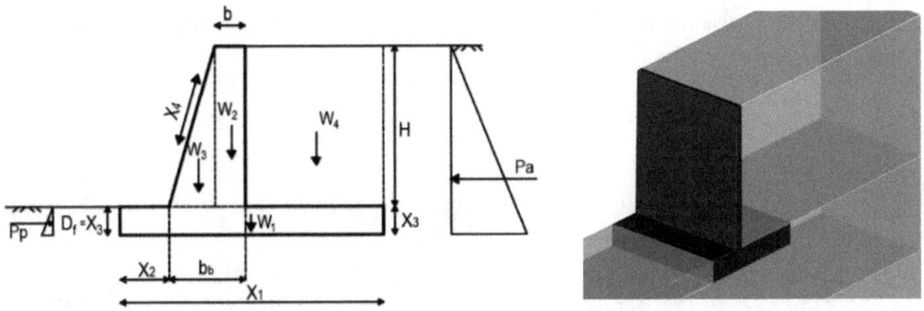

Fig. 10.3 Design variables and acting loads and a GEO 5 model of the CRW [25, 47]

Table 10.1 Discrete design parameters and definition of limits [25]

Design parameters	Lower limit	Upper limit	Interval
X_1: the base length (H)	0.25	1.0	0.05
X_2: the toe extension	0.15 × 1	0.60 × 1	0.05 × 1
X_3: the base thickness (H)	0.06	0.15	0.015
X_4 (%): the front face angle	0	4	1

$$g_2(x) = 1 - F_o/1.50 \tag{10.3}$$

$$g_3(x) = 1 - F_{ss}/1.50 \tag{10.4}$$

The constraints for the safety factor of sliding (F_s) and the safety factor of overturning (F_o) are formulated by Eqs. 10.5 and 10.6, respectively.

$$F_{ss} = \frac{\left[(X_1X_3 + bH + 0.50H^2X_4)\gamma_c + (X_1 - X_2 - b + HX_4)H\gamma_s\right]\tan\delta + P_p}{P_a} \tag{10.5}$$

$$\begin{aligned}
F_{so} &= \frac{M_{W1} + M_{W2} + M_{W3} + M_{W4} + 0.33X_3P_p}{0.33HP_a} \\
M_{W1} &= 0.50X_1^2X_3\gamma_c \\
M_{W2} &= bH\gamma_c(X_2 + 0.50b + HX_4) \\
M_{W3} &= 0.50H^2X_4\gamma_c[0.67HX_4 + X_2] \\
M_{W4} &= 0.50(X_1 - X_2 - b - HX_4)H\gamma_s(X_1 + X_2 + HX_4)
\end{aligned} \tag{10.6}$$

Here, γ_s is the unit volume weight of soil; δ is the coefficient of friction between wall and soil and is taken as $2/3\varnothing$. The $\varnothing$ is the angle of internal friction the soil.

The P_a and P_p are active and passive soil pressure forces and their mathematical expressions are given by Eqs. 10.7 and 10.8.

$$P_a = 0.50H^2\gamma_s \tan^2(45 - \phi/2) \quad (10.7)$$

$$P_p = 0.50X_3^2\gamma_s \tan^2(45 + \phi/2) \quad (10.8)$$

The mathematical model, which was statistically improved by Uray et al. [25], has been utilized in the determination of the slope stability safety factor (F_{ss}). It is a troublesome process to calculate the slope stability safety factor by utilizing theoretical formulations like safety factors of sliding and overturning. Therefore, the mathematical model given by Eq. 10.9 has been used to include checking of slope stability for the optimum CRW design.

$$\begin{aligned} F_{ss} &= \sqrt{\frac{1}{10^{-\lambda/10}}} \\ \lambda &= -0.9481(X_1/H)^3 + 1.104(X_1/H)^2 + 3.1679(X_1/H) \\ &- 0.0165(X_2/X_1)^3 - 1.1675(X_2/X_1)^2 - 1.80(X_2/X_1) \\ &- 2336.4(d/H)^3 + 702.1(X_1/H)^2 - 48.723(X_1/H) \\ &+ 43197(X_4)^3 - 2498.5(X_4)^3 + 37.666(X_4) \\ &+ 14.299(\tan\phi)^3 - 38.059(\tan\phi)^2 + 45.098(\tan\phi) - 11.0541 \end{aligned} \quad (10.9)$$

The checking of bearing capacity of the soil which provides safely carrying loads come from the wall has been realized by the constraint given in Eq. 10.10.

$$g_4(x) = 1 - F_{bc}/3.0 \quad (10.10)$$

The allowable the bearing capacity safety factor has been taken as 3.0, considering the suggested value in the literature [49]. The safety factor bearing capacity (F_{bc}) is calculated by Eq. 10.11.

$$F_{bc} = q_u/q_{max} \quad (10.11)$$

General bearing capacity expression suggested by Meyerhof [50] have been utilized for the calculation of the ultimate bearing capacity (q_u) [48].

The minimum base pressure (q_{min}) in the interface between the soil and the wall base must be greater than the zero ($e < X_1/6$) in order not to occur tensile stress in the soil. The constraint provided this condition is given by Eq. 10.12.

$$g_5(x) = X_1/(6e) \quad (10.12)$$

The mathematical formulations of the minimum (q_{min}) and the maximum (q_{max}) base pressures and eccentricity (e) are given in Eqs. 10.13 and 10.14, respectively.

$$q_{\substack{min\\max}} = \frac{(X_1X_3 + bH + 0.50H^2X_4)\gamma_c + (X_1 - X_2 - b + HX_4)H\gamma_s}{X_1}\left(1 \pm \frac{6e}{X_1}\right) \tag{10.13}$$

$$e = \frac{X_1}{2} - \frac{M_{W1} + M_{W2} + M_{W3} + M_{W4} + 0.33X_3P_p - 0.33HP_a}{(X_1X_3 + bH + 0.50H^2X_4)\gamma_c + (X_1 - X_2 - b + HX_4)H\gamma_s} \tag{10.14}$$

Equations 10.15 and 10.16 are presented the geometrical constraints due to wall dimensions for the CRW design.

$$g_6(x) = (X_2 + HX_4 + b)/X_1 - 1 \tag{10.15}$$

$$g_7(x) = b/(H * X_4 + b) - 1 \tag{10.16}$$

10.3 Taguchi Method and Implementation of the SS Algorithm to the CRW Design

10.3.1 Taguchi Method

Taguchi method is a technique, which given the effects of the parameters on the results with a small number of experiment or trial [24]. This technique is a robust and an alternative improved method to reduce the cost in the studies of optimization and parametric analysis, to reach results in less time, and to determine the effects of the parameters on the result. According to Taguchi method, the parameters affecting the results and the process are divided into two groups, controllable and uncontrollable factors.

In the investigation of controllable factors affect on the uncontrollable factors, signal/noise (*S/N*) ratio, which is defined as control criteria is employed. The *S/N* ratio has three states according to the desired goal; (i) the larger is the better, (ii) the smaller is the better, and (iii) the nominal is better. Equations 10.17–10.19 give the mathematical expressions of *S/N* ratios, respectively.

$$S/N = -10\log\left(\frac{1}{n}\sum_{i=1}^{n}\frac{1}{Y_i^2}\right) \tag{10.17}$$

$$S/N = -10\log\left(\frac{1}{n}\sum_{i=1}^{n}Y_i^2\right) \tag{10.18}$$

Table 10.2 *L*16 orthogonal array

Design no	P_1	P_2	P_3	P_4
1	1	1	1	1
2	1	2	2	2
3	1	3	3	3
4	1	4	4	4
5	2	1	2	3
6	2	2	1	4
7	2	3	4	1
8	2	4	3	2
9	3	1	3	4
10	3	2	4	3
11	3	3	1	2
12	3	4	2	1
13	4	1	4	2
14	4	2	3	1
15	4	3	2	4
16	4	4	1	3

P design parameter

$$S/N = -10\log\left(\frac{1}{n}\sum_{i=1}^{n}\frac{\bar{Y}}{\sigma^2}\right) \tag{10.19}$$

Here, Y_i is defined as the response value; n is the number of repetitions; $\overline{Y}_i$ is the arithmetic mean of the sequence; σ is the standard deviation.

Taguchi has presented the orthogonal arrays that the general representation is $Ld^k_{(a)}$ or Ld. Here, d is the total number of experiments, a is the level of the parameters, k is the number of the parameters, and L is the orthogonal array. For the investigation of the algorithm parameters effects, $L16$ and $L9$ orthogonal arrays have been taken into consideration in this chapter. $L16$ and $L9$ orthogonal array are tabulated in Tables 10.2 and 10.3, respectively.

10.3.2 Implementation of SS Algorithm to the CRW Design

In this chapter, the SS method was applied to the CRW problem in order to carry out the performance and the statistical analyses. The wall height (H) is taken 6 m, the angle of internal friction (∅) is taken 30°, and the unit volume weight of soil is taken 20 kN/m^3 for CRW optimum design.

Table 10.3 *L*9 orthogonal array

Design no	P_1	P_2	P_3
1	1	1	1
2	1	2	2
3	1	3	3
4	2	1	2
5	2	2	3
6	2	3	1
7	3	1	3
8	3	2	1
9	3	3	2

The minimum wall weight has been investigated for 5600 design combinations (16 × 10 × 5 × 7) by considering different discrete values of the design parameters. When the implementation of methods to the optimization problem, the representation of the problem has a pivotal role due to the effect on the search strategy. In the optimum design of the CRW problem, four decision variables (X_1, X_2, X_3, and X_4) have been utilized to minimize the concrete weight of the wall. Although the values of these decision variables for the CRW design problem are floating number like 0.2, 0.3, etc., its search space is not considered as a continuous form. As considering the search domain is continuous, the decision variables can take all values in the specified range. The values that X_3 variable are restricted within the specified design as 0.36, 0.45, 0.54, 0.63, 0.72, 0.81, and 0.90. Therefore, even the variables' values are floating-point number; its search space can be evaluated as discrete.

Table 10.4 shows the values of variables for the CRW design problem in the specified design explained detail in Sect. 10.2. In this design, while the parameter X_1 takes sixteen different values, the X_2 ten, the X_3 seven, and the X_4 have five different values. In order to ease computation, index numbers are used instead of values in the representation of solutions. So, Table 10.4 also indicates the transformation of values to indexes. Design variables and value ranges are limited to certain numbers in order to complete the problem analysis quickly.

A sample individual and its real values are given in Fig. 10.4. Index form is used in the population, and in search strategy for ease computation, the index form is transformed into real form for the calculation of fitness.

The initial population is randomly generated with index form, and fitness values for each individual are evaluated by objective function using real values. The objective function of CRW also contains seven constraint functions, so it should check whether the candidate solution is in the infeasible region or satisfies the constraints. Generally, a penalty value is used in order to handle with the constraints for evolutionary-based algorithms [51]. In this chapter, overflows of all constraint equations are summed as a penalty value. Besides, Deb's rules [52] are implemented to the SS algorithm as a constraint handling method for comparing the solution quality. On the selection of the best solutions, the population is firstly sorted with penalty values

Table 10.4 Values of variables for the optimum CRW design problem

	Variables			
Number	$X_1(m)$	$X_2(X_1)$	$X_3(m)$	$X_4(\%)$
1	1.50	0.15	0.36	0.00
2	1.80	0.20	0.45	0.01
3	2.10	0.25	0.54	0.02
4	2.40	0.30	0.63	0.03
5	2.70	0.35	0.72	0.04
6	3.00	0.40	0.81	
7	3.30	0.45	0.90	
8	3.60	0.50		
9	3.90	0.55		
10	4.20	0.60		
11	4.50			
12	4.80			
13	5.10			
14	5.40			
15	5.70			
16	6.00			

Fig. 10.4 Sample individual

Index Form

Indv	1	2	3	4	Index
	13	9	2	5	Value

Real Form

Indv	1	2	3	4	Index
	5.10	0.55	0.45	0.04	Real Value

and then sorted considering the fitness values. If more than one solution has the same penalty value or no penalty, these solutions are considered by fitness values. Improvement phase of the SS is optional, and this phase was not used on this implementation. Suppose that b represents the size of *RefSet*, then firstly b_1 individuals of *RefSet* are selected from the best solutions in the population. Secondly, the remaining part of *RefSet* is determined with the most different individuals from the best-selected ones [33]. So, *RefSet* is constructed considering the quality and diversity of individuals.

In the SS algorithm, individuals in the *RefSet* are used to generate new solutions by the subset generation method. Each subset consists of two-individual in this study, so $b*(b - 1)/2$ different subsets are operated. These subsets are considered as parents in the solution combination methods; in other words, crossover techniques. The crossover techniques have an essential role for evolutionary-based algorithms, and there are various crossover techniques such as one-point, two-point, uniform, and inversion [31]. Due to the representation of the CRW design problem, crossover techniques of one-point, two-point, uniform, and ensemble are used in this chapter. Ensemble crossover technique is a mixed form which is composed from one-point, two-point, and uniform techniques. These three techniques are applied with changing only the values, not the positional moves. In the CRW design problem, the values of decision variables are different from each other. Therefore, the values of the variables can be undefined when using the crossover techniques changing the position of variables such as inversion and segregation. In the one-point crossover technique, one crossover point is randomly selected, and the dimensions from after this point are copied from *Parent2*, while the others are copied from *Parent1*. Two crossover points are randomly determined, and dimensions between these points are copied from *Parent1*, and the rest one is completed from *Parent2* for the two-point crossover technique. In the uniform crossover technique, dimension is selected from one parent with an equal probability according to randomly generated number in [0,1]. Also, all of these techniques are performed as an ensemble form thanks to the flexible structure of the SS algorithm. The implementation of crossover techniques used in this chapter are given in Fig. 10.5 for one-point, in Fig. 10.6 for two-point, and in Fig. 10.7 for the uniform.

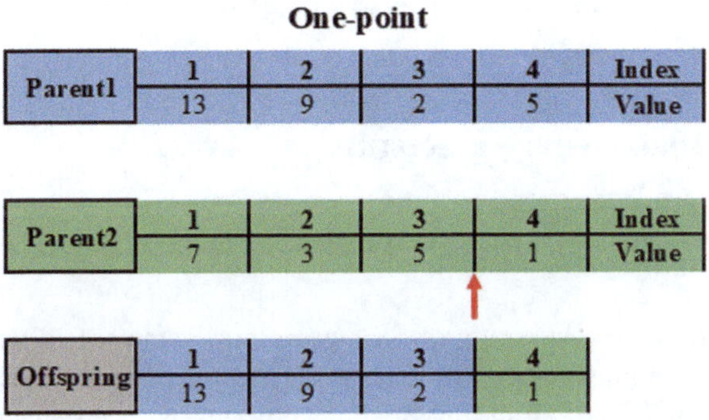

Fig. 10.5 Implementation of crossover techniques—one-point

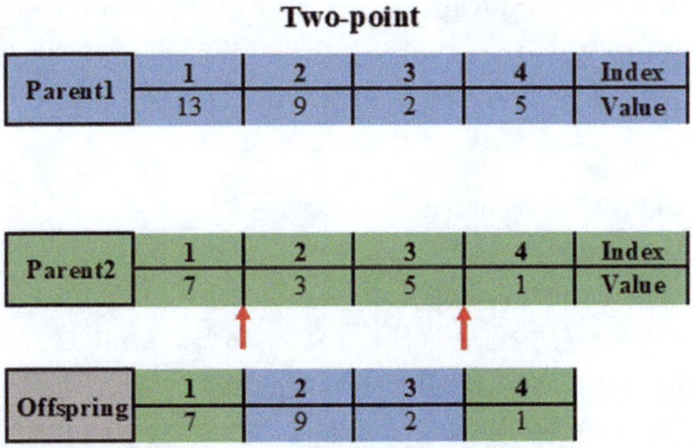

Fig. 10.6 Implementation of crossover techniques—two-point

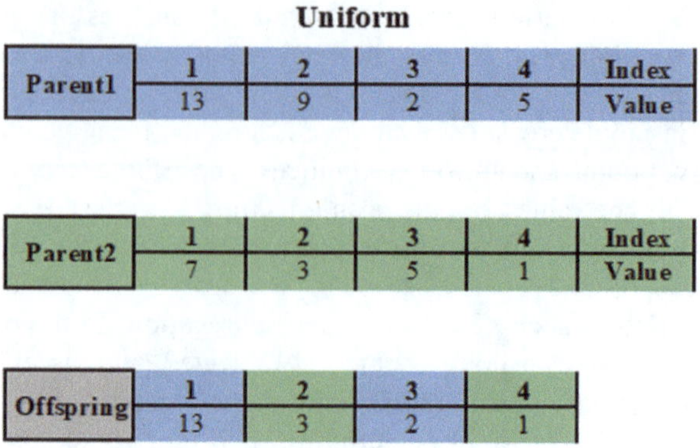

Fig. 10.7 Implementation of crossover techniques—uniform

The novel individuals generated by solution combination method are collected in a solution pool. If there is a duplicated individual in this pool, this individual is extracted from the pool in order to use efficiently of several functions evaluations (*FEs*). The objective function calculates the fitness values of individuals in this pool, and the pool is sorted in terms of fitness value. The *RefSet* is updated by selecting the best $b = b_1 + b_2$ individuals in the pool. After the update, if new RefSet is equal to the old *RefSet*, a new population is generated, and the selected b_2 individuals from this population are added to the *RefSet* until the termination criterion is satisfied [31].

10.4 Analysis Results

10.4.1 Statistical Analysis via L16 *Design Table*

In the investigation of SS algorithm parameters effect on the optimization process, the number of the population (*NP*), the reference set (*RefSet*), the maximum iteration number (*Iter*), and crossover technique have been selected as controllable factors. The different parameter values have been determined to observe which algorithm parameters values are available proper for the optimum value of the design. The controllable factors with four levels of the algorithm parameters are given in Table 10.5.

Table 10.5 Controllable factors

Parameter	Level 1	Level 2	Level 3	Level 4
NP: the number of the population	20	30	40	50
RefSet: the reference set	6	8	10	12
Iter: the maximum iteration number	125	250	375	500
Crossover technique	One-point	Two-point	Uniform	Ensemble

Table 10.6 SS algorithm parameter design table for *L*16 orthogonal array

Design no	*NP*	*RefSet*	*Iter*	Crossover technique
1	20	6	125	One-point
2	20	8	250	Two-point
3	20	10	375	Uniform
4	20	12	500	Ensemble
5	30	6	250	Uniform
6	30	8	125	Ensemble
7	30	10	500	One-point
8	30	12	375	Two-point
9	40	6	375	Ensemble
10	40	8	500	Uniform
11	40	10	125	Two-point
12	40	12	250	One-point
13	50	6	500	Two-point
14	50	8	375	One-point
15	50	10	250	Ensemble
16	50	12	125	Uniform

The SS algorithm parameters design table is shown in Table 10.6, which is formed by utilizing design parameters with their levels given in Table 10.5 and *L*16 orthogonal array in Table 10.2.

By utilizing the SS algorithm, which is given detailed in Sect. 10.2, sixteen CRW designs have been performed according to values of parameter presented in Table 10.6. The optimum wall weights of sixteen CRW designs have been obtained by operating 30 times for the specified iterations. And then, the *S/N* ratios have been calculated by employing Eq. 10.18 given for the case of the smaller is the better. The optimum wall weights (W_{wall}) and the *S/N* ratios for sixteen wall designs are shown in Table 10.7. The graphical representation of the average *S/N* ratio for the parameters and levels parameter is demonstrated in Fig. 10.8.

It is obviously seen that the most influential parameter is the crossover technique when the altering of the average *S/N* ratios in Fig. 10.8 have been examined.

The variance, which is the distribution around the arithmetic mean of the number sequence, is defined as the sum of squares from the difference between the arithmetic mean and current number value. In the investigation of the effect of the SS algorithm parameters, variance analysis has been performed by utilizing the average *S/N* ratios. Similar to the average *S/N* change graph, it has been conferred that the parameter with the most variance is the crossover technique according to the result of the variance analysis tabulated in Table 10.8.

Parameter levels which are provided with the optimum wall weight have been estimated by using the Taguchi method. The parameters levels are given in Table 10.9.

Table 10.7 Optimum wall weights and the *S/N* ratios for sixteen wall designs

Design no	W_{wall}	*S/N*
1	119.31	−41.534
2	101.16	−40.100
3	91.41	−39.220
4	90.07	−39.092
5	90.81	−39.163
6	90.94	−39.175
7	128.31	−42.165
8	95.16	−39.569
9	89.34	−39.021
10	91.5	−39.228
11	97.605	−39.789
12	119.88	−41.575
13	97.5	−39.780
14	122.49	−41.762
15	90.06	−39.091
16	89.34	−39.021

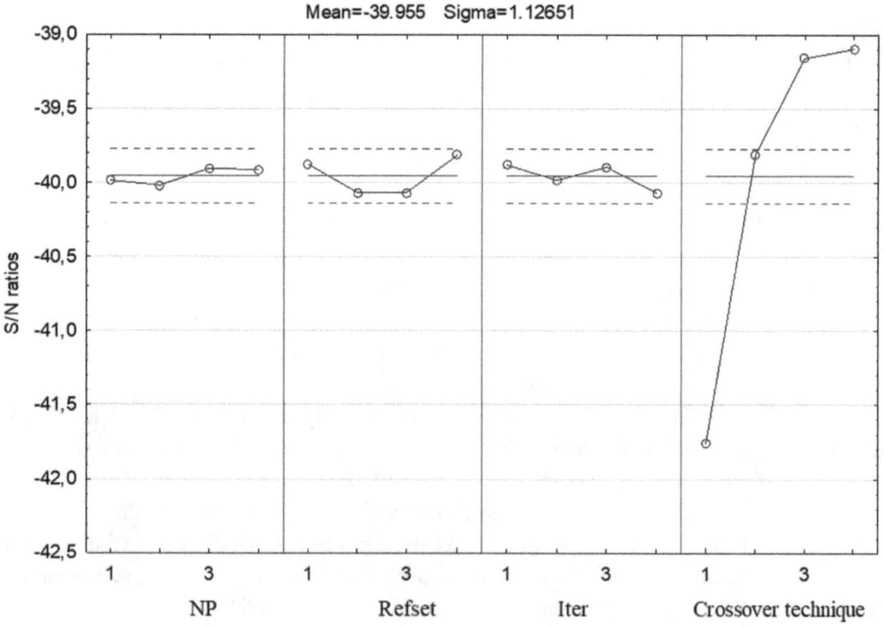

Fig. 10.8 Average *S/N* ratio for parameter levels

Table 10.8 Result of the variance analyses for CRW

Parameter	Sum of squares (SS)	Variance MS	Effect rate (P) (%)	Rank
NP: the number of the population	0.03744	0.0125	0.198	4
RefSet: the reference set	0.20457	0.0682	1.080	2
Iter: the maximum iteration number	0.09059	0.0302	0.478	3
Crossover technique	19	6.2009	98.243	1

Table 10.9 Result of the optimization analyses for the minimum wall weights

Parameter	Level	Level description	Contribution (%)
NP: the number of the population	3	40	4.60
RefSet: the reference set	4	12	12.50
Iter: the maximum iteration number	1	125	6.69
Crossover technique	4	Ensemble	76.22

By Taguchi method, the minimum wall weight has been estimated as 87.36 kN/m among 256 cases for four algorithm parameters with four levels (Table 10.5) with just 16 analysis. When results of the verification analysis which mean the optimization analysis by SS algorithm have been carried out with parameters' values (NP = 40, $RefSet$ = 12, $Iter$ = 125, Crossover technique = Ensemble) in Table 10.9, the minimum wall weight has been gained as 88.88 kN/m with the relative error of 1.7%. It is evident from Table 10.9 that the most effective algorithm parameter is the crossover technique of ensemble of the SS for the minimum wall weights alike results of variance analysis. The minimum wall weight has been estimated when the type of crossover technique was the ensemble. As the effect rate of the crossover technique is quite remarkable, it is not possible to observe the effect rate of the other algorithm parameters judiciously. For this reason, the same process of analysis has been repeated by employing the *L9* orthogonal array without considering the crossover technique.

10.4.2 Statistical Analysis via **L9** *Design Table*

In *L9* design table, algorithm parameters of the crossover technique for ensemble have been taken as an uncontrollable factor due to abovementioned results. Three different cases for the variable values of the controllable parameter designs are given in Table 10.10.

Table 10.10 Controllable factors of L9 for different cases

Case	Parameter	Level 1	Level 2	Level 3
1	*NP*	20	30	40
	RefSet	6	8	10
	Iter	100	200	300
2	*NP*	20	30	40
	RefSet	8	10	12
	Iter	100	200	300
3	*NP*	20	30	40
	RefSet	8	10	12
	Iter	125	250	375

Table 10.11 SS algorithm parameter design table for *L9* orthogonal array

	Case 1			Case 2			Case 3		
Design no	*NP*	*RefSet*	*Iter*	*NP*	*RefSet*	*Iter*	*NP*	*RefSet*	*Iter*
1	20	6	100	20	8	100	20	8	125
2	20	8	200	20	10	200	20	10	250
3	20	10	300	20	12	300	20	12	375
4	30	6	200	30	8	200	30	8	250
5	30	8	300	30	10	300	30	10	375
6	30	10	100	30	12	100	30	12	125
7	40	6	300	40	8	300	40	8	375
8	40	8	100	40	10	100	40	10	125
9	40	10	200	40	12	200	40	12	250

The revised *L9* orthogonal array considering parameter levels given in Table 10.10 and *L9* orthogonal array in Table 10.3 is shown in Table 10.11 for SS algorithm parameter design table.

The optimum wall weights (W_{wall}) and the *S/N* ratios for nine different wall designs are shown in Table 10.12.

The average *S/N* ratios change according to the abovementioned cases in Table 10.12 are depicted in Fig. 10.9 for each parameter and their levels.

Different behaviors have been observed in change between the *S/N* ratios, which corresponds to obtained optimum wall weights and parameter levels when Fig. 10.9 is examined generally. While the operation of the algorithm with more iteration is not vital in obtaining the less minimum wall weight, the increase in the population number is significant. It has been concluded from the graph that performing optimization analysis with a large number of iterations not only consume the time but also cannot guarantee the optimum result. It can be said that the increase in the population number

Table 10.12 Optimum wall weights and the *S/N* ratios for sixteen wall designs

	Case 1		Case 2		Case 3	
Design no	W_{wall}	*S/N*	W_{wall}	*S/N*	W_{wall}	*S/N*
1	93.55	−39.421	92.82	−39.353	92.49	−39.322
2	90.29	−39.113	91.31	−39.210	90.78	−39.160
3	93.11	−39.380	90.3	−39.114	90.24	−39.108
4	92.16	−39.291	89.61	−39.047	91.68	−39.245
5	90.06	−39.091	93.42	−39.409	90.06	−39.091
6	90.60	−39.143	90.51	−39.134	89.25	−39.012
7	91.20	−39.200	89.16	−39.003	89.43	−39.030
8	90.06	−39.091	90.76	−39.157	89.79	−39.065
9	89.43	−39.030	89.76	−39.062	89.61	−39.047

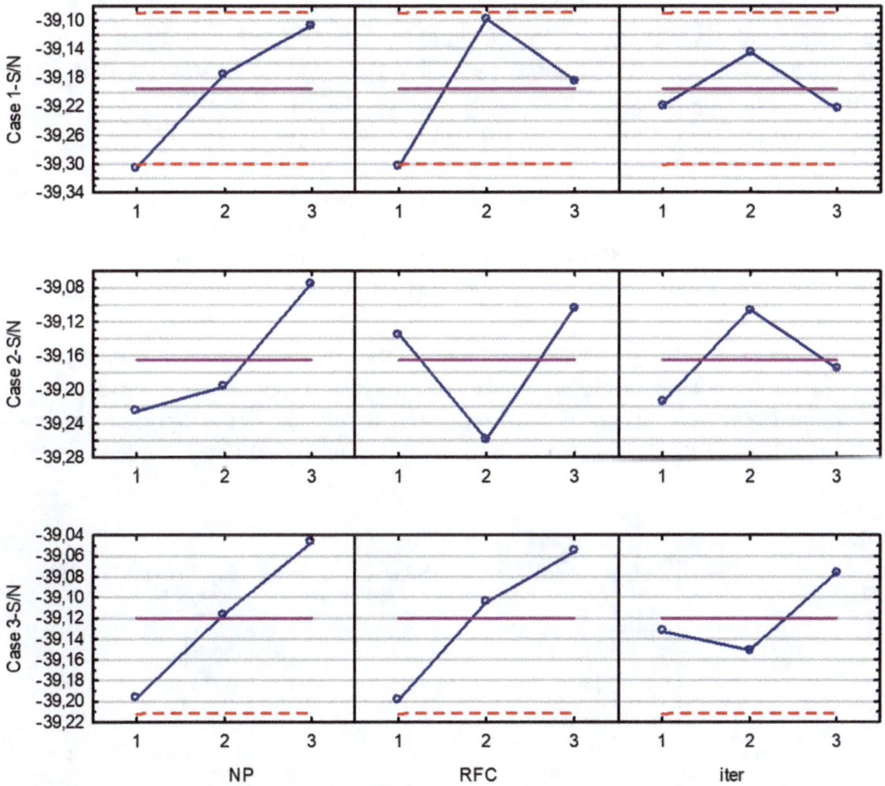

Fig. 10.9 Average *S/N* ratio of cases for parameter levels

of the algorithm is effective in obtaining fewer wall weights. It is due to an increase in individual diversity of the algorithm.

RefSet is defined as a cluster which includes the most different individuals from the best individuals in a population. In the SS algorithm, *RefSet* must have an appropriate value. In other words, having a very large or minimal value affects the quality and robustness of the algorithm. If the small value for the *RefSet* parameter is taken into account, it causes a decrease in the diversity of new solutions. On the other hand, the fact that this parameter has a great value affects the generation of many solutions and the process of reaching a solution. When the given graphs are examined in terms of *RefSet* parameter, appropriate and real-like value has been gained for case 1 by considering abovementioned characteristic of *RefSet*. For this reason, case 1 can be a decisive election in the effect of the algorithm parameters on the success of the SS algorithm, which means obtaining optimum result with reasonable algorithm parameters.

By taking the crossover technique as an uncontrollable parameter in Taguchi statistical analysis, the effect rate of change of the algorithm parameters of *NP*, *RefSet*, and *Iter* has been acquired logically. The results of variance analysis the algorithm parameters for different cases are tabulated in Table 10.13.

The order of effect ratios of parameters from large to small was obtained the same for case 1 and case 2; as *RefSet*, *NP*, and *Iter*, respectively. Since the average values of all cases are 43.8% for *RefSet* and 43.1%, respectively, it can be concluded that the effect of *NP* and *RefSet* on finding the optimum value is almost the same.

Parameter levels which are provided with the minimum wall weight have been estimated by using the Taguchi method. The parameters levels and their values are presented in Table 10.14.

It is seen from Table 10.14 that the values of the contribution for algorithm parameters of cases are compatible with variance analysis except for case 2. Besides, the minimum CRW weights have been predicted for cases by the optimum analysis of Taguchi method. By utilizing estimated algorithm parameters values given in

Table 10.13 Result of the variance analyses for CRW

Case	Parameter	Sum of squares (SS)	Variance MS	Effect rate (P) (%)	Rank
1	*NP*	0.060713	0.0304	44.443	2
	RefSet	0.064171	0.0321	46.974	1
	Iter	0.011726	0.0059	8.584	3
2	*NP*	0.038755	0.0194	39.749	2
	RefSet	0.040667	0.0203	41.711	1
	Iter	0.018076	0.0090	18.540	3
3	*NP*	0.033610	0.016805	45.067	1
	RefSet	0.031833	0.015917	42.684	2
	Iter	0.009135	0.004568	12.249	3

Table 10.14 Estimated algorithm parameter values of cases for the minimum wall weights

Case	Parameter	Level	Level description	Contribution (%)
1	*NP*	3	40	37.41
	RefSet	2	8	41.09
	Iter	2	200	21.50
2	*NP*	3	40	42.90
	RefSet	2	10	29.30
	Iter	2	200	27.80
3	*NP*	3	40	40.26
	RefSet	3	12	35.49
	Iter	3	375	24.25

Table 10.14, the verification analysis has been performed with the SS algorithm. Between estimated wall weights (*Westimated*) and wall weights obtained from the verification analysis (*Wverification*) have been compared by calculating the relative errors. All results are shown in Fig. 10.10.

It is seen from Fig. 10.10 that the minimum relative error is obtained for case 1, and the relative error is decreased comparing with *L*16 Taguchi *method* analysis. For CRW optimum design which has discrete design parameters with the SS algorithm, reasonable algorithm parameters values have been achieved in case 1 (*NP* = 20, *RefSet* = 6, *Iter* = 100).

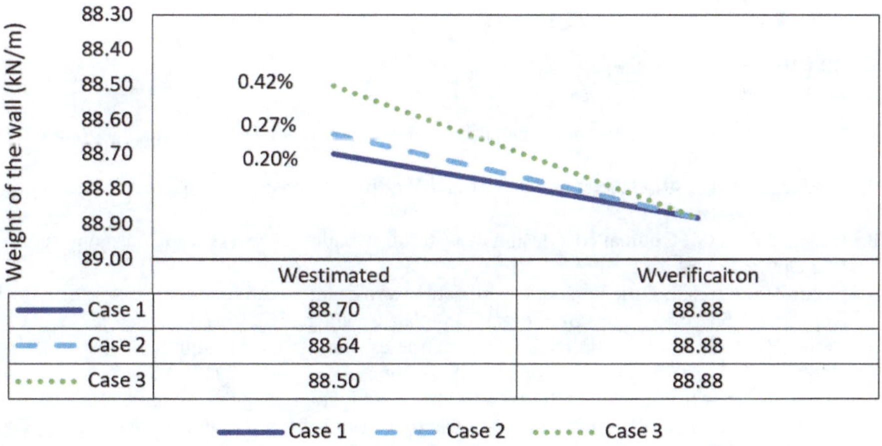

Fig. 10.10 Results of Taguchi method analysis

10.5 Conclusions

In this chapter, the effect of the algorithm parameters of the scatter search (SS) algorithm which are the number of the population (*NP*), the reference set (*RefSet*) and the maximum iteration number (*Iter*), and the type of the crossover technique on the optimum value has been investigated. The cantilever retaining wall (CRW) which is widely studied for the optimum design has been taken into consideration as the optimum design problem with discrete design variables and constraints of stability conditions. The parametric analyses have been performed for different algorithm parameters by using the Taguchi method, which includes statistical evaluation of all combinations on the response value with fewer trials. The most effective algorithm parameter has been obtained as the crossover technique by utilizing the *L*16 orthogonal array. The most effective algorithm parameter has been gained as the *RefSet* in parametric study repeated via *L*9 orthogonal array by taking the crossover technique as an uncontrollable factor. Thus, reasonable values of the SS algorithm parameters, which make the algorithm robust have been acquired with the effect rate on the optimum result in a short time.

If all the results obtained are evaluated only in terms of the number of iterations, it is an advantage to have the knowledge that the global solution will be reached with the appropriate number of iterations because many iterations which take much time to fulfill are defined in optimization algorithms to achieve optimum results.

Consequently, the Taguchi method can be used reliably and effectively in the investigation of available values for design parameters of the optimization algorithms.

References

1. Rao S Engineering optimization: theory and practice—Singiresu—Google Kitaplar. John Wiley Sons
2. Goldberg DE (1989) Genetic algorithms and Walsh functions: part I, a gentle introduction. Complex Syst 3:129–152
3. Geem ZW (2006) Optimal cost design of water distribution networks using harmony search. Eng Optim 38:259–277
4. Geem ZW, Lee KS, Park Y (2005) Application of harmony search to vehicle routing. Am J Appl Sci 2:1552–1557. https://doi.org/10.3844/ajassp.2005.1552.1557
5. Carbas S (2016) Design optimization of steel frames using an enhanced firefly algorithm. Eng Optim 48:2007–2025. https://doi.org/10.1080/0305215X.2016.1145217
6. Uray E, Çarbaş S, Erkan İH, Olgun M (2020) Investigation of optimal designs for concrete cantilever retaining walls in different soils. Chall J Concr Res Lett 11. https://doi.org/10.20528/cjcrl.2020.02.003
7. Cheng YM, Li L, Fang SS (2011) Improved harmony search methods to replace variational principle in geotechnical problems. J Mech 27:107–119. https://doi.org/10.1017/jmech.2011.12
8. Niu M, Wan C, Xu Z (2014) A review on applications of heuristic optimization algorithms for optimal power flow in modern power systems. J Mod Power Syst Clean Energy 2:289–297. https://doi.org/10.1007/s40565-014-0089-4

9. Hatamlou A (2013) Black hole: a new heuristic optimization approach for data clustering. Inf Sci (Ny) 222:175–184. https://doi.org/10.1016/j.ins.2012.08.023
10. Kirkpatrick S, Gelatt CD, Vecchi MP (1983) Optimization by simulated annealing. Science 220(80):671–680. https://doi.org/10.1126/science.220.4598.671
11. Glover F (1989) Tabu search—part I. ORSA J Comput 1:190–206. https://doi.org/10.1287/ijoc.1.3.190
12. Glover F (1990) Tabu search—part II. ORSA J Comput 2:4–32. https://doi.org/10.1287/ijoc.2.1.4
13. Dorigo M, Gambardella LM (1997) Ant colonies for the travelling salesman problem. BioSystems 43:73–81. https://doi.org/10.1016/S0303-2647(97)01708-5
14. Geem ZW, Kim JH, Loganathan GV (2001) A new heuristic optimization algorithm: harmony search. Simulation 76:60–68. https://doi.org/10.1177/003754970107600201
15. Kennedy J, Eberhart R Particle swarm optimization. In: Proceedings of ICNN'95—international conference on neural networks. IEEE, pp 1942–1948
16. Karaboga D (2005) An idea based on honey bee swarm for numerical optimization
17. Yang XS (2009) Firefly algorithms for multimodal optimization. In: International symposium on stochastic algorithms. pp 169–178
18. Rajabioun R (2011) Cuckoo optimization algorithm. Appl Soft Comput J 11:5508–5518. https://doi.org/10.1016/j.asoc.2011.05.008
19. Perez RE, Behdinan K (2007) Particle swarm approach for structural design optimization. Comput Struct 85:1579–1588. https://doi.org/10.1016/j.compstruc.2006.10.013
20. Akay B, Karaboga D (2012) A modified artificial bee colony algorithm for real-parameter optimization. Inf Sci (Ny) 192:120–142. https://doi.org/10.1016/j.ins.2010.07.015
21. Carbas S, Saka MP (2013) Efficiency of improved harmony search algorithm for solving engineering optimization problems. Iran Univ Sci Technol 3:99–114
22. Bossek J Smoof: single-and multi-objective optimization test functions
23. Molga M, Smutnicki C (2005) Test functions for optimization needs
24. Taguchi G (1986) Introduction to quality engineering: designing quality into products and processes
25. Uray E, Tan Ö, Çarbaş S, Erkan H (2021) Metaheuristics-based pre-design guide for cantilever retaining walls. Tek Dergi 32. https://doi.org/10.18400/tekderg.561956
26. Cheng BW, Chang CL (2007) A study on flowshop scheduling problem combining Taguchi experimental design and genetic algorithm. Expert Syst Appl 32:415–421. https://doi.org/10.1016/j.eswa.2005.12.002
27. Yildirim T, Kalayci CB, Mutlu Ö, İşlem Daire Başkanlığı B, Üniversitesi P, Makalesi A, Article Öz R (2016) Gezgin satıcı problemi için yeni bir meta-sezgisel: kör fare algoritması A novel metaheuristic for traveling salesman problem: blind mole-rat algorithm. dergipark.org.tr 22:64–70. https://doi.org/10.5505/pajes.2015.38981
28. Russell RA, Chiang WC (2006) Scatter search for the vehicle routing problem with time windows. Eur J Oper Res 169:606–622. https://doi.org/10.1016/j.ejor.2004.08.018
29. Pendharkar PC (2013) Scatter search based interactive multi-criteria optimization of fuzzy objectives for coal production planning. Eng Appl Artif Intell 26:1503–1511. https://doi.org/10.1016/j.engappai.2013.01.001
30. Naderi B, Ruiz R (2014) A scatter search algorithm for the distributed permutation flowshop scheduling problem. Eur J Oper Res 239:323–334. https://doi.org/10.1016/j.ejor.2014.05.024
31. Hakli H, Ortacay Z (2019) An improved scatter search algorithm for the uncapacitated facility location problem. Comput Ind Eng 135:855–867. https://doi.org/10.1016/j.cie.2019.06.060
32. Duman E, Ozcelik MH (2011) Detecting credit card fraud by genetic algorithm and scatter search. Expert Syst Appl 38:13057–13063. https://doi.org/10.1016/j.eswa.2011.04.110
33. Hakli H (2020) A performance evaluation and two new implementations of evolutionary algorithms for land partitioning problem. Arab J Sci Eng 45:2545–2558. https://doi.org/10.1007/s13369-019-04203-z
34. Khooban Z, Farahani RZ, Miandoabchi E, Szeto WY (2015) Mixed network design using hybrid scatter search. Eur J Oper Res 247:699–710. https://doi.org/10.1016/j.ejor.2015.06.025

35. Keskin BB, Üster H (2007) A scatter search-based heuristic to locate capacitated transshipment points. Comput Oper Res 34:3112–3125. https://doi.org/10.1016/j.cor.2005.11.020
36. Pinol H, Beasley JE (2006) Scatter search and bionomic algorithms for the aircraft landing problem. Eur J Oper Res 171:439–462. https://doi.org/10.1016/j.ejor.2004.09.040
37. İstinat Duvarlarıyla İlgili Yapılan Hatalar - Serbest Cihangir. http://www.serbestcihangir.com/istinat-duvarlariyla-ilgili-yapilan-hatalar/. Accessed 24 Oct 2020
38. Chau KW, Albermani F (2003) Knowledge-based system on optimum design of liquid retaining structures with genetic algorithms. J Struct Eng 129:1312
39. Ceranic B, Fryer C, Baines R (2001) An application of simulated annealing to the optimum design of reinforced concrete retaining structures. Comput Struct
40. Gandomi AH, Kashani AR, Roke DA, Mousavi M (2015) Optimization of retaining wall design using recent swarm intelligence techniques. Eng Struct 103:72–84. https://doi.org/10.1016/j.engstruct.2015.08.034
41. Camp C, Akin A (2012) Design of retaining walls using big bang-big crunch optimization. J Struct Eng 138:438–448
42. Sheikholeslami R, Khalili B, Zahrai S (2014) Optimum cost design of reinforced concrete retaining walls using hybrid firefly algorithm. ijetch.org. https://doi.org/10.7763/IJET.2014.V6.742
43. Talatahari S, Sheikholeslami R (2014) Optimum design of gravity and reinforced retaining walls using enhanced charged system search algorithm. KSCE J Civ Eng
44. Temür R, Bekdas G (2016) Teaching learning-based optimization for design of cantilever retaining walls. Struct Eng Mech 57:763–783
45. Glover F (1977) Heuristics for integer programming using surrogate constraints. Decis Sci 8:156–166. https://doi.org/10.1111/j.1540-5915.1977.tb01074.x
46. Martí R, Laguna M, Glover F (2006) Principles of scatter search. Eur J Oper Res 169:359–372. https://doi.org/10.1016/j.ejor.2004.08.004
47. GEO5 Geotechnical software https://www.finesoftware.eu/geotechnical-software/
48. Das BM (2017) Principles of foundation engineering, 9th edn. Cengage Learning
49. Das BM (2002) Principles of foundation engineering
50. Meyerhof GG (1963) Some recent research on the bearing capacity of foundations. Can Geotech J 1:16–26. https://doi.org/10.1139/t63-003
51. Lin C (2013) A rough penalty genetic algorithm for constrained optimization. Inf Sci (Ny) 241. https://doi.org/10.1016/j.ins.2013.04.001
52. Deb K (2000) An efficient constraint handling method for genetic algorithms. Comput Methods Appl Mech Eng 186. https://doi.org/10.1016/S0045-7825(99)00389-88

Chapter 11
Optimum Design of Beams with Varying Cross-Section by Using Application Interface

Osman Tunca

11.1 Introduction

Optimization is a numerical solution method frequently used in engineering designs. It has specific objective or objectives, design variables, and constraints [1]. Most of design problems comply with these requirements in engineering. If an example from civil engineering is given, it can be taken as an objective to reduce the cost of a building to be designed. So, the geometric and material properties of the building may be considered as design variables and relevant specifications may contain boundary conditions required for the designed building [2–4]. In such a design problem, many design variations may need to be tried one by one. This generally becomes impossible. Thus, optimization is becoming more and more widely used in engineering design.

The title of optimization is examined under three sub-headings: topology optimization, shape optimization, and size optimization. In topology optimization, particles are removed from a certain volume and the optimum geometry is obtained [5–7]. The optimum shape of an object is decided in shape optimization [8–10]. However, shape of the object is certain in size optimization. This type of optimization is only aimed to achieve optimum dimensions of designed objects [11–13]. Also, optimization methods can be categorized as deterministic and probabilistic. The deterministic methods contains quadratic complex mathematical functions [14]. The other optimization methods are based on probabilistic calculation. They do not need complex mathematical expressions [15]. However, since these methods are based on probabilistic calculation, the obtained results cannot be claimed to be exact optimum. It would not be wrong to say that the probabilistic methods having such an important place in engineering, still cannot complete its development.

There are various types of stochastic optimization methods such as Symbiotic Organisms Search (SOS) algorithm [16], Cuckoo Search Optimization (CSO) [17],

---

O. Tunca (✉)
Department of Civil Engineering, Karamanoglu Mehmetbey University, 70200 Karaman, Turkey
e-mail: osmantunca@kmu.edu.tr

S. Carbas et al. (eds.), *Nature-Inspired Metaheuristic Algorithms for Engineering Optimization Applications*, Springer Tracts in Nature-Inspired Computing,
https://doi.org/10.1007/978-981-33-6773-9_11

Grey Wolf Optimizer (GWO) [18], and so forth. New ones continue to be added to these constantly [19, 20]. Among these, Harmony Search Algorithm (HSA) was developed by Lee and Geem in 2004 [21]. This algorithm, which was designed by inspiring the harmony between music instruments played in an orchestra, is used by researchers in many studies and it is well accepted [22–25]. Moreover, in 2013, Civicioglu enhanced a new stochastic optimization algorithm named Backtracking Search Optimization Algorithm (BSA) [26]. It is encoded by inspiring from Differential Evolution (DE) algorithm [27]. It has two selection operators and population-based structure. BSA comes forward with its convergence performance with the help of them [28]. There are less studies taken place in the literature conducted using BSA [29–31].

Another calculation method frequently used by engineers is Finite Element Analysis (FEA). FEA method is based on assumption of dividing objects into a finite number of sub-structures. Then, each sub-part are calculated and the all results are combined. Thanks to this method, many problems encountered in engineering can be simulated easily [32–34]. Even the most suitable designs can be achieved with parametric studies [35]. With the advancing computer technology, this method is developing rapidly. Some companies have been established in the software industry in this field [36–39]. These companies offer package softwares to researchers. And, they are constantly improving them with new updates. Some of these presents to its users Application Programming Interface (API) property. Also, researchers can be reach to application interface of some packaged softwares via programming languages.

Most of the researchers especially studying on optimization, determine the mathematical expression of the optimization problem manually [40, 41]. Doing so can be very costly in terms of time and effort. Additionally, possible attention errors can also be made when specifying complex mathematical expressions. Some researchers even provide the obtained results after the optimization process with finite element packaged softwares. Instead, optimization algorithms developed in various programming languages can be executed with the mentioned packaged softwares.

In this study, the algorithmic performances of HSA and BSA are investigated on Three-Bar Truss Design Problem. In this problem, the weight of the structure is minimized by specifying cross-section area of truss elements. There are three constraint functions in this basic benchmark problem. Thus, optimization algorithm coded in Microsoft Visual Basic for Application (MS VBA 6.0) are tested on this problem. At the end, it was revealed that BSA shows a better convergence performance to the optimum design for this problem. After that, two different beams with varying cross-section are modeled via Ansys Workbench v18.1 (finite element packaged software) [36]. Finally, obtained simulation models are executed with optimization algorithms by using Ironpython script file. With this study, BSA is used for the first time in optimum structural design of beams with varying cross-sections. So, a weak area in the literature is tried to be strengthened by including API in the field of size optimizations. General structure of this study is as follows:

1. In the first section, the aim and general concept of the study are mentioned.
2. In second section, the methods used in the study were examined.
3. Detailed explanations of the optimum design examples discussed are given in Sect. 11.3. Also, the obtained results of all study are given with detail in this section.
4. Finally, the conclusions are explained in Sect. 11.4.

11.2 Optimization

Optimization can be defined finding most useful result or results [42]. This has specific purpose and limits. Finally, it is aimed to find the optimum values of the variables that affect the outcomes of the problem. Two different stochastic optimization algorithms, HSA and BSA, are used in this study.

11.2.1 Harmony Search Algorithm (HSA)

There are many different instruments used in a music orchestra. Each instrument is played by a different musician. There must be a harmony between the musical instruments used for the mentioned orchestra to produce quality music. During music development, each artist performs within the possible sound ranges by tuning own instrument. Thus, they together create a harmony.

Lee and Geem are developed the Harmony Search Algorithm (HSA) by inspiring this phenomena in 2004 [21]. Researchers notice that harmony searching by orchestra is similar to research optimum results in optimization problems. For instances, each of musical instruments in orchestra is similar with variables in optimization problems These are restricted by specific bounds. Additionally, orchestra and optimization problems have a goal. All of them aim to achieve the most suitable result. Relationship between music improvement and engineering optimization is shows in Fig. 11.1.

In orchestras, musicians change the notes. If this provides a better harmony, the harmony is memorized. Likewise, there are changes for each design variable in optimization problems. If better solution is obtained by using new solution vector, this is saved to harmony memory matrices. The following process is applied in the HSA, respectively.

A. Initialize HSA parameters

HSA includes some specific constants. First of all, some constant included in the algorithm should be determined. For example, number of iteration and number of row of the harmony memory matrices should be determined before running of algorithm. So, Harmony memory size (HMS), harmony memory consideration ratio (HMCR),

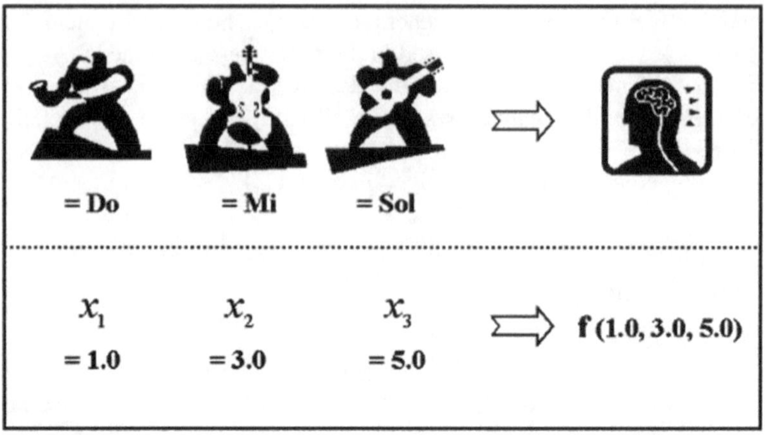

Fig. 11.1 Relationship between music improvement and engineering optimization [21]

pitch adjustment ratio (PAR) etc. are defined to algorithm at the beginning. Additionally, each of design variables has some boundaries. A range of values is defined for each design variables in the optimum design problem.

B. Initialize harmony memory matrices

Each row of the harmony memory matrices (HMM) is a solution vector. Design variables are defined randomly by considering boundary conditions. After that, these are assigned to HMM. This shown in Eq. 11.1.

$$[\text{HMM}] = \begin{bmatrix} x_{1,1} & x_{1,2} & \cdots\cdots & x_{1,n-1} & x_{1,n} \\ x_{2,1} & x_{2,2} & \cdots\cdots & x_{2,n-1} & x_{2,n} \\ \cdots & \cdots & \cdots\cdots & \cdots & \cdots \\ \cdots & \cdots & \cdots\cdots & \cdots & \cdots \\ x_{\text{HMS}-1,1} & x_{\text{HMS}-1,2} & \cdots\cdots & x_{\text{HMS}-1,n-1} & x_{\text{HMS}-1,n} \\ x_{\text{HMS},1} & x_{\text{HMS},2} & \cdots\cdots & x_{\text{HMS},n-1} & x_{\text{HMS},n} \end{bmatrix} \quad (11.1)$$

Here, column size of the matrices is symbolized with n and row size of the matrices equal to harmony memory size (HMS). Each element of the HMM is generated by Eq. 11.2.

$$x_{i,j} = x_{\text{min}} + (x_{\text{max}} - x_{\text{min}}) * \text{Rnd} \quad (11.2)$$

x_{max} and x_{min} represent the maximum and minimum values $x_{i,j}$ can take. And, Rnd is a random number between 0 and 1. Each row of the matrice is named as solution vector. Objective and fitness values are calculated by using elements of solution vectors. Then, all row is shorted by considering fitness values.

C. Improvise New Harmony

After the generation of HMM, algorithm generates new solution vector for each iteration. In this step, memory consideration, pitch adjustment and random selection rules became part of activity. Which rule to use is determined by comparing a chosen random number with the harmony memory consideration rate (HMCR) and pitch adjustment ratio (PAR). In cycle, if randomly generated Rnd is bigger than the HMCR, elements of new harmony vector is selected from HMM randomly (Eq. 11.3). Otherwise it is generated by considering Eq. 11.2.

$$x_i = \begin{cases} x_i \in \{x_i^1, x_i^2, x_i^3, \ldots, x_i^{\text{HMS}}\}, & \text{if Rnd} < \text{HMCR} \\ x_i \text{ is randomly generated} & \text{otherwise.} \end{cases} \tag{11.3}$$

Afterward, when the HMCR is bigger than the Rnd, PAR is considered. On condition that Rnd can not exceed the PAR, Eq. 11.4 is activated. In the contrary case, the value generated from Eq. 11.4 is taken exactly.

$$x_i = \begin{cases} x_i = x_i \pm \text{Rnd} * b_w, & \text{if Rnd} < \text{PAR} \\ x_i = x_i & \text{otherwise.} \end{cases} \tag{11.4}$$

D. Update Harmony Memory

Obtained new solution vector which is named new harmony is used to generate fitness value. Fitness value is compared to other values from previous step. If new fitness is bigger than the previous one, the new harmony saved to last row of HMM. If this situation is not provided, generated new harmony is deleted.

E. Check cycling criteria

Step C and D is repeated in algorithm. In the last step of the algorithm, iteration criteria is checked. When required criteria is established, the cycle is stopped.

11.2.2 Backtracking Search Optimization Algorithm (BSA)

Although there are numerous optimization methods, they are still developed continuously. New metaheuristic optimization algorithms are emerging day by day. Among these, Civicioglu studies on evolutionary optimization algorithm which is named Backtracking Search Optimization Algorithm (BSA) in 2013 [26]. BSA consist of five main steps. These are initialization, selection-1, mutation, crossover and selection-2. General process of BSA is indicated in Fig. 11.2.

A. Initialization

In this process, specific constants of algorithm such as population size (N), problem dimension (D) is assigned. Thus, algorithm is set. After the setting process of algorithm, initial population matrice (P) is generated by considered uniform distribution as shown in Eq. 11.5.

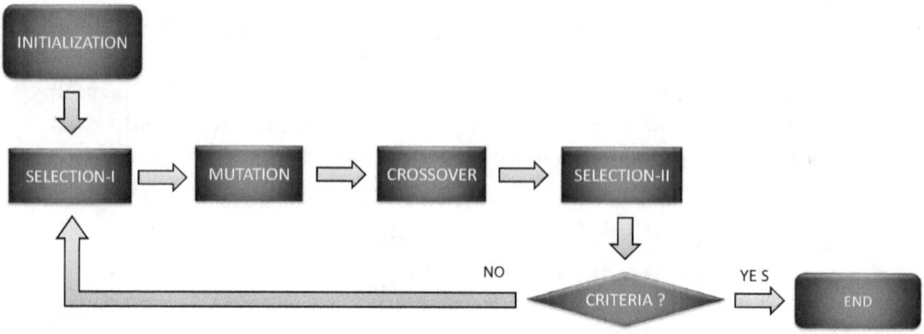

Fig. 11.2 Process of BSA

$$P = \begin{bmatrix} x_{1,1} & x_{1,2} & \cdots\cdots & x_{1,D-1} & x_{1,D} \\ x_{2,1} & x_{2,2} & \cdots\cdots & x_{2,D-1} & x_{2,D} \\ \cdots & \cdots & \cdots\cdots & \cdots & \cdots \\ \cdots & \cdots & \cdots\cdots & \cdots & \cdots \\ x_{N-1,1} & x_{N-1,2} & \cdots\cdots & x_{N-1,D-1} & x_{N-1,D} \\ x_{N,1} & x_{N,2} & \cdots\cdots & x_{N,D-1} & x_{N,D} \end{bmatrix} \quad (11.5)$$

B. Selection-1

To calculate search direction, second matrice is created. Another purpose of doing this is to give the algorithm a memory. Same process is used to generate historical population (P_{old}). In each iteration, P_{old} may be regenerated. This stage is determined by the equation below.

$$P_{\text{old}} = \begin{cases} P_{\text{old}} = P_i, & \text{if } a < b \\ P_{\text{old}} = P_{\text{old}} & \text{otherwise.} \end{cases} \quad (11.6)$$

Equation 11.6 determines the randomly selected initial matrices as historical matrice. Thus, it is saved as historical population until it is changed. Here, a and b is a random number between 0 and 1. P_i is the any element of P. Then, shuffling function is randomly used to permute P_{old}.

C. Mutation

Trial population matrices (T) is generated in mutation step. Initial form of this is named mutant population (M). It is calculated as follows.

$$M = P + F * (P_{\text{old}} - P) \quad (11.7)$$

In Eq. 11.8, F is a control parameter. It decides the size of the searching attempt and is randomly produced with Eq. 11.9.

$$F = 3 * \text{Rnd} \quad (11.8)$$

D. Crossover

The crossover stage of procedure is used to reach the final state of trial population (T). This is composed of two steps. First step includes generation of a binary integer-valued matrices (map). Its size is equal to size of P and T. map is used for decision of manipulation of T with P. Crossover strategy is given in Algorithm 1.

Algorithm 1 Crossover Strategy of BSA

```
map(1 to N,1 to D) = 1
if a < b then
  for i = 1 to N do
    map_{i,u(1,α)} = 0   'u is the random sequence of numbers from 1 through D
  end for
else
  for i = 1 to N do
    map_{i,rand(D)} = 0   'rand(D) is the random selected number from 1 through D
  end for
end if
T=Mutant
for i = 1 to N do
  for j = 1 to D do
    if map_{i,j} = 1 then
      T_{i,j} = P_{i,j}
    end if
  end for
end for
```

It is observed that Algorithm 1 use two randomization strategy for update map. In the first of these, u matrice which is the random sequence of numbers 1 through D, is used. Here, it is decided by using the following equation to choose which element is selected from u.

$$\alpha = \text{Int}(\text{mixrate} * \text{Rnd} * D) \tag{11.9}$$

In Eq. 11.9, mixrate may vary depending on the type of optimization problem. In the second strategy, the random selected number from 1 through D is used for update map.

At the end of crossover, some elements of T may be exceed the search space boundaries ($up_{i,j}$, $\text{low}_{i,j}$). The following algorithm is activated to sure.

Algorithm 2 Boundaries Control Mechanism

```
for i = 1 to N do
  for j = 1 to D do
    if T_{i,j} < low_{i,j} or T_{i,j} > up_{i,j} then
      T_{i,j} = low_{i,j} + Rnd * (up_{i,j} − low_{i,j})
    end if
  end for
end for
```

D. Selection-II

Fitness values of T and P are compared in the end of the process. On condition that row of T has better fitness than P, corresponding row of P is replaced by row of T. After that, end form of the P is shorted from less to more. Thus, $P_{\text{best fitness}}$ value is unearthed. Finally, Algorithm 3 is used to save best solution of current iteration.

Algorithm 3 Boundaries Control Mechanism

if $P_{bestfitness} < global\ mimimum$ **then**
  $global\ minimum = fitness\ of\ P\ row$
  $global\ minimizer = row\ of\ the\ P$
end if

Here, obtained best fitness value saved to variable of $global\ minimum$ and corresponding row of the matrice is saved to $global\ minimizer$ which is a array, respectively. Steps from B to D form a loop. And, Iteration continues until the initial satisfactory conditions are met.

11.2.3 Constraint Handling

The obtained design must meet some initial conditions in most engineering design problems. These limits can sometimes be related to physical conditions, while others are related to design provision which are issued by governments. In the optimization process, the final design must meet these conditions. Equation 11.10 is used to penalize the objective function in designs that do not meet the required conditions.

$$\text{Fit} = \text{Obj} * \left(1 + \sum g_i\right)^2 \tag{11.10}$$

Penalized objective value is Fit. And, Objective value is symbolized with Obj in here. $\sum g_i$ is calculated by using Eqs. 11.13 and 11.14 in Sect. 11.3.

11.2.4 Discrete Design Variables

In engineering design, design values are selected by considering production conditions. For example, the diameters of the rebars to be used in a reinforced concrete structure are selected from the values offered by the steel factories. Or, the strength of the concrete material to be used is selected from certain fixed values. Therefore, it is important for optimization algorithms developed for engineering problems to run with discrete variables in terms of the applicability of the optimum results. This can be applicable with Eq. 11.11

$$x_i = \mathrm{Int}\left(\mathrm{Rnd} * \frac{x_{\max} - x_{\min}}{x_{\mathrm{inc}}}\right) * x_{\mathrm{inc}} + x_{\min} \tag{11.11}$$

Here, $x_{\max}$ and $x_{\min}$ are the limit values that x_i can take. And, x_{inc} is the increment value. Int() is also a rounding function.

11.2.5 Programming Application Interfaces

Nowadays, FEA based packaged softwares are commonly used is many research process. These continues to develop in parallel with computer technology. Software companies increase their knowledge from generation to generation. They compete with each other over their programs. Among these, some package program presents application interface to their users. So, researchers can benefit from this feature in solving problems whose mathematical expressions are difficult and complex. In this way, the knowledge of the companies mentioned is effectively used.

In this study, Ansys Workbench v18.1 is selected as a FEA packaged software. Ansys presents own control with using Ironpython scripts to users. Ironpython programming language is able to work with MS Excel. In the light of this information, Ansys Workbench v18.1 is controlled via MS Excel macro codes which has Microsoft Visual Basic for Application 6.0 programming language. The schematic representation of this process is given in Fig. 11.3.

11.3 Problem Definition and Results

Firstly, the performance of the optimization algorithms are tested on the benchmark problem. By this way, both the expectations on the results to be obtained from the study were shaped and it is verified that the algorithms run correctly. Then, size optimization of beams with varying cross-section is investigated by using Application Programming Interface (API).

11.3.1 Three-Bar Truss Design Problem

Benchmark problems are very useful for checking of encoded optimization algorithms. These basic problems give us information about the performance of the algorithms. Three-Bar Truss Design Problem (TBT) is emerged by Ray and Saini in 2007 [43], is selected as a benchmark problem of this study. In this engineering problem, three rods with a hinge at one end are connected to each other at the other ends as shown in Fig. 11.4. Two design variables are cross-section area of bars. The objective of this problem is to minimize all system weight.

Fig. 11.3 Interactive relations of Ansys Workbench, Ironpython and MS Visual Basic

Fig. 11.4 Three-bar truss design problem [43]

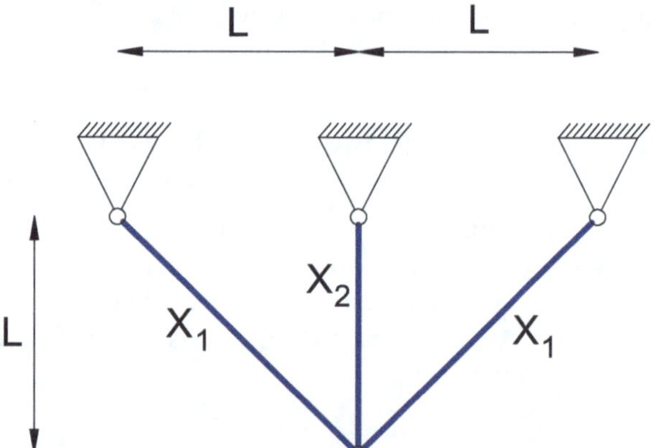

Table 11.1 Results of circular section with detail

Algorithm	x_1 (cm^2)	x_2 (cm^2)	Weight (kg)
HSA	0.789	0.408	263.90
BSA	0.789	0.408	263.90

The weight of the whole system $f(x)$ can be calculated with Eq. 11.12.

$$f(x) = L * (2\sqrt{2} * x_1 + x_2) \tag{11.12}$$

L is the horizontal and vertical distance between nodes of the bars. It is equal to 100 cm. There are two design variable that can take values between 0 and 1. Design variables of the problem are continuous. This shows that the problem has an infinite number of solution sets. Additionally, there are three constraints shown in Eq. 11.13.

$$\begin{aligned} g_1(x) &= \frac{\sqrt{2} * x_1 + x_2}{\sqrt{2} * x_1^2 + 2 * x_1 * x_2} * P - \sigma \leq 0 \\ g_2(x) &= \frac{x_2}{\sqrt{2} * x_1^2 + 2 * x_1 * x_2} * P - \sigma \leq 0 \\ g_3(x) &= \frac{1}{\sqrt{2} * x_2 + x_1} * P - \sigma \leq 0 \end{aligned} \tag{11.13}$$

In Eq. 11.13, P and σ are two constants that take values 2 kN and 2 kN/cm^2, respectively.

As stated before TBT is a simple benchmark problem aimed to minimize the structural weight. The HSA and BSA are conducted on this problem. In this way, the comparison is made for the performance of algorithms. The obtained optimal designs are given in Table 11.1.

Both HSA and BSA find the optimum result in the first 2500 iterations as 263.90 kg. However, there is a clear difference between convergence performances of optimization algorithms. The BSA converges to the minimum weight much earlier than HSA. The design history graph is shown in Fig. 11.5.

11.3.2 Beams with Varying Cross-Section

Optimization problems that can be easily solved by the computer were selected due to the iterative structure of the stochastic optimization algorithms used. However, in selected problems, both nonlinear materials and second order effects are considered. So, it is also predicted that it is very difficult to model them with the matrix displacement method. There are two optimization problem relating with beams with varying cross-section. One of these has varying rectangular cross-section and the other has circular.

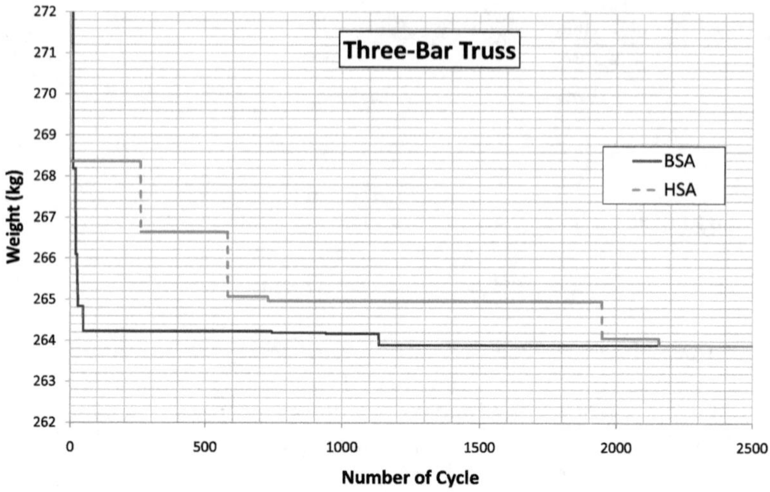

Fig. 11.5 Performance of HSA and BSA on TBT design problem

In this section, two recent metaheuristic algorithms are united with Finite Element Analysis (FEA) packaged software beams with varying cross-section by using Application Programming Interface (API). In previous section, Harmony Search Optimization Algorithm (HSA) and Backtracking Search Optimization Algorithm (BSA) is tested on three-bar truss (TBT) design problem via Microsoft Visual Basic Application 6.0 (MS VBA 6.0). After making sure of the encoding, rectangular and circular beams with varying cross-section modeled on Ansys Workbench v18.1 (FEA packaged software) in this section. The optimization problem, which aims to find the minimum weight design under certain constraints, has been realized by connecting the Ansys Workbench v18.1 to MS VBA 6.0 via Ironpython (programming language).

11.3.2.1 Circular Beams with Varying Cross-Section

Circular Beams with Varying Cross-Section has 250 cm length. It is a thin-walled cantilever beam. In the other words, one end of each beam is fixed while the other is free. Additionally, there is a load of 22.50 kN at the free ends of the beam. Static structural of the circular beam is shown in Fig. 11.6.

ANSYS Workbench v18.1 has a rich library of materials. This library is used when assigning materials to the model. The *Structural Steel NL* is selected as material. The modulus of elasticity, yield strength and tangent modulus are 200 GPa, 250 MPa and 1450 MPa, respectively (Fig. 11.7).

There are infinite element in any object on nature. Finite element analysis (FEA) is based on dividing objects in nature into a finite number of sub-parts. Then, the solutions of each sub-part created are yielded in itself and the results are combined.

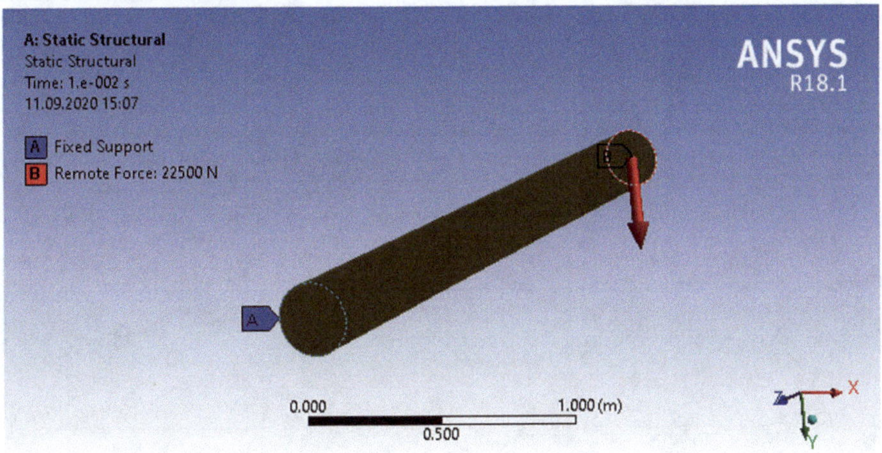

Fig. 11.6 Static structural of circular beam with varying cross-section

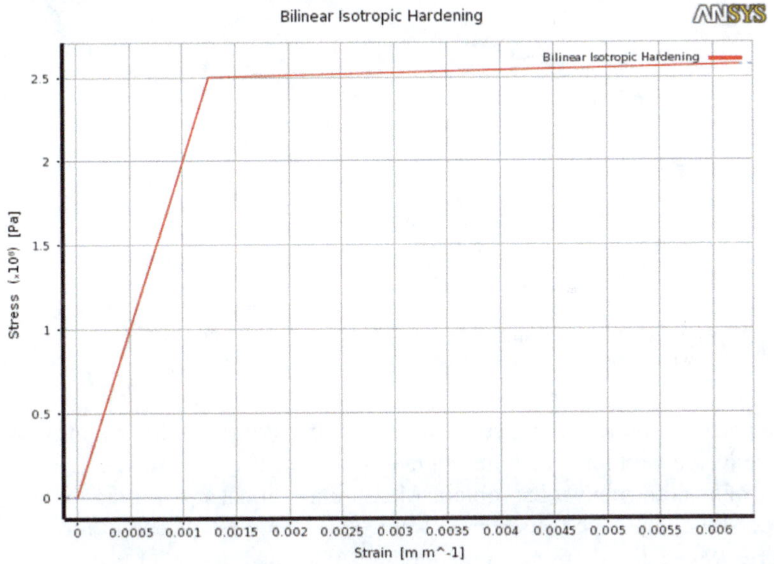

Fig. 11.7 Stress–strain curve of *Structural Steel NL*

This separation process is called meshing. Each separated part is termed an element and, each point on element called node. At this point, the number of elements and nodes are very important. If this number increases, the accuracy of the results obtained from the analysis will increase, yet the solution time will be longer. However, after the number of nodes and elements exceed a limit value, the solution values obtained from the analysis change at very low rates.

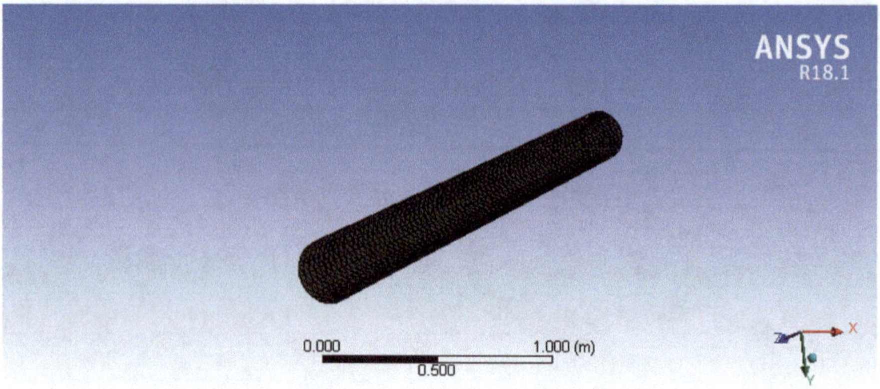

Fig. 11.8 Mesh distribution on circular beams

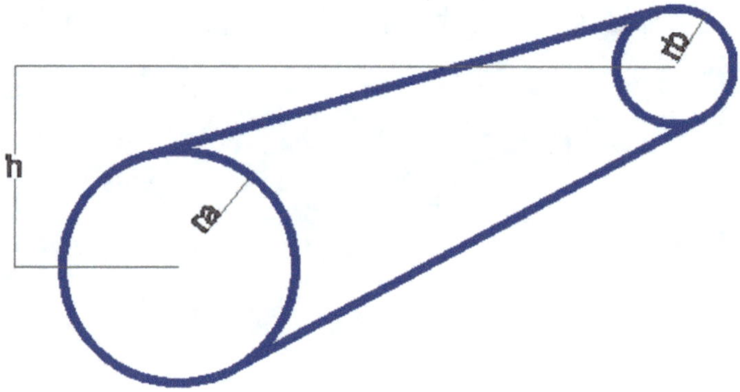

Fig. 11.9 Design variables of beams with varying cross-section problems

Geometry of models changes through iteration process. This situation may be cause converge problem. In finite element models, triangle surface mesh is used because of avoiding converge problem. Maximum element size is assigned as 30 mm. *SHELL181* element type is used by default. This element type useful for thin wallet structures. Mesh distributions on the model is indicated in Fig. 11.8.

Main objective of the circular beam with varying cross-section problem is the minimizing structural weight. The design variables in the circular cross-section problem are the radius of the circles at the ends of section, the vertical distance between the centers of the end of sections and the thickness, respectively. The considered geometric design variables are shown in Fig. 11.9.

Moreover, deformation in the direction of force, lateral deformation of the beam and Von Mises stress on the beam are limited. All restrains are given in Table 11.2.

Table 11.2 Limits of optimization problems

Variables	Min. value	Max. value	Increment
r_a	0.1 m	0.3 m	0.001 m
r_b	0.1 m	0.3 m	0.001 m
h	0.1 m	1 m	0.1 m
t	0.005 m	0.01 m	0.001 m
VM	–	260 MPa	–
$\mathrm{Def}_{\mathrm{ver}}$	–	16.67 mm	–
$\mathrm{Def}_{\mathrm{lat}}$	–	1 mm	–

Table 11.3 Results of circular section with detail

Algorithm	r_a (m)	r_b (m)	h (m)	t (m)	$\mathrm{VM}_{\mathrm{max}}$ (Pa)	$\mathrm{Def}_{\mathrm{lat}}$ (m)	$\mathrm{Def}_{\mathrm{ver}}$ (m)	Weight (kg)
HSA	0.106	0.141	0.1	0.005	1.8e+08	2.9e−05	0.0166	76.17
BSA	0.116	0.137	0.1	0.005	1.9e+08	2.9e−05	0.0167	78.02

In Table 11.2, r_a and r_b represents the radius of the circles at the end sections for circular steel beam. h is the distance between two beam ends in force direction. t is the thickness of the thin walled beams. VM, $\mathrm{Def}_{\mathrm{ver}}$ and $\mathrm{Def}_{\mathrm{lat}}$ are Von Mises stress, vertical deformation (in force direction) and lateral deformation, respectively. Three restriction functions given in Eq. 11.14.

$$
\begin{aligned}
g_1(x) &= \frac{\mathrm{VM}}{\mathrm{MaxVM}} - 1 \leq \epsilon \\
g_2(x) &= \frac{\mathrm{Def}_{\mathrm{ver}}}{\mathrm{MaxDef}_{\mathrm{ver}}} - 1 \leq \epsilon \\
g_3(x) &= \frac{\mathrm{Def}_{\mathrm{lat}}}{\mathrm{MaxDef}_{\mathrm{lat}}} - 1 \leq \epsilon
\end{aligned}
\tag{11.14}
$$

Here, MaxVM, $\mathrm{MaxDef}_{\mathrm{ver}}$ and $\mathrm{MaxDef}_{\mathrm{lat}}$ is determined by considering Table 11.2. 0.1% of error is not important for such a problems in engineering. So, ϵ is considered as 0.001 in study. This will help the algorithm not to catch local optimum points during the solution.

The obtained results are given in Table 11.3. h and t design values are minimal as expected. Additionally, obtained results from both algorithm actually very close to each other.

The Circular beams with varying cross-section problem includes four design variables and three constrains (Table 11.3). The objective of this optimization is the obtaining minimum weight. Performance comparison of the HSA and BSA are shown in Fig. 11.10.

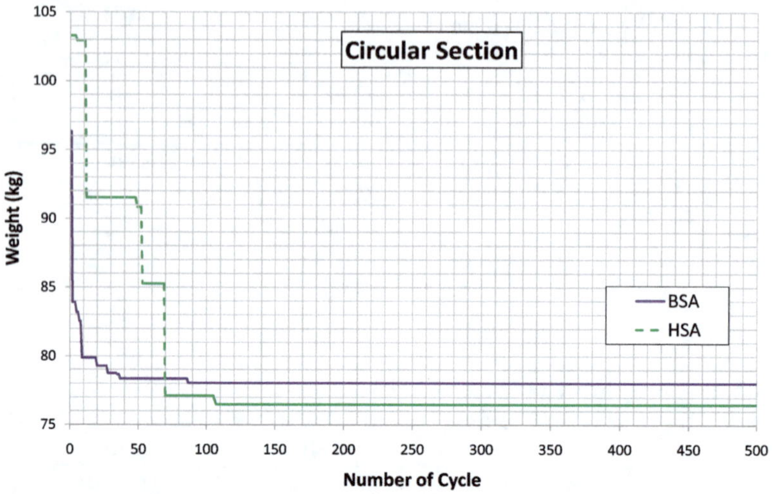

Fig. 11.10 Performance of HSA and BSA on circular beam design problem

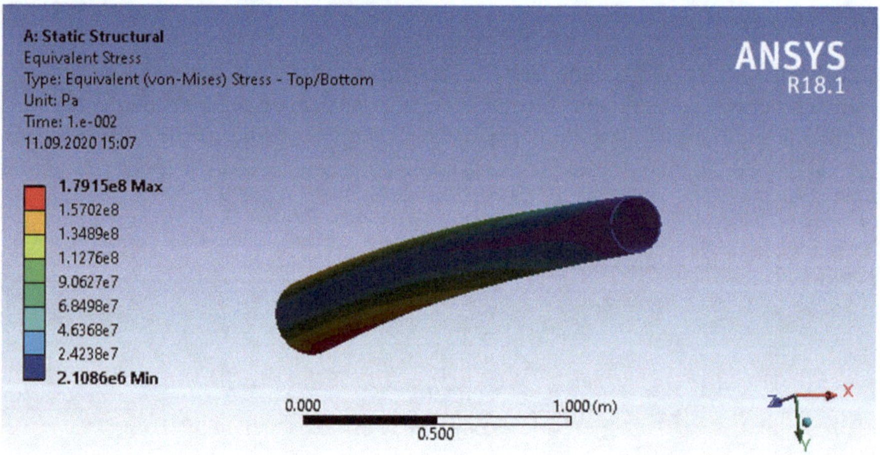

Fig. 11.11 Von Mises stress distribution on beams

It is observed that converge performance of BSA is higher than HSA. But, HSA is much more effective in finding minimum weight on circular beams with varying cross-section at the end of optimization process. When minimum weight is calculated as 78.02 kg by BSA, HSA finds this as 76.17 kg by 2.4%. Maximum Von Mises stress as 179.15 MPa occurs on the end of the circular beam as shown in Fig. 11.11.

It is clearly observed that, maximum deformation in *y* axes exceed to 0.0166 m at the end of the beam (Fig. 11.12). Minimum value is zero because of that other end of the beam has fixed support.

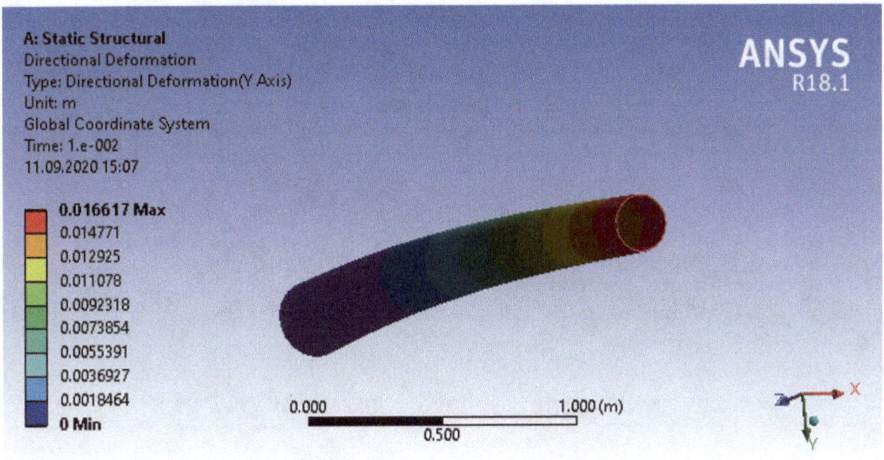

Fig. 11.12 Vertical deformation on beams

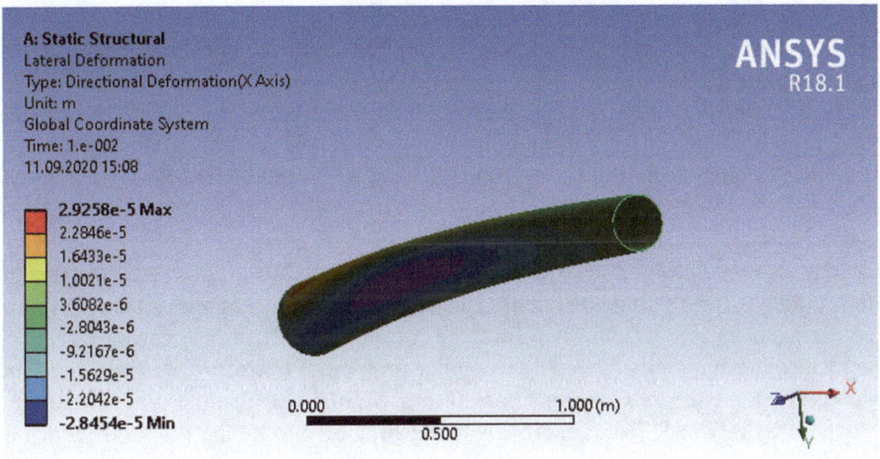

Fig. 11.13 Lateral deformation on beams

Finally, maximum lateral deformation is 2.9×10^{-5} m as given in Fig. 11.13. This shows also that large deflection is active.

11.3.2.2 Rectangular Beam with Varying Cross-Section

Rectangular beams with varying cross-section design problem has similar with circular one in terms of some geometrical quantity. It has 250 cm length. Additionally, static structural of this is same with circular beam with varying cross-section design

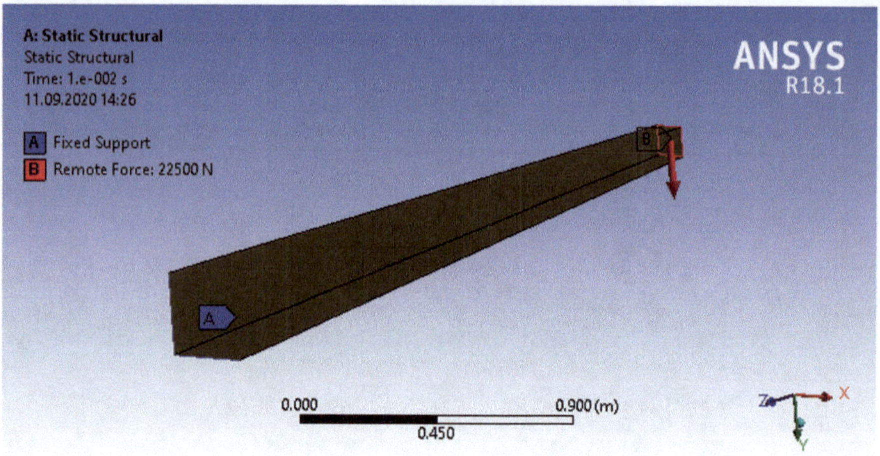

Fig. 11.14 Static structural of rectangular beam with varying cross-section

problem. There is a load of 22.50 kN at the free ends of the beams. Its static structural are shown in Fig. 11.14.

The *Structural Steel NL* is selected from materials library of ANSYS Workbench v18.1 for rectangular beams as material. As previous, the modulus of elasticity, yield strength and tangent modulus are 200 GPa, 250 MPa and 1450 MPa, respectively (Fig. 11.7).

Meshing process has an important place in FEA. If this process is not done properly, convergence problems may occur during the FEA. Meshing is also effective on the accuracy of the obtained results. During optimization, algorithms try different geometries in each iteration. Therefore, the converge problems may cause. The triangle surface mesh is safe to avoid convergence problems. Therefore, it is selected. The maximum element size is assigned as 30 mm. *SHELL181* element type is assigned as default by ANSYS Workbench v18.1. Meshing display of the model is shown in Fig. 11.15.

The objective of the rectangular beams with varying cross-section is the minimizing structural weight. The design variables in the square cross-section problem are the length of one side of the squares at the ends of sections, the vertical distance between the centers of the end of sections and the thickness, respectively. The geometric design variables of rectangular beams with varying cross-section optimization problem are shown in Fig. 11.16.

The maximum deformation in the direction of force, maximum lateral deformation of the beam and maximum Von Mises stress which is occurred on the beam are given in Table 11.4. These are same with previous design problem.

In Table 11.2, a_a and a_b represents the length of one side of the squares for beams with rectangular cross section. h is the distance between two center of square at the ends of beam in force direction. t is the thickness of this. VM, $\mathrm{Def}_{\mathrm{ver}}$ and $\mathrm{Def}_{\mathrm{lat}}$ are Von Mises stress, vertical deformation (in force direction) and lateral deformation,

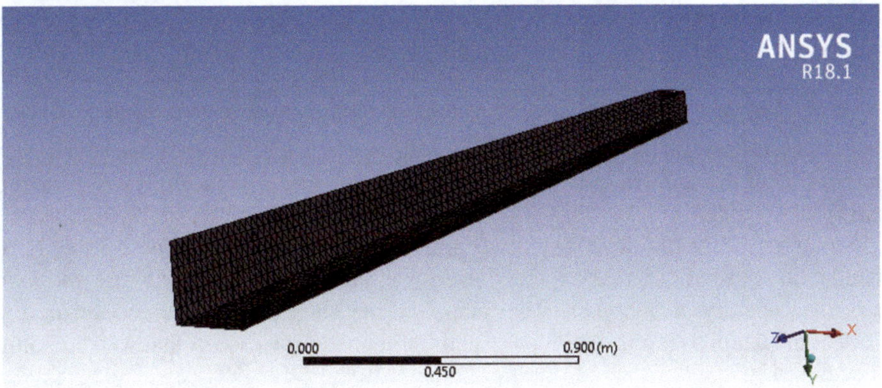

Fig. 11.15 Mesh distribution on beams

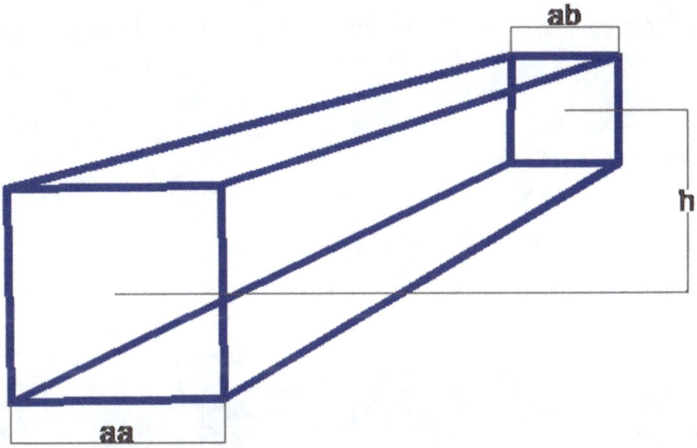

Fig. 11.16 Design variable of rectangular beam with varying cross-section problems

Table 11.4 Limits of optimization problems

Variables	Min. value	Max. value	Increment
a_a	0.1 m	0.3 m	0.001 m
a_b	0.1 m	0.3 m	0.001 m
h	0.1 m	1 m	0.1 m
t	0.005 m	0.01 m	0.001 m
VM	–	260 MPa	–
$\mathrm{Def}_{\mathrm{ver}}$	–	16.67 mm	–
$\mathrm{Def}_{\mathrm{lat}}$	–	1 mm	–

respectively. Three restriction functions are same with given in Eq. 11.14. MaxVM, $MaxDef_{ver}$ and $MaxDef_{lat}$ is determined by considering Table 11.2 and, the ϵ is considered as 0.001 in study.

Rectangular beams with varying cross-section design problem is similar to circular one. There is only one difference in first two design variable. They are used the length of one side of the squares instead of the radius of the circles at the ends as the design variable.

Summary of the obtained results are given in Table 11.5. As similar to Table 11.3, *h* and *t* value are minimal as expected. Because, the value of *h* slightly effect to load carrying capacity of the beam. The local web-buckling is not observed during the FEA of the beam. It is expected that *t* is minimal unless the case of that web buckling is occurred.

BSA shows superior algorithmic performance than HSA both in finding minimum weight and convergence speed for rectangular beams with varying cross-section design problem. There is a 2.0% difference between them in minimum weight. The design history graph representing minimum weight versus number of iteration is given in Fig. 11.17.

Table 11.5 Results of rectangular section with detail

Algorithm	a_a (m)	a_b (m)	h (m)	t (m)	VM_{max} (Pa)	Def_{lat} (m)	Def_{ver} (m)	Weight (kg)
HSA	0.122	0.263	0.1	0.005	1.3e+08	2.6e−05	0.0166	75.62
BSA	0.100	0.278	0.1	0.005	1.2e+08	2.5e−05	0.0165	74.26

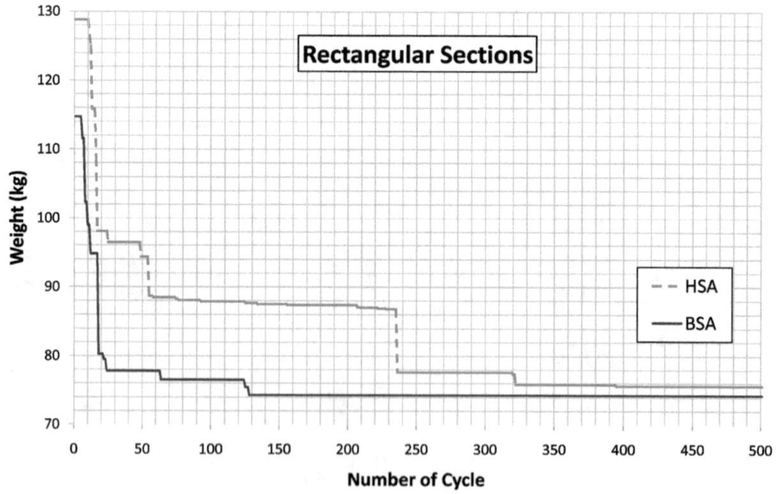

Fig. 11.17 Performance of HSA and BSA on rectangular beam design problem

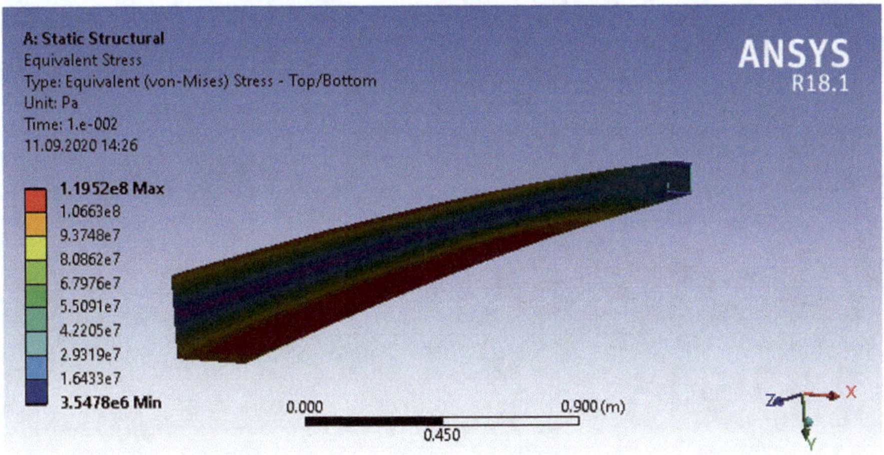

Fig. 11.18 Von Mises stress distribution on beams

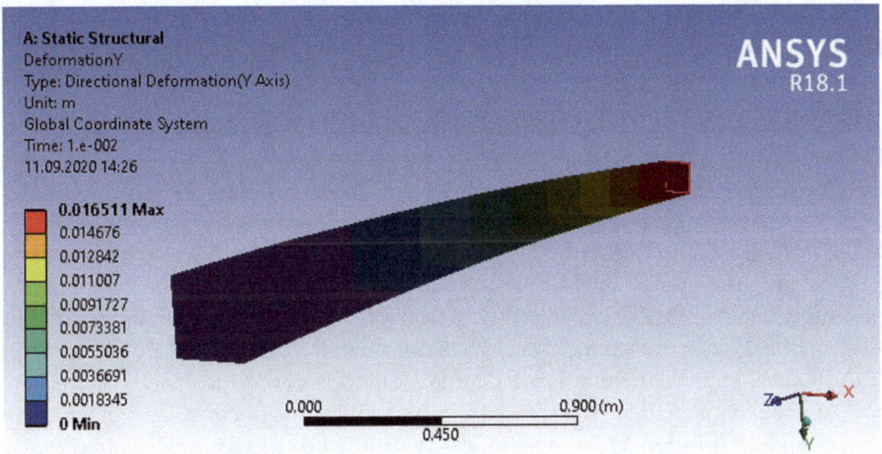

Fig. 11.19 Vertical deformation on beams

Maximum Von Mises stress occurs on the end of the rectangular beam as shown in Fig. 11.18. Thus, maximum stress 119.52 MPa occurs in the region where maximum strain occurred (Table 11.5).

It is clearly observed that, maximum deformation in y axes exceed to 0.0165 m at the end of the beam (Fig. 11.19). This value too close to the displacement limit in y direction.

Finally, maximum lateral deformation is 2.5×10^{-5} m as given in Fig. 11.20. Lateral deflection limit is quite bigger than this.

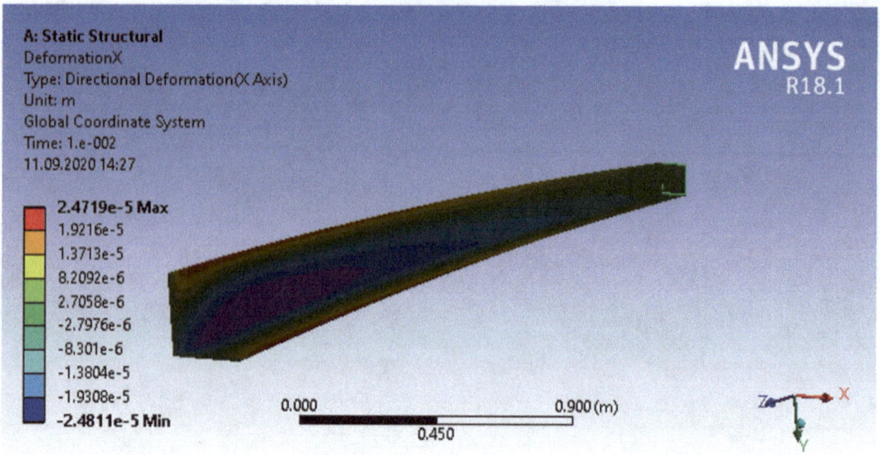

Fig. 11.20 Lateral deformation on beams

11.4 Conclusions

In this study, firstly, performances of Harmony Search Algorithm (HSA) and Backtracking Search Optimization Algorithm (BSA) are compared on Three-Bar Truss Design Problem (TBT), which is treated as a benchmark problem. Considered benchmark problem has two design variables and three constraints. In this phase of the study, Both algorithm find same optimum result by 263.90 kg. This result coincides with the previous study which is presented by Wang et al. [29]. It is proved that the used algorithms are encoded correctly. However, while HSA reached the optimum result in the 2163th iteration, BSA achieved the same result in the 1127th iteration. BSA indicate superior performance than HSA in time cost.

Then, rectangular and circular beams with varying cross-sections design problems are optimized. It is considered that it is very difficult to model this design problem manually with the matrix displacement method in the optimum design problem selection stage of this study. Geometrically two different thin-walled steel beam simulated by Ansys Workbench v18.1 packaged software. Nonlinear material is used for both model. Large displacement in other words second order effect is activated in analysis setting. By this way, the design problems are conducted more realistic and, the difficulty levels are increased. After that, HSA and BSA coded in Microsoft Visual Basic for Application v6.0 (MS VBA 6.0) programming language are connected to Ansys Workbench v18.1 by using Ironpython (programming language) script file.

Both algorithm which are encoded in Microsoft Office Excel (MS Excel) show well integration to Ansys Workbench v18.1. BSA shows superior performance in circular beams with varying cross-sections design problems. But, HSA is achieved a lower design weight that also provides the necessary constraints. It is thought that BSA may be trapped a local optimum during the optimization process of this problem. Nevertheless, there is a slight difference between obtained design weights as 2.4%. On the rectangular beams with varying cross-sections design problems, BSA both converges faster than HSA to the optimum design by providing a lower weight design. There is 2.00% difference between optimal design weights.

h value are proved minimal value as expected. Because, the value of *h* slightly effect to load carrying capacity of the both beams with varying cross-sections. And, it can increase the weight of the structure. The local web-buckling is not observed during the FEA of the optimum designed beams. It is predicted that *t* is minimal unless the case of that web buckling is occurred. There are two design variables and three constraints in both beams with varying cross-sections design problems. Among these, deformation constraints in force direction are dominant alone during the analysis (Tables 11.3 and 11.5).

If the obtained gains from the study are listed:

1. The first indicator of success in the study is that the algorithms established using the application programming interface work efficiently.
2. The solution of optimization problems whose mathematical model is difficult and complex can be easily constituted with this method.
3. Usage of the Application Programming Interface (API) may slightly extend the solution time passing to reach the optimum design.
4. Meshing is very important both in terms of both the solution time and the accuracy of the optimum result obtained.
5. BSA shows a good algorithmic performance and it can be used effectively in structural engineering design problems.

As a future work, the solutions of much more complex real-sized structural engineering problems can be realized by using the metaheuristics and solution methods used in this study. Also, the performance of BSA can be enhanced and/or improved much more effectively with small mathematical adjustments.

References

1. Tunca O, Carbas S (2016) Biogeography-based optimization algorithm for designing of planar steel frames. Int J Intell Syst Appl Eng 4:53–57. https://doi.org/10.18201/ijisae.266128
2. Erdal F, Tunca O, Dogan E (2017) Optimum design of composite corrugated web beams using hunting search algorithm. Int J Eng Appl Sci 9(2):156. https://doi.org/10.24107/ijeas.323633
3. Tunca O, Aydogdu I, Carbas S (2018) Symbiotic organisms search based optimal design of rigid frames. In: International congress on computational mechanics (9 GRACM), Chania/GREECE
4. Dogan E (2014) Solving design optimization problems via hunting search algorithm with Levy flights. Struct Eng Mech 52:351–368. https://doi.org/10.12989/sem.2014.52.2.351
5. Zhang L, Song B, Fu JJ, Weia SS, Yang L, Yan CZ, Li H, Gao L, Shia YS (2020) Topology-optimized lattice structures with simultaneously high stiffness and light weight fabricated by selective laser melting: design, manufacturing and characterization. J Manuf Process 56:1166–1177. https://doi.org/10.1016/j.jmapro.2020.06.005
6. Niu C, Zhang W, Gao T (2020) Topology optimization of elastic contact problems with friction using efficient adjoint sensitivity analysis with load increment reduction. Comput Struct 238:1–22. https://doi.org/10.1016/j.compstruc.2020.106296

7. Mortazavi A (2020) A new fuzzy strategy for size and topology optimization of truss structures. Appl Soft Comput 93:106412. https://doi.org/10.1016/j.asoc.2020.106412
8. Sun D, Dong C (2020) Shape optimization of heterogeneous materials based on isogeometric boundary element method. Comput Methods Appl Mech Eng 370:113279. https://doi.org/10.1016/j.cma.2020.113279
9. Zhu S, Ohsaki M, Guo X, Zeng Q (2020) Shape optimization for non-linear buckling load of aluminum alloy reticulated shells with gusset joints. Thin-Walled Struct 154:106830. https://doi.org/10.1016/j.tws.2020.106830
10. Mam K, Douthe C, Roy RL, Consigny F (2020) Shape optimization of braced frames for tall timber buildings: influence of semi-rigid connections on design and optimization process. Eng Struct 216:110692. https://doi.org/10.1016/j.engstruct.2020.110692
11. Dede T, Ayvaz Y (2015) Combined size and shape optimization of structures with a new meta-heuristic algorithm. Appl Soft Comput 28:250–258. https://doi.org/10.1016/j.asoc.2014.12.007
12. Khatibini M, Yazdani H (2018) Accelerated multi-gravitational search algorithm for size optimization of truss structures. Swarm Evol Comput 38:109–119. https://doi.org/10.1016/j.swevo.2017.07.001
13. Bekdaş G, Nigdeli SM, Yang XS (2015) Sizing optimization of truss structures using flower pollination algorithm. Appl Soft Comput 37:322–331. https://doi.org/10.1016/j.asoc.2015.08.037
14. Ziadi R, Bencherif-Madani A, Ellaia R (2020) A deterministic method for continuous global optimization using a dense curve. Math Comput Simul 178:62–91. https://doi.org/10.1016/j.matcom.2020.05.029
15. Kvasov DE, Mukhametzhanov MS (2018) Metaheuristic vs. deterministic global optimization algorithms: the univariate case. Appl Math Comput 318:245–259. https://doi.org/10.1016/j.amc.2017.05.014
16. Cheng MY, Prayogo D (2004) Symbiotic Organisms Search: a new metaheuristic optimization algorithm. Comput Struct 139:98–112. https://doi.org/10.1016/j.compstruc.2014.03.007
17. Yang XS, Deb S (2013) Multiobjective cuckoo search for design optimization. Comput Oper Res 40(6):1616–1624. https://doi.org/10.1016/j.cor.2011.09.026
18. Mirjalili S, Mirjalili SM, Lewis A (2014) Grey wolf optimizer. Adv Eng Softw 69:46–61. https://doi.org/10.1016/j.advengsoft.2013.12.007
19. Moazzeni AR, Khamehchi E (2020) Rain optimization algorithm (ROA): a new metaheuristic method for drilling optimization solutions. J Pet Sci Eng 195:1–13. https://doi.org/10.1016/j.petrol.2020.107512
20. Houssein EH, Saad MR, Hashim FA, Shaban H, Hassaballah M (2020) Levy flight distribution: a new metaheuristic algorithm for solving engineering optimization problems. Eng Appl Artif Intell 94:103731. https://doi.org/10.1016/j.engappai.2020.103731
21. Lee KS, Geem ZW (2004) A new structural optimization method based on the harmony search algorithm. Comput Struct 82(1):781–798
22. Huang YF, Chen PH (2020) Fake news detection using an ensemble learning model based on Self-Adaptive Harmony Search algorithms. Expert Syst Appl 159:113584. https://doi.org/10.1016/j.eswa.2020.113584
23. Alshammari N, Asumadu J (2020) Optimum unit sizing of hybrid renewable energy system utilizing harmony search. Jaya and particle swarm optimization algorithms. Sustain Cities Soc 60:102255. https://doi.org/10.1016/j.scs.2020.102255
24. Almeida FS (2019) Optimization of laminated composite structures using harmony search algorithm. Compos Struct 221:110852. https://doi.org/10.1016/j.compstruct.2019.04.024
25. Mansourzadeh F, Gh Khamseh AA, Safdari J, Norouzi A (2019) Utilization of harmony search algorithm to optimize a cascade for separating multicomponent mixtures. Prog Nucl Energy 111:165–173. https://doi.org/10.1016/j.pnucene.2018.11.005
26. Civicioglu P (2013) Backtracking search optimization algorithm for numerical optimization problems. Appl Math Comput 219:8121–8144. https://doi.org/10.1016/j.amc.2013.02.017

27. Storn R, Price K (1997) Differential evolution—a simple and efficient heuristic for global optimization over continuous spaces. J Global Optim 11(4):341–359. https://doi.org/10.1023/A:1008202821328
28. Zhang C, Lin Q, Gao L, Li X (2015) Backtracking Search Algorithm with three constraint handling methods for constrained optimization problems. Expert Syst Appl 42(21):7831–7845. https://doi.org/10.1016/j.eswa.2015.05.050
29. Wang H, Hu Z, Sun Y, Su Q, Xia X (2018) Modified backtracking search optimization algorithm inspired by simulated annealing for constrained engineering optimization problems. Comput Intell Neurosci 2018:27. https://doi.org/10.1155/2018/9167414
30. El-Fergany A (2015) Optimal allocation of multi-type distributed generators using backtracking search optimization algorithm. Int J Electr Power Energy Syst 64:1197–1205. https://doi.org/10.1016/j.ijepes.2014.09.020
31. Akhtar M, Hannan MA, Begum RA, Basri H, Scavino E (2017) Backtracking search algorithm in CVRP models for efficient solid waste collection and route optimization. Waste Manag 61:117–128. https://doi.org/10.1016/j.wasman.2017.01.022
32. Noraphaiphipaksa N, Poapongsakorn P, Hasap A, Kanchanomai C (2020) Failure analysis of pressure vessel with sight ports using finite element analysis. Eng Fail Anal 117:104791. https://doi.org/10.1016/j.engfailanal.2020.104791
33. Baragetti S, Arcieri EV (2020) Study of impact phenomena for the design of a mobile anti-terror barrier: experiments and finite element analyses. Eng Fail Anal 113:104564. https://doi.org/10.1016/j.engfailanal.2020.104564
34. Wisanuyotin T, Sirichativapee W, Paholpak P, Kosuwan W, Kasai Y (2020) Optimal configuration of a dual locking plate for femoral allograft or recycled autograft bone fixation: a finite element and biomechanical analysis. Clin Biomech 80:105156. https://doi.org/10.1016/j.clinbiomech.2020.105156
35. Guneyisi EM, Tunca O, Azez I (2015) Nonlinear dynamic response of reinforced concrete building retrofitted with buckling restrained braces. Earthq Struct 8:1349–1362. https://doi.org/10.12989/eas.2015.8.6.1349
36. ANSYS Inc (2017) ANSYS Mechanical Theory Reference: Release 18.1. Canonsburg, PA, USA
37. Smith M (2009) Abaqus, standard user's manual, version 6.9. Dassault Systèmes Simulia Corp., Providence, RI
38. SAP2000 (2000) Structural analysis program, nonlinear version 7.12. Computer and structures, Inc., Berkeley, CA, USA
39. Plaxis introductory: finite element code for soil and rock analyses (2000). Plaxis, Delft
40. Aydogdu I (2017) Cost optimization of reinforced concrete cantilever retaining walls under seismic loading using a biogeography-based optimization algorithm with Levy flights. Eng Optim 3:381–400. https://doi.org/10.1080/0305215X.2016.1191837
41. Idels O, Lavan O (2020) Performance based formal optimized seismic design of steel moment resisting frames. Comput Struct 235:106269. https://doi.org/10.1016/j.compstruc.2020.106269
42. Liu Q, Li X, Liu H, Guo Z (2020) Multi-objective metaheuristics for discrete optimization problems: a review of the state-of-the-art. Appl Soft Comput 93:106382. https://doi.org/10.1016/j.asoc.2020.106382
43. Ray T, Saini P (2007) Engineering design optimization using a swarm with an intelligent information sharing among individuals. J Eng Optim 33:735–748

Chapter 12
Metaheuristic-Based Structural Control Methods and Comparison of Applications

Serdar Ulusoy, Aylin Ece Kayabekir, Sinan Melih Nigdeli, and Gebrail Bekdaş

12.1 Introduction

All structures exposed to earthquake or wind have energy absorption capacity. The higher energy absorption capacity of the structure resulted with the lower amplitude of the vibrations occurring in the structures. For this purpose, different control systems are used within the structures. These structural control systems are divided into three groups: (a) passive systems, (b) active systems, and (c) semi-active and hybrid systems. In these three structural control systems, the passive systems are more widely used compared to other systems.

Passive structural control systems are provided by structural elements such as reinforced concrete wall and diagonal steel brace or mechanical system parts such as base isolation and tuned mass damper. These control systems are widely used in the retrofitting of historical structures. Passive systems are economical and more suitable than other control systems. However, these systems may be insufficient in the near fault ground motion with high peak velocities because of more ductility requirement for the structures.

S. Ulusoy
Department of Civil Engineering, Turkish-German University, Sahinkaya Cad 86, 34820 Istanbul, Turkey
e-mail: serdar.ulusoy@tau.edu.tr

A. E. Kayabekir · S. M. Nigdeli (✉) · G. Bekdaş
Department of Civil Engineering, Istanbul University-Cerrahpaşa, 34320 Avcılar/Istanbul, Turkey
e-mail: melihnig@istanbul.edu.tr

A. E. Kayabekir
e-mail: ecekayabekir@gmail.com

G. Bekdaş
e-mail: bekdas@istanbul.edu.tr

S. Carbas et al. (eds.), *Nature-Inspired Metaheuristic Algorithms for Engineering Optimization Applications*, Springer Tracts in Nature-Inspired Computing,
https://doi.org/10.1007/978-981-33-6773-9_12

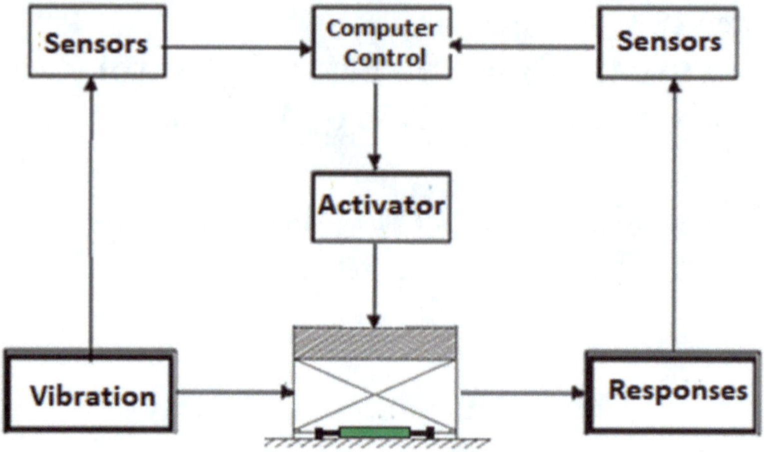

Fig. 12.1 Working scheme of active control systems in structures

Active control systems keep the structural responses caused by the effects of earthquake or wind within appropriate limits with the help of the energy source inside the structures. The working scheme of active control systems in structures is shown in Fig. 12.1.

The main task of the sensors inside the structures is to measure the structural responses such as displacement, velocity, or acceleration. These measured values are evaluated by the computer. After the evaluation, the required control forces for the structures are generated by the activators and applied to the structures. Active structural control generally occurs in two ways. One of them is active mass damper. Another one is active tendon systems consisting of prestressed cables.

Hybrid systems consist of a combination of a passive control system and an active control system. The working scheme of hybrid systems in structures is shown in Fig. 12.2. Since this structural control includes a passive control system, the amount of energy required in the building is less in comparison with the active controlled structures. For this reason, it is an important advantage of these systems that will not be affected by power cuts during earthquakes because these continue to protect the building with passive control systems.

Semi-active control system is formed by controlling the passive control system with a low energy source active control system. The working scheme of semi-active control systems in structures is shown in Fig. 12.3. No additional energy is applied to the structures in semi-active systems. Therefore, its performance to reduce structural responses is low compared to active systems, but it provides more effective protection in the structures compared to passive systems. In addition to their low power needs, they provide great advantages against power cuts in the event of an earthquake, as in hybrid systems, since they can operate even with small batteries. The most important semi-active control element is the magnetorheological (MR) damper.

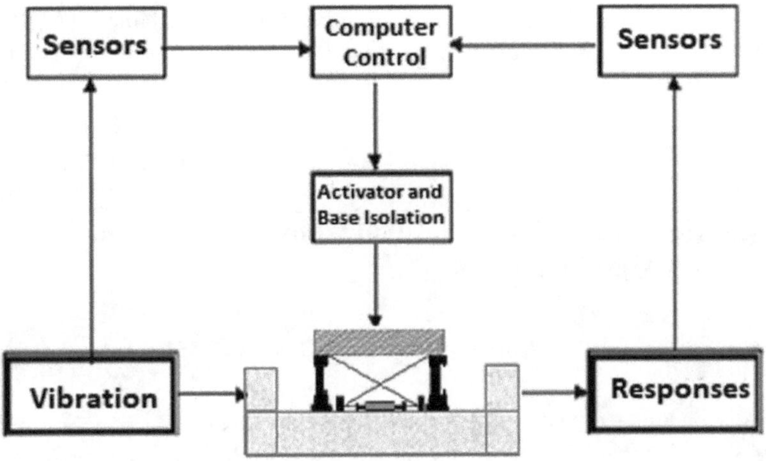

Fig. 12.2 Working scheme of hybrid control systems in structures

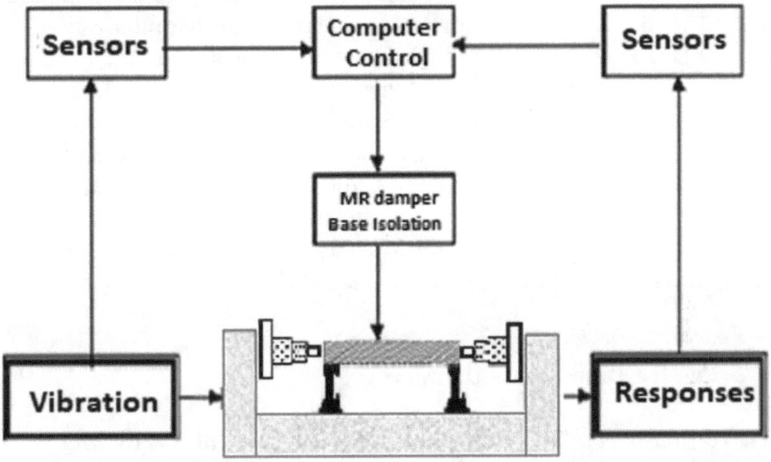

Fig. 12.3 Working scheme of semi-active control systems in structures

In the literature, two types of control systems are mentioned, namely open loop and closed loop. The most important difference of the open-loop control system from the closed-loop control system is that the desired output has no effect on the control action. For this reason, the control system, which has an important place in terms of use in structural engineering, is a closed-loop control system, because feedback takes place in structural responses. In other words, the output (structural responses) is measured with the sensors and compared with the reference value. Then, the control signal is obtained, and the generated control force is applied to the structure.

In this study, two active control models active tuned mass damper (ATMD) and active tendon control (ATC) are compared via metaheuristic algorithms and investigated to control the structures, actively. It is determined that structural reactions are significantly reduced in both active control systems for specific cases.

12.2 Review of Recent Structural Control Applications Using Metaheuristics

12.2.1 Tuned Mass Dampers

Tuned mass damper (TMD) is an invention that absorbs undesired vibrations. The parameters of TMDs must be tuned properties, and these parameters include its mechanical components in passive systems. In active systems, the controller parameters are also needed to be tuned. The optimum tuning of TMDs is subjected to the structural properties that define the dynamic behavior like period and damping of the structure. Also, the content of excitation is another factor in optimization and efficiency of TMD. The usage area of TMDs is not limited to structures and includes the vibrating dynamic systems such as machines, all types of moving vehicles including automotive, trains and ships, robotic devices, and all types of civil structures exposed with dynamic excitations such as earthquake, strong winds, vehicle or pedestrian traffics, and water. In structural engineering, the practical examples can be shown as the previously or newly constructed high-rise buildings, towers, critically important vibration-sensitive structures like nuclear plants, offshore structures and motorway bridges, viaducts and footbridges that may easily affected due to resonance.

12.2.1.1 Passive Tuned Mass Dampers

The first version of TMDs is accepted as the device invented by Frahm [1] that contains a mass attached to a spring-like stiffness element. Ormondroyd and Den Hartog [2] generated the classical form of TMD by adding additional damping devices. Due to this modification, to absorb vibrations resulting from random frequency, excitations became possible.

As the most well-known application of passive TMDs, the sphere-shaped TMD of the Taipei 101 building in Taipei, Taiwan can be shown. It was installed during the construction for the aim to reduce both wind and earthquake indicated vibrations. Differently, TMDs have been installed to retrofit the structures after their construction to improve seismic protection and comfort against strong winds. One comfort application against strong winds is the usage of a 1.5 t TMD suspended by three cables and four hydraulic telescopic shock absorbers in TV Tower in Berlin, Germany shown in Fig. 12.4. As a seismic retrofit example, the theme building found in The Los Angeles International Airport and called as Lax Theme Building was renovated by adding a

Fig. 12.4 Berlin TV tower

TMD positioned on the top of the concrete core. The TMD is constructed as a slab-like structure supported by eight fluid viscous dampers and isolation systems. By the advantage of TMD, a response reduction between 30 and 40% is found according to the numerical analyses for the TMD having a mass with 20% mass of the main tower structure [3].

For single-degree-of-freedom (SDOF) structures, the basic working principle of TMDs is to include a secondary mode that is tuned with a frequency close to the natural frequency of the main structural system. In that reason, the combined system of main structure and TMD have two frequencies that are a little lower and higher than SDOF frequency. Due to that, the resonance case that occurs when the excitation frequency and the SDOF natural frequency are equal is prevented. Similarly, the excitation has also more effect on the system when the excitation frequency is too close to the natural frequency of the system. The purpose in structures is to reduce these increasing effects. The damping of TMDs plays an important role in the reduction of vibrations resulting from excitations with various frequencies and reduce the movement of TMD to keep the control application in applicable range.

The optimum estimation of TMD parameters such as stiffness and damping has been investigated by proposing a lot of methods that assume or model the structural system differently. Currently, it is still an active research area because of zero possibility of deriving an exact formulation due to the response of damped structures exposed to excitations including random frequencies.

The basic formulations that are obtained according to several assumptions are presented in Table 12.1. These formulations are the best known optimum tuning equations of TMD that are given by Den Hartog [4]. The equations of Den Hartog are used to find two optimum values. One of them is the optimum frequency ratio (f_{opt}) that is the ratio of frequencies of TMD ($\omega_{\text{d,opt}}$) and main structure (ω_{s}). The other ones are the optimum damping ratio of TMD ($\xi_{\text{d,opt}}$) that is the ratio of the damping coefficient of TMD (c_{d}) and two times of mass of the damper (m_{d}) multiplied with $\omega_{\text{d,opt}}$. The equations of Den Hartog are derived for undamped SDOF main systems, and the effect of inherent damping of main system is also considered in several studies [5–8] via numerical analyses. Another well-known formula was found by Warburton that suggested equations for different excitation like white noise [9]. Sadek et al. conducted numerical algorithms and investigations to find equations for main systems with inherent damping (ξ) by using a curve fitting method [10]. The application of these equations derived for SDOF structures can be approximately done for multiple-degree-of-freedom (MDOF) systems by considering a single critical mode [11].

To consider all vibration modes, numerical algorithms and metaheuristic-based methods must be used. In the documented methods, the employed metaheuristic algorithms that imitate a natural phenomenon in numerical iterations include genetic algorithm (GA) [12–15], particle swarm optimization (PSO) [16, 17] (close form expressions for this algorithm are also given in Table 12.1), harmony search (HS)

Table 12.1 Basic frequency and damping ratio expressions of optimum TMD tuning application

Method	$f_{\text{opt}} = \frac{w_{\text{d,opt}}}{w_{\text{s}}}$	$\xi_{\text{d,opt}} = \frac{c_{\text{d,opt}}}{2m_{\text{d}}w_{\text{d,opt}}}$
Den Hartog [4]	$\frac{1}{1+\mu}$	$\sqrt{\frac{3\mu}{8(1+\mu)}}$
Warburton [9]	$\frac{\sqrt{1-(\mu/2)}}{1+\mu}$	$\sqrt{\frac{\mu(1-\mu/4)}{4(1+\mu)(1-\mu/2)}}$
Sadek et al. [10]	$\frac{1}{1+\mu}\left[1-\xi\sqrt{\frac{\mu}{1+\mu}}\right]$	$\frac{\xi}{1+\mu}+\sqrt{\frac{\mu}{1+\mu}}$
Leung and Zhang [17]	$\frac{\sqrt{1-(\mu/2)}}{1+\mu} + (-4.9453+20.2319\sqrt{\mu}-37.9419\mu)\sqrt{\mu}\xi + (-4.8287+25.0000\sqrt{\mu})\sqrt{\mu}\xi^2$	$\sqrt{\frac{\mu(1-\mu/4)}{4(1+\mu)(1-\mu/2)}} -5.3024\xi^2\mu$

[18–22], bat algorithm (BA) [23], ant colony optimization (ACO) [24], artificial bee colony algorithm (ABC) [25], flower pollination algorithm (FPA) [21, 22, 26, 27], teaching–learning-based optimization (TLBO) [21, 22, 28], and Jaya algorithm (JA) [22]. Yucel et al. [29] used optimum TMD parameters found via metaheuristics in machine learning to develop an artificial neural network (ANN) model that estimates the TMD parameters.

12.2.1.2 Active, Semi-Active, and Hybrid Tuned Mass Dampers

Active tuned mass damper (ATMD) systems have been successfully applied to many structures. As practical examples of ATMDs, ORC 2000 Symbol tower (188 m) in Osaka, Japan; Shanghai World Financial Center (492 m) in Shanghai, China; Shinjuku Park Tower (227 m) in Tokyo, Japan; Air Traffic Control Tower (57 m) in Edinburgh, UK and Incheon International Airport Control Tower (100.4 m) in Incheon, Korea can be given as examples [30].

In this section, studies related to ATMD system are summarized chronologically. Ankireddi and Yang [31] presented a method using full feedback (all of displacement, velocity and acceleration feedback) control algorithm for the design of an ATMD. In the study, it is aimed to mitigate first-mode responses due to wind loads in tall buildings. For the similar purpose, Mackriell et al. [32] suggested a control algorithm using acceleration feedback. The analytical expressions derived by Ankireddi and Yang [31] to determine the optimal feedback gains were examined by Yan et al. [33] in order to test accuracy of expressions under more sophisticated wind loadings. According to results obtained by a norm control technique, it was concluded that although the expressions are sufficient under along-wind loadings, the expression related with the damper frequency ratio needs to be modified in the case of across-wind loadings. Qu et al. [34] have stated that reduced-order control is required for the multi-degree-of-freedom systems. For that reason, in the study, a type of system reduction scheme named dynamic condensation method was introduced and tested on a tall building with TMD and ATMD. The tests showed that the proposed method provides efficient reduced-order control and gives very close result to exact. The performance of a five-story benchmark model with ATMD having fuzzy logic controller (FLC) under the earthquake excitations was investigated by Samali and Al-Dawod [35]. In the numerical example comparing FLC with linear quadratic regulator (LQR) controllers, similar building response, active control force, and the stroke of the actuator were obtained with both controllers, whereas it was seen that FLC gives better results considering number of sensors, computational resources, and the required control power. Samali et al. [36] used ATMD system with FLC for 76-story, 306-m tall a benchmark building under cross-wind excitation and did performance comparation of FLC and linear quadratic Gaussian (LQG) controller. From the analysis results, it was concluded that the FLC is also better than LQG in terms of required control power and number of sensors. In 2006, Han and Li [37] proposed to multiple ATMDs (AMTMDs) consisting of several ATMD to control the oscillations of structures exposed to ground acceleration. The study showed that

AMTMDs are more successful than single ATMDs and multiple ATMDs in the frequency domain. The multiple ATMDs were also used by Li and Xiong [38] to suppress translational and torsional responses due to earthquake excitations in asymmetric structures. Pourzeynali et al. [39] in 2007 combined FLC and genetic algorithm (GA), a type of the metaheuristic algorithms, for the design of ATMD systems applied to high-rise building structures under earthquake excitation. Next year, Guclu and Yazici [40] designed FLC and proportional derivative (PD) controller using two types of actuators installed on the first and fifteenth story of a fifteen-degree-of-freedom structural system with ATMD and observed that FLC provides better active control performance against earthquake effects. Then, fuzzy proportional–integral–derivative (PID) was implemented to ATMD, attached to a nonlinear structural system, to suppress seismic vibrations [41]. For the similar purpose, in another study, self-tuning fuzzy logic controllers (STFLC) were applied [42]. Li et al. [43] introduced a design methodology for the optimum control performance of the ATMD in order to control the translational and the torsional responses caused by earthquake in asymmetric structures. To determine optimum ATMD parameters, minimum translational and torsional displacement minimization criteria with the gradient search method were utilized. Li [44] also investigated optimum AMTMD parameters for asymmetric structures considering soil structure interaction (SSI) and determined the optimum parameter criteria according to minimizing the translational and torsional displacement mean square responses. In order to obtain optimal control forces applied by ATMD to structure, a method using three algorithm, i.e., discrete wavelet transform (DWT), particle swarm optimization (PSO), and linear quadratic regulator (LQR) was developed by Amini et al. [45]. In another study, for the optimal control of tall structures under along-wind excitations, You et al. [46] proposed the use of a linear quadratic Gaussian (LQG) controller, and the numerical optimization method developed by Ayorinde and Warburton [11] to find tuning frequency and damping ratio of ATMD was used. In the same year, Shariatmadar and Razavi [47] used a fuzzy logic controller (FLC) together with particle swarm optimization (PSO) to reduce earthquake resulted vibrations in a structure with ATMD. Shariatmadar et al. [48] investigated the interval type-2 FLC (IT2FLC) for an ATMD system and concluded that the presented method is quite effective in suppressing structural vibrations. An approach using a multi-objective adaptive genetic fuzzy controller was presented by Soleymani and Khodadadi [49] to reduce destructive effects due to earthquakes and strong winds in tall buildings. According to the investigation, the approach could not be adequate and the ATMD designed against earthquakes in the case of wind loadings and vice versa. Heidari et al. [50] suggested using a hybrid controller combining LQR and PID controllers and employed cuckoo search (CS) algorithm, a metaheuristic algorithm, in the optimum design process of the controller. Kayabekir et al. [51] optimized ATMD for seismic structures via proposing a modified HS that have adaptive strategies on algorithm parameter tuning. In that study, both mechanical properties of ATMD and the parameters of PID controller are taken as design variables.

As hybrid application, Li and Cao [52] developed a newly hybrid active tuned mass dampers (HATMD) that consist of two different ATMDs and combine the negative normalized acceleration feedback gain factors (NAFGF) with positive NAFGF schemes, in order to mitigate undesirable oscillations of structures exposed seismic excitations. Then, a dashpot was added between the structure and ATMD masses to provide stroke limitation to the HATMD and this system was named as enhanced HATMD (EHATMD) [53]. Djedoui et al. [54] proposed combination of base isolation and LQR-controlled ATMD in a single rigid building. Mamat et al. [55] investigated the placement of control device in structure for HATMDs using FLC. Chesne et al. [56] proposed a new dual transducer and its associated control law that is used in HTMDs. TMDs and magnetorheological (MR) dampers are proposed to use together in the hybrid control application proposed by Bhaiya et al. [57].

As semi-active applications, Demetriou and Nikitas [58] proposed a hybrid semi-active TMD that is efficient in cost and energy. Maślanka [59] measured the performance of TMDs with MR damper using acceleration feedback as a semi-active system. Zhang et al. [60] proposed a TMD with tunable stiffness mechanism changing the number of active coils of springs in parallel and in series configurations. Magnetorheological elastomer was proposed by Sun et al. [61] to generate a semi-active TMD that especially eliminates the magnetic circuit gap in a design. Moutinho et al. [62] described the work done for reduction of vibrations in a slender footbridge located at FEUP campus that use passive and semi-active TMDs. Karami et al. [63] proposed a semi-active TMD by using the damage detection algorithm based on identified system Markov parameters (DDA/ISMP) method for estimation of the structural parameters of the bare structure. Zelleke and Matsagar [64] developed an energy-based predictive algorithm for semi-active tuned mass dampers. A semi-active tuned mass damper [65] working according to acceleration and relative motion feedbacks was optimized in the frequency domain to provide the same vibration damping efficiency with TMD having seven times larger mass than the optimized via Den Hartog's formula [4].

12.2.2 Active Tendon Control

In recent years, many studies including different control algorithms and different structural models, different ground motions (near or far fault ground motions), and various feedback strategies are conducted in order to control the structures with active tendons. A single-degree-of-freedom system with active tendons was investigated by Chung et al. experimentally and analytically considering the time delay effect [66]. By increasing the degree of the same single free degree system, the behavior of two control algorithms (new algorithms based on instantaneous optimal control) in this system was compared [67]. A frame building and a shear wall building with active tendons are examined to compute the required control force [68]. The performance of a full-scale dedicated structure controlled was evaluated under different earthquake records using two modified control algorithms which was discussed by Lin et al. [69]. Artificial neural networks algorithm was proposed by Ghaboussi and Joghataie to solve especially nonlinear problems in active tendon-controlled structures [70]. The feasibility and efficiency of an SDOF system with active tendons were investigated by considering varying soil properties as long-span bridges [71]. The multi-degree-of-freedom system with active tendons using acceleration feedback strategy was experimentally demonstrated to achieve more robust and effective control [72]. The success of a modified predictive control algorithm on the structural responses of active tendon-controlled systems was verified according to the results obtained the eigenvalue analysis, time, and frequency domain analysis [73]. Fuzzy sliding-mode control was used in active tendon-controlled structures to overcome the chattering effect which is the most important disadvantage of conventional sliding-mode controller [74]. The superiority of LQR with a new energy-based technique over LQR is presented for active tendon structures [75]. The important reduction in structural responses and applicable control forces was achieved under various earthquake records using the combination of sliding-mode control and artificial neural network techniques [76]. Torsionally irregular structures with active tendons were controlled considering soil structure interaction using $H_{\inf}$ controller [77]. PID controller was implemented to decrease the effect of the impulsive motion on torsionally irregular structures with active tendons placed on different floors [78]. LQR algorithm was tested on the irregular active tendon-controlled structures under near fault ground motions considering soil structure interaction [79]. Six different metaheuristic algorithms such as differential evolution, HS, imperialist competitive algorithm, firefly algorithms, BA, and bees algorithm are separately employed for active tendon-controlled structures using wavelet-based LQR controller to determine the optimum feedback gains [80]. TLBO, FPA, and JA were used discretely to calculate the optimum parameters of PID controller of active tendon-controlled structures considering soil structure interaction [81]. The parameters of PID controller of active tendon-controlled structures with different cases of time delay and different control force limits were optimized via metaheuristic algorithms to enhance the robustness of structures [82].

12.3 Equations of Motion and Optimization Methodologies

12.3.1 TMD and ATMD

Dynamic analysis of the structures is required to analyze the structural systems subject to seismic ground motion. In order to perform dynamic analysis, it is necessary to write the equations of motion of the structural model and solve these equations. Figure 12.5 shows a shear building (uncontrolled structure) and structures controlled with tuned mass damper (TMD) and active tuned mass damper (ATMD). The equations of motion based on these shear building models are presented in this section.

In Fig. 12.5, m_i, c_i, k_i, and x_i are the mass, damping coefficient, stiffness coefficient of the ist floor in a N-storey building and relative displacement of the ist floor relative to the ground. Similarly, the mass, stiffness coefficient, damping coefficient, and

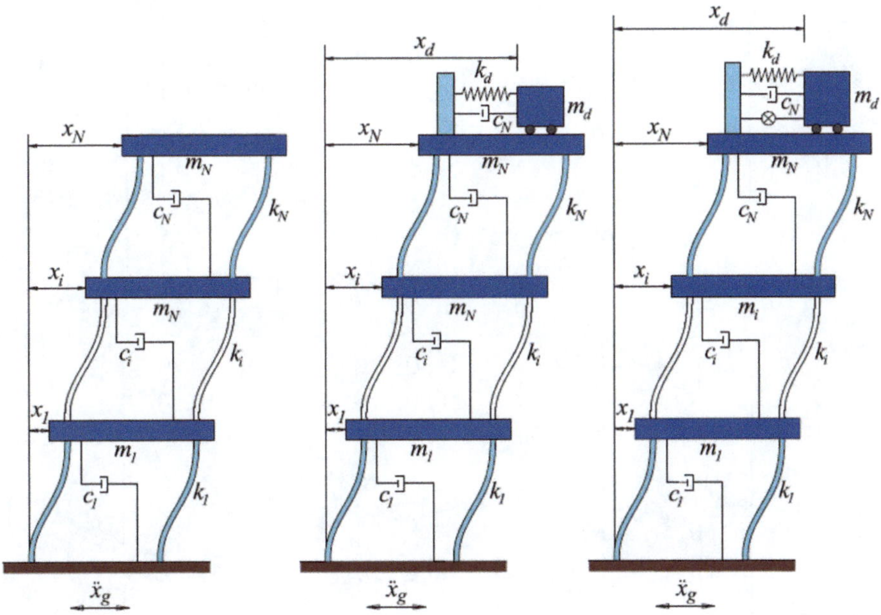

Fig. 12.5 N-story building models without structural control **a**, with TMD **b** and **c** with ATMD

relative displacement of the control systems correspond to the expressions of m_{d}, c_{d}, k_{d}, and x_{d}, respectively. The $\ddot{x}_{\mathrm{g}}$ in the figure shows the ground acceleration.

Equation of motion for the uncontrolled shear building under seismic effects can be written as

$$[M] \cdot \ddot{x}(t) + [C] \cdot \dot{x}(t) + [K] \cdot x(t) = -[M] \cdot \{1\} \cdot \ddot{x}_{\mathrm{g}}(t) \tag{12.1}$$

where $[M]$, $[C]$, and $[K]$ represent mass, damping, and stiffness coefficient matrices, respectively, and these matrices can be symbolized by Eqs. (12.2–12.4).

$$[M] = \begin{bmatrix} m_1 & 0 & 0 & 0 \\ 0 & m_2 & 0 & 0 \\ 0 & 0 & \ddots & 0 \\ 0 & 0 & 0 & m_N \end{bmatrix} \tag{12.2}$$

$$[C] = \begin{bmatrix} c_1+c_2 & -c_2 & 0 & 0 & 0 & 0 & 0 & 0 \\ -c_2 & c_2+c_3 & -c_3 & 0 & 0 & 0 & 0 & 0 \\ 0 & -c_3 & c_3+c_4 & -c_4 & 0 & 0 & 0 & 0 \\ 0 & 0 & \vdots & \vdots & \vdots & 0 & 0 & 0 \\ 0 & 0 & 0 & \vdots & \vdots & \vdots & 0 & 0 \\ 0 & 0 & 0 & 0 & \vdots & \vdots & \vdots & 0 \\ 0 & 0 & 0 & 0 & 0 & -c_{N-1} & c_{N-1}+c_N & -c_N \\ 0 & 0 & 0 & 0 & 0 & 0 & -c_N & c_N \end{bmatrix} \quad (12.3)$$

$$[K] = \begin{bmatrix} k_1+k_2 & -k_2 & 0 & 0 & 0 & 0 & 0 & 0 \\ -k_2 & k_2+k_3 & -k_3 & 0 & 0 & 0 & 0 & 0 \\ 0 & -k_3 & k_3+k_4 & -k_4 & 0 & 0 & 0 & 0 \\ 0 & 0 & \vdots & \vdots & \vdots & 0 & 0 & 0 \\ 0 & 0 & 0 & \vdots & \vdots & \vdots & 0 & 0 \\ 0 & 0 & 0 & 0 & \vdots & \vdots & \vdots & 0 \\ 0 & 0 & 0 & 0 & 0 & -k_{N-1} & k_{N-1}+k_N & -k_N \\ 0 & 0 & 0 & 0 & 0 & 0 & -k_N & k_N \end{bmatrix} \quad (12.4)$$

As seen from Eq. (12.2), these matrices are multiplied by the time-varying structural response vectors $\ddot{x}(t)$ (acceleration), $\dot{x}(t)$ (velocity) ve $x(t)$ (displacement). The displacement vector can be defined by Eq. (12.5) and the velocity and acceleration vectors can be obtained by deriving the displacement vector.

$$x(t) = \begin{bmatrix} x_1 \\ x_2 \\ \vdots \\ x_{N-1} \\ x_N \end{bmatrix} \quad (12.5)$$

The right side of Eq. (12.1) corresponds to the force acting on the structure due to ground acceleration. The force is equal to the multiplication of the mass matrix of the structure, the unit column vector $\{1\}$, and the time-varying ground acceleration $\left(\ddot{x}_g(t)\right)$. The unit column vector $(1 \times N)$ provides matrix size compatibility in multiplication.

If the equation of motion is written for a TMD-controlled shear building, arrangements depending on TMD parameters must be done in the $[M]$, $[C]$, and $[K]$ matrices shown in Eq. (12.1). For the structural model attached TMD to the top story as in Fig. 12.5, the matrices are given in Eqs. (12.6)–(12.8).

$$[M] = \begin{bmatrix} m_1 & 0 & 0 & 0 & 0 \\ 0 & m_2 & 0 & 0 & 0 \\ 0 & 0 & \ddots & 0 & 0 \\ 0 & 0 & 0 & m_N & 0 \\ 0 & 0 & 0 & 0 & m_d \end{bmatrix} \quad (12.6)$$

$$[C] = \begin{bmatrix} c_1+c_2 & -c_2 & 0 & 0 & 0 & 0 & 0 & 0 \\ -c_2 & c_2+c_3 & -c_3 & 0 & 0 & 0 & 0 & 0 \\ 0 & -c_3 & c_3+c_4 & -c_4 & 0 & 0 & 0 & 0 \\ 0 & 0 & \vdots & \vdots & \vdots & 0 & 0 & 0 \\ 0 & 0 & 0 & \vdots & \vdots & \vdots & 0 & 0 \\ 0 & 0 & 0 & 0 & \vdots & \vdots & \vdots & 0 \\ 0 & 0 & 0 & 0 & 0 & -c_N & c_N+c_d & -c_d \\ 0 & 0 & 0 & 0 & 0 & 0 & -c_d & c_d \end{bmatrix} \quad (12.7)$$

$$[K] = \begin{bmatrix} k_1+k_2 & -k_2 & 0 & 0 & 0 & 0 & 0 & 0 \\ -k_2 & k_2+k_3 & -k_3 & 0 & 0 & 0 & 0 & 0 \\ 0 & -k_3 & k_3+k_4 & -k_4 & 0 & 0 & 0 & 0 \\ 0 & 0 & \vdots & \vdots & \vdots & 0 & 0 & 0 \\ 0 & 0 & 0 & \vdots & \vdots & \vdots & 0 & 0 \\ 0 & 0 & 0 & 0 & \vdots & \vdots & \vdots & 0 \\ 0 & 0 & 0 & 0 & 0 & -k_N & k_N+k_d & -k_d \\ 0 & 0 & 0 & 0 & 0 & 0 & -k_d & k_d \end{bmatrix} \quad (12.8)$$

Similarly, TMD responses need to be added to structural response vectors. In this case, the displacement vector in the equation of motion can be express by Eq. (12.9)

$$x(t) = \begin{bmatrix} x_1 \\ x_2 \\ \vdots \\ x_N \\ x_d \end{bmatrix} \quad (12.9)$$

For the equation of motion of the shear building attached to the ATMD system, the control force must be added to the equation, unlike the uncontrolled and TMD-controlled systems. Accordingly, the equation of motion for the ATMD system can be written as

$$[M] \cdot \ddot{x}(t) + [C] \cdot \dot{x}(t) + [K] \cdot x(t) = -[M] \cdot \{1\} \cdot \ddot{x}_g(t) + F(t) \quad (12.10)$$

ATMD systems are systems derived by adding an active control mechanism (including sensor, controller and actuator) to the TMD system. Therefore, the matrices $[M]$, $[C]$, and $[K]$ and the structural response vectors ($\ddot{x}(t)$, $\dot{x}(t)$, and x(t)) of the system controlled with ATMD are similar to system controlled with TMD. The control force $F(t)$ of the ATMD system in the equation of motion is a vector with $N + 1$ elements and can be found by

$$F(t) = \begin{bmatrix} 0 \\ 0 \\ \vdots \\ F_u \\ -F_u \end{bmatrix} \quad (12.11)$$

where F_u represents the active control force produced by ATMD and can be obtained by multiplying the trust constant (K_f) with the current of armature coil (i_{ATMD}) as given in Eq. (12.12).

$$F_u = K_f \cdot i_{\text{ATMD}} \quad (12.12)$$

The current of armature coil (i_{ATMD}) can be calculated utilizing Eq. (12.13).

$$R \cdot i_{\text{ATMD}} + K_e \cdot (\dot{x}_{\text{d}} - \dot{x}_N) = u \quad (12.13)$$

In Eq. (12.13), expressions show R; the resistance value—K_e; induced voltage constant of armature coil—u; control signal generated according to the control algorithm, in other words, the current of the linear motor. $\dot{x}_{\text{d}}$ and $\dot{x}_N$ are the velocity values of ATMD and the top story of the building, respectively.

12.3.2 Active Tendon Control

The active control of a single-degree-of-freedom system (SDOF) with active tendons is shown in Fig. 12.6. It is given as m_1—the mass of the structure, c_1—the damping coefficient of the structure, k_1—the stiffness coefficient of the structure, k_{c}—the stiffness coefficient of the cable, α—the angle of the cables according to the ground, and F_1 and F_2—the forces of the cables under dynamic state that vary according to the direction of the earthquake. If there is an F force in the cables under static state and the direction of the dynamic load is in the $+$ x-direction indicated in Fig. 12.6, the resulting F_2 force is equal to $F + k_{\text{c}}u(t)$, while the F_1 force is equal to F-$k_{\text{c}}u(t)$. In this case, the equation of motion of the active tendon structure model is obtained as in Eq. 12.14 by adding the horizontal force $4k_{\text{c}}u_1(t)\cos\alpha$ in the opposite direction of the movement of the mass of the uncontrolled structure model.

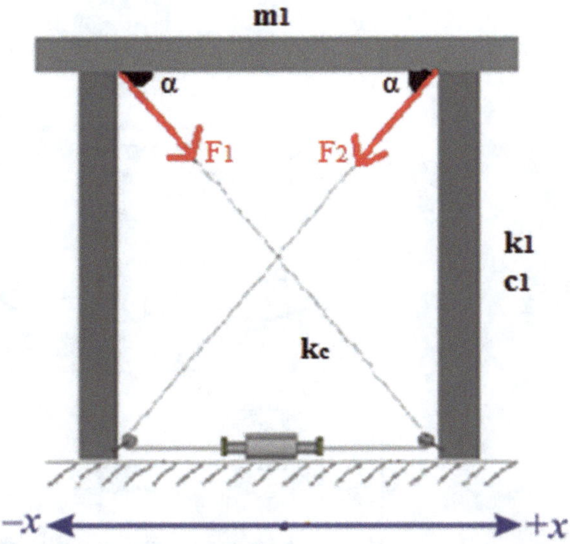

Fig. 12.6 Active control of a single degree of freedom system with active tendons

$$m_1 a_1(t) + c_1 v_1(t) + k_1 x_1(t) = -m_1 a_g(t) - 4k_c u_1(t) \cos \alpha \tag{12.14}$$

$a_1(t)$, $v_1(t)$, $x_1(t)$, $a_g(t)$, and $u_1(t)$ are the time-dependent acceleration, velocity, displacement, ground acceleration, and control signal of SDOF system, respectively.

12.3.3 Proportional–Integral–Derivative Controller

In this chapter, the proportional–integral–derivative (PID) controller is used to obtain the control signal. PID is a feedback controller that converts the error signals into control signals by controlling three separate actions: proportion (P), integration (I), and derivation (D). According to the PID controller equation, the control signal is calculated by Eq. (12.15).

$$u = K_p \cdot \left(e(t) + \frac{1}{T_i} \cdot \int e(t)\mathrm{d}t + T_d \cdot \frac{\mathrm{d}e(t)}{\mathrm{d}t} \right) \tag{12.15}$$

In Eq. (12.15), K_p, T_d, and T_i are the PID controller parameters and show the proportional gain, derivative time, and integral time, respectively. $e(t)$ is the error signal showing the difference between the desired and response of system. PID controller block diagram generated via MATLAB with Simulink [83] is given as Fig. 12.7.

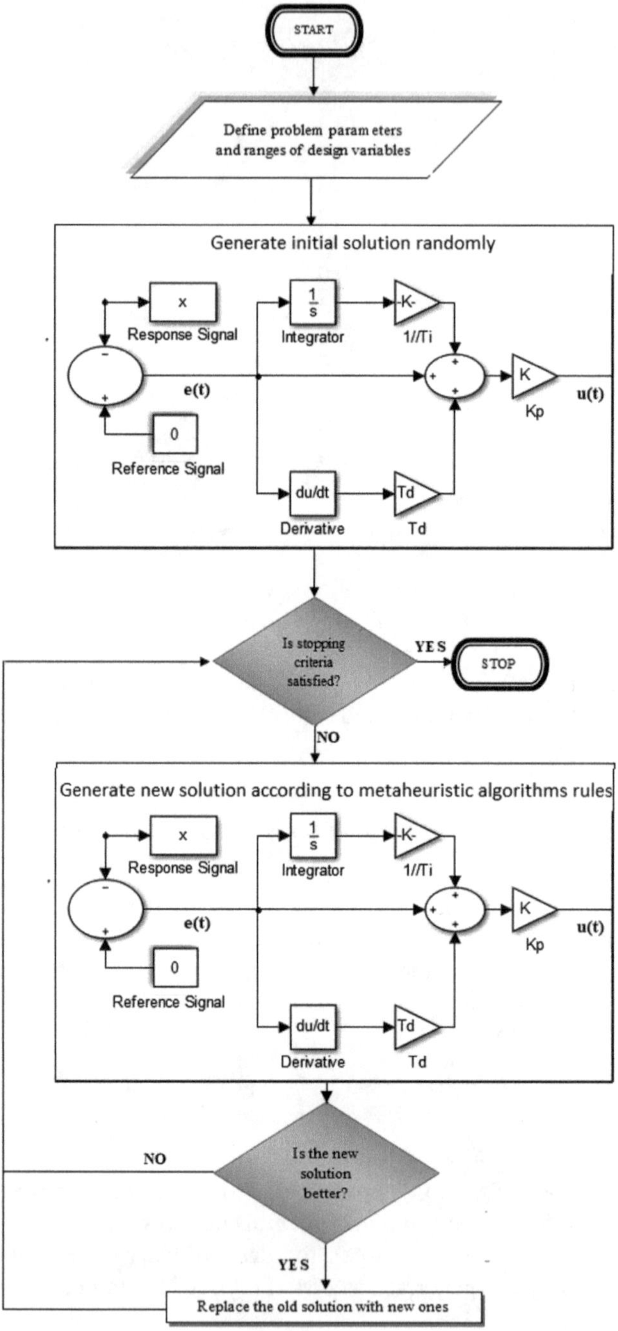

Fig. 12.7 PID control block diagram generated via MATLAB with Simulink

12.3.4 Metaheuristic-Based Optimization

In this section, the metaheuristic algorithm-based optimization process is presented for structural control system. The general methodology of the optimization process is based on generating candidate solutions and, according to objective of optimization, finding of the best solution by modifying existing candidate solutions. Considering the operation of this process, it is possible to summarize optimization in 5 steps as given below.

Step 1: The design constants, lower and upper limits of the design variables, the population number (pn), the user-defined parameters of the algorithm, and the stopping criteria of the optimization problem are defined. In this chapter, stopping criteria is defined as maximum iteration number.

Step 2: For each design variable, a value is randomly generated within the its lower and upper limits as shown in Eq. (12.16). This process is repeated until the providing population number defined in Step 1 matrix.

$$X_{i,j} = X_{i(\text{low})} + \text{rand} \cdot \left(X_{i(\text{up})} - X_{i(\text{low})}\right) \tag{12.16}$$

In Eq. (12.16), i and j symbolize number of design variable and solution vector, respectively. $X_{i,j}$, $X_{i(\text{low})}$, and $X_{i(\text{up})}$ show the value of design variable, lower and upper limit of the design variable, respectively. rand is a command that generates random numbers between 0 and 1. Then, these generated values stored in an initial solution matrix such as CL matrix (Eq. 12.17).

$$\text{CL} = \begin{bmatrix} X_{1,1} & X_{1,2} & \cdots & X_{1,\text{pn}} \\ X_{2,1} & X_{2,2} & \cdots & X_{2,\text{pn}} \\ \vdots & \vdots & \cdots & \vdots \\ X_{N-1,1} & X_{N-1,2} & \cdots & X_{N,\text{pn}} \\ X_{N,1} & X_{N,2} & \cdots & X_{N-1,\text{pn}} \end{bmatrix} \tag{12.17}$$

Each column of the CL matrix is a candidate solution vector including values of design variables. In the matrix, there are N design variables and pn solution vectors.

Step 3: For each solution vector, the objective function is calculated, and the design constraints of the optimization problem are checked. If a design constraint is not provided by a solution vector, the objective function of the related solution is given as a penalized value. The objective function is the minimization of a structural response.

Step 4: In this step, new solutions are generated and stored in a new solution matrix. Since the generation is done according to the algorithm rules, this step is specific to algorithm.

Two metaheuristics are presented in this chapter. These are harmony search (HS) developed according to musical performances [84] and teaching–learning-based optimization (TLBO) imitating education process [85].

In the optimization process performed with the modified HS [51], Eqs. 12.8 or 12.19 is used for this step. According to harmony memory considering rate (HMCR), which is a user-defined algorithm parameter, it is decided that which equation will be used in generation.

$$X_{\text{new}} = X_{\text{low}} + \text{rand} \cdot \left(X_{\text{up}} - X_{\text{low}}\right) \tag{12.18}$$

$$X_{\text{new}} = X_{\text{sel}} + \text{rand} \cdot \text{PAR} \cdot \left(X_{\text{up}} - X_{\text{low}}\right) \tag{12.19}$$

In Eqs. (12.18 and 12.19), X_{new}, X_{sel}, X_{low}, and X_{up} are a new solution, a selected existing solution, lower and upper limits of design variables, respectively. PAR indicates the pitch adjusting rate and is a user-defined algorithm parameter. PAR calculated with

$$\text{PAR} = \text{PAR}_{\text{in}}\left(1 - \frac{\text{IN}}{\text{MI}}\right) \tag{12.20}$$

where PAR_{in}, IN, and MI represent the initial value of PAR at the start of the optimization process, current iteration number, and maximum iteration number, respectively. Also, calculation of HMCR is done according to

$$\text{HMCR} = \text{HMCR}_{\text{in}}\left(1 - \frac{\text{IN}}{\text{MI}}\right) \tag{12.21}$$

in which HMCR_{in} is initial value of HMCR at the start of the optimization process.

In TLBO, two phases are consequently performed for the fourth step of the optimization. In the first phase, new solutions are generated according to Eq. (12.22) as teaching phase.

$$X_{\text{new}} = X_{\text{old}} + \text{rand} \cdot (X_{\text{best}} - T_F \cdot X_{\text{mean}}) \tag{12.22}$$

In Eq. (12.22), X_{best}, X_{old}, T_F, and X_{mean} show best solution, existing (old) solution, and teaching factor randomly chosen as 1 or 2 and mean value of existing solutions, respectively.

In the second phase called learner phase, two solutions (X_a and X_b) are randomly chosen from the existing solutions and new solutions are generated by considering objective functions of the existing solutions ($f(X_a)$ and $f(X_b)$) as given in Eq. (12.23).

$$X_{\text{new}} = \begin{cases} X_{\text{old}} + \text{rand} \cdot (X_a - X_b) & \text{if } f(X_a) > f(X_b) \\ X_{\text{old}} + \text{rand} \cdot (X_b - X_a) & \text{if } f(X_a) < f(X_b) \end{cases} \tag{12.23}$$

Step 5: In the last step, the existing (old) solution matrix is compared with the new solution matrix. If the objective function of new solutions is better than the existing ones, the existing solution matrix is updated.

Steps 4 and 5 are continued until the stopping criteria defined as maximum iteration number are provided.

The optimum values of PID controller coefficients (design variables) are determined with the help of metaheuristic algorithms. For ATMD, the mechanical properties are also optimized. The general flowchart diagram of the optimization process using metaheuristic algorithms is given in Fig. 12.8.

12.4 Numerical Examples Comparing ATMD and Active Tendons

In this study, ATMD and active tendon-controlled SDOF structures with different periods (Case 1: 0.5 s period, Case 2: 1.0 s period, Case 3: 1.5 s period) are used as numerical examples. The numerical parameters are presented in Table 12.2.

The optimization was conducted under a directivity pulse. The vibration equations of Makris [86] given in Eq. 12.24 are utilized to determine the optimum parameters of three different control systems. The pulse period (T_p) and peak ground velocity (V_p) were taken as 1.5 s and 230 cm/s, respectively. In optimization process, the time delay effect as 0.2 ms and the force control limit as 30% of the total mass of the structures are considered to avoid the stability problem of structures and providing a feasible solution.

The optimization objective is the minimization of the maximum displacement (x_1) that is found according to dynamic analysis conducted via MATLAB with Simulink [83]. The objective function is formulized as Eq. 12.25.

The numerical results of optimization process according to HS and TLBO are presented in Tables 12.3 and 12.4 for ATMD and active tendons, respectively.

$$a_g(t) = \omega_p V_p \cos(\omega_p t) 0 \leq t \leq T_p \quad \text{for normal component} \tag{12.24}$$

$$\mathrm{f} = \text{minimize } (\max(|x_1|)) \tag{12.25}$$

12.5 Conclusions and Future Studies

The maximum displacement of the uncontrolled structures is 9.77 cm, 70.58 cm, and 138.41 cm for periods 0.5 s, 1 s, and 1.5 s, respectively. As seen from the results, ATMD is not effective for structure with 0.5 s period. For the same structure, active tendons are effective to reduce the maximum displacement by 25%. For the other

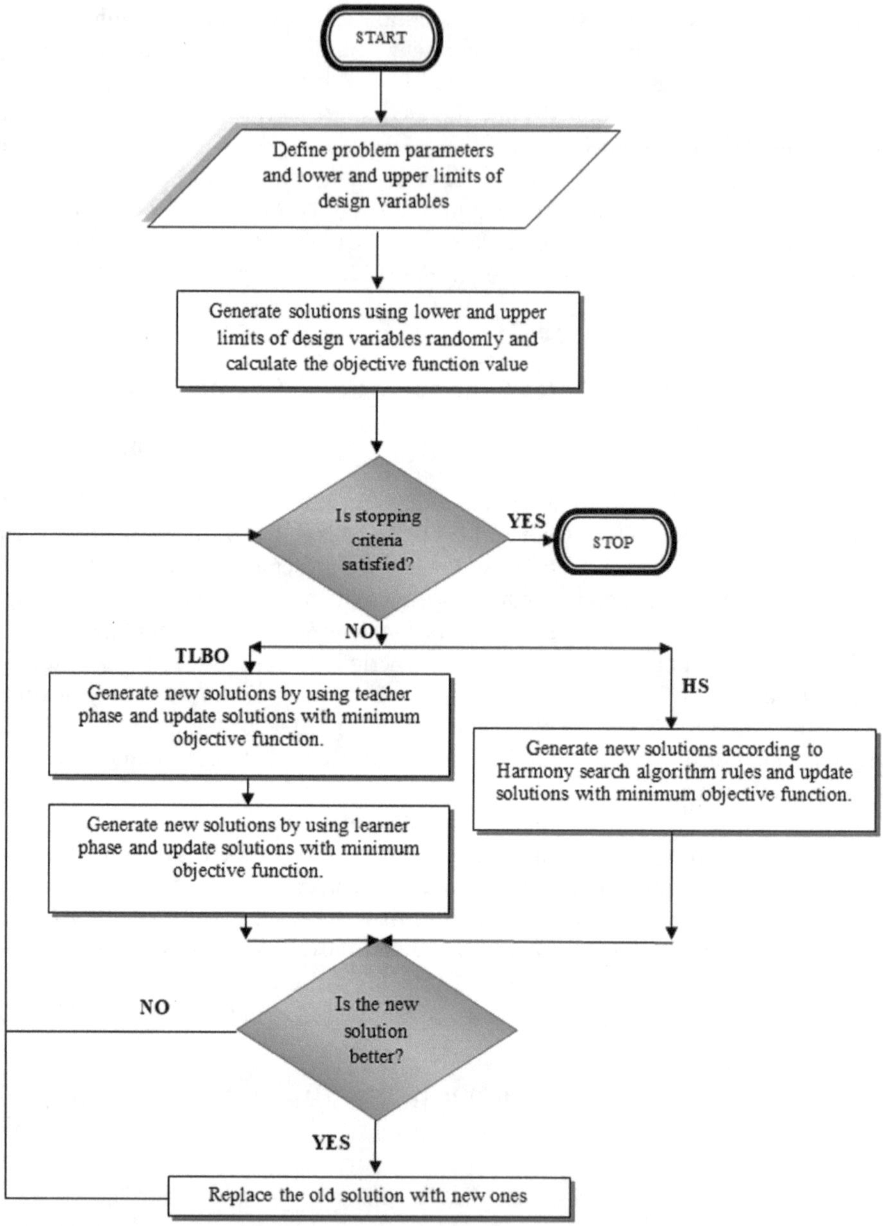

Fig. 12.8 Flowchart diagram of optimization process

Table 12.2 Numerical data of structures and control system

Parameters	Case 1	Case 2	Case 3
m_1 (kg)	100,000	100,000	100,000
k_1 (N/m)	15,791,367	3,947,842	1,754,596
c_1 (Ns/m)	125,664	62,832	41,888
R (Ω)	4.2	4.2	4.2
K_f (N/A)	2	2	2
K_e (V)	2	2	2
α	36°	36°	36°
k_c (N/m)	372,100	372,100	372,100

structures, ATMD is more effective than active tendons. Especially, 37% reduction is achieved with ATMD, while active tendons are effective to reduce it by 26%. For algorithms, TLBO has a small advantage comparing to HS in minimization of the structural displacement.

As a conclusion, ATMDs are more effective on structures that are subjected to excitation that is close to the period of the structure. As future studies, more cases of time delay and control force limitations can be investigated to understand the effectiveness of both control systems. Also, multiple degrees of freedom structures can be investigated, and ATMDs may have significant advantage since the control force needed for active tendons increases by the stories.

Table 12.3 Numerical data of optimization process of ATMD-controlled structures according to HS and TLBO

Definition	Range	Unit	TLBO case 1	HS case 1	TLBO case 2	HS case 2	TLBO case 3	HS case 3
Mass, m_d	10,000	kg	10,000	10,000	10,000	10,000	10,000	10,000
Period, T_d	(0.5)–(1.5) times of main structure period	s	0.4131	0.4887	1.1382	1.1202	1.2010	1.1880
Damping ratio, ξ_d	(0.01)–(0.5)	–	0.2754	0.4809	0.0239	0.0722	0.1207	0.1218
Proportional gain, K_p	(−10,000)–(10,000)	–	879.59	−433.51	153.45	−1858.6	967.88	−353.18
Derivative gain, T_d	(−10,000)–(10,000)	s	−57.47	−19.95	−282.80	23.93	−46.04	126.41
Integral gain, T_i	(−10,000)–(10,000)	s	4096.5	4220.8	9992.8	5931	−1288.8	−7385.4
Objective function, f	–	cm	9.61	9.77	41.97	42.55	87.47	87.52

Table 12.4 Numerical data of optimization process of active tendon-controlled structures according to HS and TLBO

Definition	Range	Unit	TLBO case 1	HS case 1	TLBO case 2	HS case 2	TLBO case 3	HS case 3
Proportional gain, K_p	(−5)–(5)	–	0.1429	0.1384	−0.2530	−0.2496	0.0530	0.0378
Derivative gain, T_d	(−5)–(5)	s	−2.9762	−3.0689	0.5482	0.5380	−0.5161	−0.7755
Integral gain, T_i	(−5)–(5)	s	−0.0114	−0.0013	0.1755	0.1851	0.1470	0.1137
Objective function, f	–	cm	7.37	7.37	46.21	46.48	101.98	102.21

References

1. Frahm H (1911) Device for damping of bodies. U.S. Patent No: 989,958
2. Ormondroyd J, Den Hartog JP (1928) The theory of dynamic vibration absorber. T ASME 50:922
3. Miyamoto HK, Gilani ASJ, Gündoğdu YZ (2011) Innovative seismic retroit of an iconic building. In: Seventh national conference on earthquake engineering, Istanbul, Turkey, 30 May–3 June
4. Den Hartog JP (1947) Mechanical vibrations. McGraw-Hill, New York
5. Bishop RED, Welbourn DB (1952) The problem of the dynamic vibration absorber. Engineering (London), 174 and 769
6. Snowdon JC (1959) Steady-state behavior of the dynamic absorber. J Acoust Soc Am 31:1096–1103
7. Falcon KC, Stone BJ, Simcock WD, Andrew C (1967) Optimization of vibration absorbers: a graphical method for use on idealized systems with restricted damping. J Mech Eng Sci 9:374–381
8. Ioi T, Ikeda K (1978) On the dynamic vibration damped absorber of the vibration system. Bull JSME 21:64–71
9. Warburton GB (1982) Optimum absorber parameters for various combination of response and excitation parameters. Earthq Eng Struct Dynam 10:381–401
10. Sadek F, Mohraz B, Taylor AW, Chung RM (1997) A method of estimating the parameters of tuned mass dampers for seismic applications. Earthq Eng Struct Dynam 26:617–635
11. Warburton GB, Ayorinde EO (1980) Optimum absorber parameters for simple systems. Earthq Eng Struct Dynam 8:197–217
12. Hadi MNS, Arfiadi Y (1998) Optimum design of absorber for MDOF structures. J Struct Eng-ASCE 124:1272–1280
13. Marano GC, Greco R, Chiaia B (2010) A comparison between different optimization criteria for tuned mass dampers design. J Sound Vib 329:4880–4890
14. Singh MP, Singh S, Moreschi LM (2002) Tuned mass dampers for response control of torsional buildings. Earthq Eng Struct Dynam 31:749–769
15. Desu NB, Deb SK, Dutta A (2006) Coupled tuned mass dampers for control of coupled vibrations in asymmetric buildings. Struct Control Health 13:897–916
16. Leung AYT, Zhang H, Cheng CC, Lee YY (2008) Particle swarm optimization of TMD by non-stationary base excitation during earthquake. Earthq Eng Struct Dynam 37:1223–1246
17. Leung AYT, Zhang H (2009) Particle swarm optimization of tuned mass dampers. Eng Struct 31:715–728

18. Bekdaş G, Nigdeli SM (2017) Metaheuristic based optimization of tuned mass dampers under earthquake excitation by considering soil-structure interaction. Soil Dyn Earthq Eng 92:443–461
19. Nigdeli SM, Bekdaş G (2017) Optimum tuned mass damper design in frequency domain for structures. KSCE J Civ Eng 21(3):912–922
20. Zhang HY, Zhang LJ (2017) Tuned mass damper system of high-rise intake towers optimized by improved harmony search algorithm. Eng Struct 138:270–282
21. Nigdeli SM, Bekdas G, Aydn A (2017) Metaheuristic based optimization of tuned mass dampers on single degree of freedom structures subjected to near fault vibrations. In: International Conference on Engineering and Natural Sciences (ICENS 2017), Budapest, Hungary, 3–7 May 2017
22. Bekdaş G, Kayabekir AE, Nigdeli SM, Toklu YC (2019) Tranfer function amplitude minimization for structures with tuned mass dampers considering soil-structure interaction. Soil Dyn Earthq Eng 116:552–562
23. Bekdaş G, Nigdeli SM, Yang XS (2018) A novel bat algorithm based optimum tuning of mass dampers for improving the seismic safety of structures. Eng Struct 159:89–98
24. Farshidianfar A, Soheili S (2013) Ant colony optimization of tuned mass dampers for earthquake oscillations of high-rise structures including soilstructure interaction. Soil Dyn Earthq Eng 51:14–22
25. Farshidianfar A, Soheili S (2013) ABC optimization of TMD parameters for tall buildings with soil structure interaction. Interact Multiscale Mech 6:339–356
26. Nigdeli SM, Bekdas G, Yang X-S (2017) Optimum tuning of mass dampers by using a hybrid method using harmony search and flower pollination algorithm. In: Del Ser J (ed) Harmony search algorithm. Advances in intelligent systems and computing, vol 514. Springer, pp 222–231
27. Nigdeli SM, Bekdaş G (2019) Optimum design of multiple positioned tuned mass dampers for structures constrained with axial force capacity. Struct Des Tall Spec Build 28(5):e1593
28. Nigdeli SM, Bekdas G (2015) Teaching-learning-based optimization for estimating tuned mass damper parameters. In: 3rd International conference on optimization techniques in engineering (OTENG '15), Rome, Italy, 7–9 Nov 2015
29. Yucel M, Bekdaş G, Nigdeli SM, Sevgen S (2019) Estimation of optimum tuned mass damper parameters via machine learning. J Build Eng 26:100847
30. Soto MG, Adeli H (2013) Tuned mass dampers. Arch Comput Methods Eng 20(4):419–431
31. Ankireddi S, Yang HTY (1996) Simple ATMD control methodology for tall buildings subject to wind loads. J Struct Eng 122(1):83–91
32. Mackriell LE, Kwok KCS, Samali B (1997) Critical mode control of a wind-loaded tall building using an active tuned mass damper. Eng Struct 19(10):834–842
33. Yan N, Wang CM, Balendra T (1999) Optimal damper characteristics of ATMD for buildings under wind loads. J Struct Eng 125(12):1376–1383
34. Qu ZQ, Shi Y, Hua H (2001) A reduced-order modeling technique for tall buildings with active tuned mass damper. Earthq Eng Struct Dynam 30(3):349–362
35. Samali B, Al-Dawod M (2003) Performance of a five-storey benchmark model using an active tuned mass damper and a fuzzy controller. Eng Struct 25(13):1597–1610
36. Samali B, Al-Dawod M, Kwok KC, Naghdy F (2004) Active control of cross wind re-sponse of 76-story tall building using a fuzzy controller. J Eng Mech 130(4):492–498
37. Han B, Li C (2006) Seismic response of controlled structures with active multiple tuned mass dampers. Earthq Eng Eng Vibr 5(2):205–213
38. Li C, Xiong X (2008) Estimation of active multiple tuned mass dampers for asymmetric structures. Struct Eng Mech 29(5):505–530
39. Pourzeynali S, Lavasani HH, Modarayi AH (2007) Active control of high rise building structures using fuzzy logic and genetic algorithms. Eng Struct 29(3):346–357
40. Guclu R, Yazici H (2008) Vibration control of a structure with ATMD against earthquake using fuzzy logic controllers. J Sound Vib 318(1–2):36–49

41. Guclu R, Yazici H (2009) Seismic-vibration mitigation of a nonlinear structural system with an ATMD through a fuzzy PID controller. Nonlinear Dyn 58(3):553
42. Guclu R, Yazici H (2009) Self-tuning fuzzy logic control of a non-linear structural system with ATMD against earthquake. Nonlinear Dyn 56(3):199
43. Li C, Li J, Qu Y (2010) An optimum design methodology of active tuned mass damper for asymmetric structures. Mech Syst Signal Process 24(3):746–765
44. Li C (2012) Effectiveness of active multiple-tuned mass dampers for asymmetric structures considering soil–structure interaction effects. Struct Des Tall Spec Build 21(8):543–565
45. Amini F, Hazaveh NK, Rad AA (2013) Wavelet PSO-based LQR algorithm for optimal structural control using active tuned mass dampers. Comput Aided Civ Infrastruct Eng 28(7):542–557
46. You KP, You JY, Kim YM (2014) LQG control of along-wind response of a tall building with an ATMD. Math Probl Eng
47. Shariatmadar H, Razavi HM (2014) Seismic control response of structures using an ATMD with fuzzy logic controller and PSO method. Struct Eng Mech 51
48. Shariatmadar H, Golnargesi S, Akbarzadeh TMR (2014) Vibration control of buildings using ATMD against earthquake excitations through interval type-2 fuzzy logic controller. Asian J Civ Eng Build Hous 15
49. Soleymani M, Khodadadi M (2014) Adaptive fuzzy controller for active tuned mass damper of a benchmark tall building subjected to seismic and wind loads. Struct Des Tall Spec Build 23(10):781–800
50. Heidari AH, Etedali S, Javaheri-Tafti MR (2018) A hybrid LQR-PID control design for seismic control of buildings equipped with ATMD. Front Struct Civ Eng 12(1):44–57
51. Kayabekir AE, Bekdaş G, Nigdeli SM, Geem ZW (2020) Optimum design of PID controlled active tuned mass damper via modified harmony search. Appl Sci 10(8):2976
52. Li C, Cao B (2015) Hybrid active tuned mass dampers for structures under the ground acceleration. Struct Control Health Monit 22(4):757–773
53. Cao L, Li C (2018) Enhanced hybrid active tuned mass dampers for structures. Struct Control Health Monit 25(2):e2067
54. Djedoui N, Ounis A, Pinelli JP, Abdeddaim M (2017) Hybrid control systems for rigid buildings structures under strong earthquakes. Asian J Civ Eng (BHRC) 18(6):893–909
55. Mamat N, Yakub F, Salim SAZS, Ali MSM, Putra SMSM (2018) Analysis of implementation control device in hybrid mass damper system. In: IEEE International conference on automatic control and intelligent systems (I2CACIS). IEEE, pp 123–127
56. Chesne S, Inquieté G, Cranga P, Legrand F, Petitjean B (2019) Innovative hybrid mass damper for dual-loop controller. Mech Syst Signal Process 115:514–523
57. Bhaiya V, Bharti SD, Shrimali MK, Datta TK (2020) Hybrid seismic control of buildings using tuned mass and magnetorheological dampers. Proc Inst Civ Eng Struct Build 173(7):471–487
58. Demetriou D, Nikitas N (2016) A novel hybrid semi-active mass damper configuration for structural applications. Appl Sci 6(12):397
59. Maślanka M (2017) Measured performance of a semi-active tuned mass damper with acceleration feedback. In: Active and passive smart structures and integrated systems 2017, vol 10164. International Society for Optics and Photonics, p 1016423
60. Zhang L, Hong L, Dhupia JS, Johnson S, Qaiser Z, Zhou Z (2018) A novel semi-active tuned mass damper with tunable stiffness. In: IEEE/ASME International conference on advanced intelligent mechatronics (AIM). IEEE, pp 274–279
61. Sun S, Yang J, Du H, Zhang S, Yan T, Nakano M, Li W (2018) Development of magnetorheological elastomers–based tuned mass damper for building protection from seismic events. J Intell Mater Syst Struct 29(8):1777–1789
62. Moutinho C, Cunha Á, Caetano E, De Carvalho JM (2018) Vibration control of a slender footbridge using passive and semiactive tuned mass dampers. Struct Control Health Monit 25(9):e2208
63. Karami K, Manie S, Ghafouri K, Nagarajaiah S (2019) Nonlinear structural control using integrated DDA/ISMP and semi-active tuned mass damper. Eng Struct 181:589–604

64. Zelleke DH, Matsagar VA (2019) Energy-based predictive algorithm for semi-active tuned mass dampers. Struct Des Tall Spec Build 28(12):e1626
65. Maślanka M (2019) Optimised semi-active tuned mass damper with acceleration and relative motion feedbacks. Mech Syst Signal Process 130:707–731
66. Chung LL, Reinhorn AM, Soong TT (1988) Experiments on active control of seismic structures
67. Chung LL, Lin RC, Reinhorn AM, Soong TT (1989) Experimental study of active control for MDOF seismic structures. J Struct Eng (ASCE) 115:1609–1627
68. López-Almansa F, Rodellar J (1989) Control systems of building structures by active cables. J Struct Eng (ASCE) 115:2897–2913
69. Lin C, Reinhorn AM, Soong TT, Wang YP (1991) Full-scale implementation of active control. 1: design and simulation. J Struct Eng (ASCE) 117:3516–3536
70. Ghaboussi J, Joghataie A (1995) Active control of structures using neural networks. J Struct Eng (ASCE) 115:2897–2913
71. Betti R, Panariello GF (1995) Active tendon control system for structures subjected to multiple support excitations. Smart Mater Struct 4:153–163
72. Dyke SJ, Spencer BF, Quast P, Sain MK, Kaspari DC, Soong TT (1996) Accelera-tion feedback control of Mdof structures. J Struct Eng (ASCE) 122:907–918
73. Chung LL (1999) Modified predictive control of structures. Eng Struct 21:1076–1085
74. Alli H, Yakut O (2005) Fuzzy sliding-mode control of structures. Eng Struct 27:277–284
75. Alavinasab A, Muharrami H (2006) Active control of structures using energy-based LQR method. Comput Aided Civ Infrastruct Eng 21:605–611
76. Yakut O, Alli H (2011) Neural based sliding-mode control with moving sliding surface for the seismic isolation of structures. J Vib Control 17:2103–2116
77. Chang CC, Lin CC, Wang JF (2010) Active control of irregular buildings considering soil-structure interaction effects. Soil Dyn Earthq Eng 30:98–109
78. Nigdeli SM, Boduroğlu BH (2013) Active tendon control of torsionally irregular structures under near-fault ground motion excogitations. Comput Aided Civ Infrastruct Eng 28:718–736
79. Nazarimofrad E, Zahrai SM (2016) Seismic control of irregular multistory buildings using active tendons considering soil-structure interaction effect. Soil Dyn Earthq Eng 89:100–115
80. Katebi J, Shoaei-parchin M, Shariati M, Trung NT, Khorami M (2019) Developed comparative analysis of metaheuristic optimization algorithms for optimal active control of structures. Eng Comput 1–20
81. Ulusoy S, Bekdas G, Nigdeli SM (2020) Active structural control via metaheuristic algorithms considering soil-structure interaction. Struct Eng Mech 75(2):175–191
82. Ulusoy S, Nigdeli SM, Bekdaş G (2020) Novel metaheuristic-based tuning of PID controllers for seismic structures and verification of robustness. J Build Eng 33:101647
83. The MathWorks Inc (2015) MATLAB R2015b. Natick, MA, USA
84. Geem ZW, Kim JH, Loganathan GV (2001) A new heuristic optimization algorithm: harmony search. Simulation 76(2):60–68
85. Rao RV, Savsani VJ, Vakharia DP (2011) Teaching–learning-based optimization: a novel method for constrained mechanical design optimization problems. Comput Aided Des 43(3):303–315
86. Makris N (1997) Rigidity-plasticity-viscosity: can electrorheological dampers protect base-isolated structures from near-source ground motions? Earthq Eng Struct Dynam 26:571–591

Chapter 13
Evolutionary Structural Optimization—A Trial Review

Fatih Mehmet Özkal

13.1 Introduction

Design process is a social activity based on the construction of social needs [1] and can be defined as a systematic series of actions that must be done to design a product, system or plan [2]. Most of the design problems in nature can be considered as optimization problems that require appropriate optimization techniques and algorithms to be addressed [3]. Difficulties in designing sophisticated engineering products are not just due to their technical complexity. It should be kept in mind that the managerial complexity required to ensure interaction between different engineering disciplines also affects the design process. Therefore, in recent years, for product development, integrated engineering has gained an increasingly significant characteristic [4, 5]. In order to provide an integration between design and planning processes, requirements of concurrent engineering should be determined, and these processes should be performed in an integrated way for the same purpose [6].

In the last few decades, there has been considerable popularity in numerical simulation and optimization for computer-aided design (CAD) and computer-aided engineering (CAE). This includes the emergence of optimization tools based on numerical simulation, as well as capabilities in multi-physics analysis and solid design. In particular, engineering disciplines such as aerospace, automotive, civil and naval have significantly benefited from advanced numerical tools for simulation-based structural design that are available in a variety of commercial packages today. Gradient-based mathematical programming techniques or metaheuristic algorithms are offered to find optimized designs [7]. Advancement of technology and the accompanying economic conditions reveal the importance of the concept of optimization. Optimization, which is the search for the most suitable solution under a certain objective and constraints,

F. M. Özkal (✉)
Department of Civil Engineering, Atatürk University, Erzurum, Turkey
e-mail: fmozkal@atauni.edu.tr

S. Carbas et al. (eds.), *Nature-Inspired Metaheuristic Algorithms for Engineering Optimization Applications*, Springer Tracts in Nature-Inspired Computing,
https://doi.org/10.1007/978-981-33-6773-9_13

is successfully applied in almost all disciplines of engineering [8]. In structural engineering, substantially, the solution of two major problems is sought. The first of these are the improvements made to reduce stress levels (safety condition) and secondly, to reduce the weight of the structure (economic condition) [9].

A design engineer working in the field of research and development often has to design completely new structures [10]. Since it is known that the most important factor affecting the construction cost is the amount of material used, one of the main issues considered during the design is the reduction of the weight of the structure [4]. Designer could try to solve the problem intuitively, based on previous experience; but this approach seems tiring for engineering problems and does not always guarantee a good result. Another option is to apply structural optimization approaches to the engineering problem described above [11].

Structural optimization approach, which could provide minimum material usage and lowest cost for a structural design, has become widespread especially in the last decades [12], and it has always been a popular research subject since it is intended to comprehend structural analysis, optimization algorithms and geometric modeling within an automated design procedure. Aim of structural optimization mainly stands on improving structural properties such as reduction of stress concentration and weight and enhancement of stiffness owing to the change of structural geometry [13].

Optimization has been classified under three titles: size optimization, shape optimization and topology optimization [14]. Topological optimization aims to achieve the best product prototype by changing the topological structure regarding the best load transfer path in a given design domain. Shape optimization is established on the optimization object by revising border shape of the initial design and considering coordinates as design variables in order to improve structural performance of the product [15, 16]. Size optimization takes individual dimensional variables into consideration for the final model to represent better performance by its new dimensional design [13].

Many researches on optimization algorithms searching feasibility, less material usage and stiffness increment have been conducted in an extensive field [17]. For the purpose of eluding the complex processes of traditional continuous and discrete methods, evolutionary structural optimization (ESO) has been suggested by Xie and Steven in 1993 [18–20].

ESO was constructed upon a native inspiration that the optimum topology can be generated by eliminating the inefficient material iteratively from the initial design region. This idea has been originated from homogenization method, which achieves the optimal topology by forming voids at the regions with low level of stress. In other words, homogenization and ESO methods have an association on the porosity concept [21, 22]. Within many optimization approaches, internal cavities are not allowed, and only the structural boundaries could be modified. These are traditionally classified as shape optimization problems [19]. Although material removal concept has been considered by also Maier [23], Rodriguez-Velazquez and Seireg [24], Atrek [25], etc, none of these studies have been recognized as a generalized method [26].

Design region is constituted by finite element method (FEM), and thereafter, load and boundary conditions are applied to the element model. Within this concept, inefficient material notion corresponds to the material that has not adequate contribution on the whole structural performance [11, 27, 28]. In addition, optimal structure notion basically corresponds to the structure with maximum stiffness and minimum weight, which has also fully stressed design characteristics. Hence, eliminating proportionally ineffective elements from the topology of the structure will yield a higher structural performance [29]. Various design objectives and design constraints, regarding design requirements, such as stress, stiffness, displacement, frequency, buckling load, moment of inertia and thermal parameters may be imposed upon a structure, and FEM is preferred for the assessment of structural behavior [19, 30].

Main goal of the evolutionary structural optimization is achieving the lightest design while satisfying the stiffness constraints and converging a fully stressed design [22]. Considering the engineering perspective, ESO has been found to have some attractive features: It is relatively easy to code the algorithm over finite element analysis (FEA) and requires a significantly less computation time. [11, 27]. Other than many experimental and heuristic design approaches, topology optimization has shown great potential for the design of advanced structures and materials, which can automatically achieve the best material layout within a design domain subject to specific constraints [31–34].

13.2 Structural Optimization Concept

Development and production processes often bring the question of which measures should be taken to increase quality and reliability without exceeding a certain cost limit, especially in the industrial area [35]. Geometry of load-carrying members is usually complex due to their strength and efficiency requirements. These requirements generally lead to an increase in production costs [36]. In the field of structural design, rather than limited mathematical purposes, real-world conditions should be considered [27]; design process should be dominated in the light of various purposes and constraints such as function, cost, aesthetics, production conditions and other technical requirements [37]. Existence of affordable high-performance computing technologies has encouraged the implementation of optimization algorithms in structural design, easing the burden of trial-and-error cycles of design optimization process. In general, optimum structural design is an approach to find a minimum weight/cost structure without violating the strength and serviceability constraints provided in structural design codes [38].

Optimization studies were triggered by the tendency of integrating the accumulated experience from natural facts with intentional experimental conclusions. Application areas of the optimization approaches and developed algorithms have consistently been improved ever since [4, 29]. Structural design and optimization have always played an important role in engineering. It is because how to find the optimal material distribution in a given design region to achieve the best structural performance is the basic problem of structural design and optimization [39, 40]. Schmit [41] proposed the idea of designing minimum cost systems by mathematical programming techniques [42]. Structural optimization within mechanics discipline has been comprehensively studied following the modern formulation of the optimal layout theory proposed by Prager and Shield [43] and Prager and Rozvany [44]. Thenceafter, generalized shape optimization has been pioneered by Bendsøe and Kikuchi [21] with homogenization and Bendsøe [45] without homogenization [46].

Therefore, structural optimization is based on the basis of obtaining a reasonable structure that meets some of the predefined requirements and enables to reach the desired goal under certain geometric and behavioral constraints [47, 48]. At its simplest, it is accepted as a fusion in engineering, mathematics, science and technology fields aiming to achieve the best achievement for a structure such as a bridge, a spacecraft or a large frame [49]. Structural optimization, although sometimes challenging at the design stage, has been identified as also the most economically rewarding process [18].

While mathematical methods for structural optimization have been developing consistently [50, 51], an overarching algorithm capable of performing simultaneous shape and topology optimization has also been one of the significant research goals [26]. For the purpose of improving structural performance, topology optimization and finite element method have commonly been preferred in order to optimize structural size, shape and topology by minimizing weight while increasing stiffness, which provides a simple and efficient process for optimum design [52, 53]. Topology optimization has a capability of responding various types of design necessities in engineering, especially in areas that require high-performance components such as the automotive, aerospace and aerospace industries [46]. Figure 13.1 demonstrates the final designs of different optimization methods for a cylindrical shell.

Size optimization considers optimum dimensional parameters such as cross section dimensions and thickness, without altering structural shape and topology. Shape optimization focuses on outer boundary of the structure to achieve the optimal shape. Topology optimization optimizes the material layout in a given design area, and this can find the optimal topology that is different from existing topologies or a specific topology. Size optimization of these three categories is a relatively simple parameter optimization [55]. In other words, cross-sectional areas of members are considered as design variables of the optimization problem in size optimization; nodal coordinates are evaluated in shape optimization, and connectivity of nodes are treated as design variables in topology optimization [38].

Traditional structural optimization algorithms focus on size optimization during the construction design phase alongside limited construction improvement [56]. However, topology optimization intends to achieve optimal distribution of material of an object in continuous space [35]. It has attracted great attention from both academics and engineers. Since it is not based on the initial configuration and the experience of an engineer, a completely unexpected and innovative configuration could be acquired [12]. The most significant characteristic of topology optimization could be asserted to attain completely a new layout which does not resemble existing design in a considerably manner. Essential factor on this concept is that modification

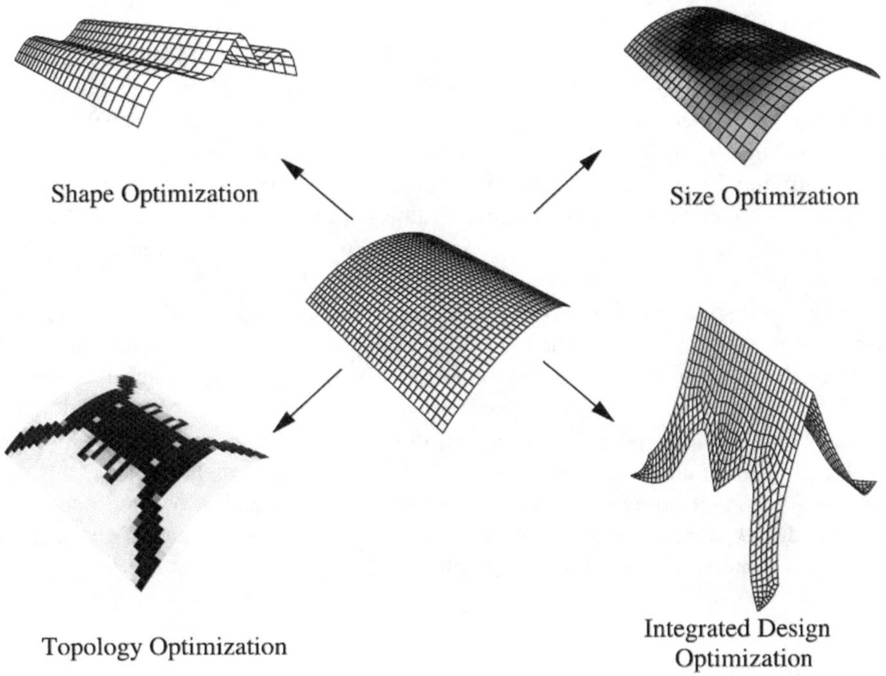

Fig. 13.1 Final designs of different optimization methods for a cylindrical shell [54]

of existing design is not the objective while seeking of the best topology among all possible solutions is performed by starting from a simplified design domain [52].

13.3 Topology Optimization Methodology

The word optimum was first used by Leibniz in the eighteenth century to mean the best of the possible ones [57]. Optimization has applicability in all fields of study and has left its mark on engineering science. This is because searching for the best solution for a problem is a fundamental and constant rule of engineering [58]. An optimization problem can be described as searching for the best solution without violating predetermined conditions and technical constraints [59].

Recent studies in the field of structural optimization, especially in truss systems, are constantly increasing. These studies are mainly grouped under three titles; reducing the weight of the structure, reducing the strain energy for a certain amount of material and having the stress distribution within the structure uniformly distributed [58].

The concept of topology can be expressed simply as the mesh pattern or the internal geometry that forms a structure [4]. Structural topology optimization has become an assistive mechanism in order to achieve designs with higher structural performance alongside less material usage since the precipitating research of Michell [60] in 1904 and also Bendsøe and Kikuchi [21] in 1988 [61]. Michell theory has typically been preferred for the verification of topology optimization results [62]. Michell truss is a type of frame with continuous distributions of members that is often called "truss-like continuum" [63]. Besides having the most effective truss structure, it is aimed to design members at almost fully stressed state. Furthermore, Bendsøe and Kikuchi [21] have emphasized research direction into the structural topology optimization by introducing the homogenization method [28].

Principal inspiration of homogenization method is constructing the design region from a finite number of cells, each of which can have individual microstructure, and additionally, each of these cells can be defined by either having material or representing a rectangular void [11]. Objective of the optimization problem is defined by searching optimal porosity of such a porous medium [18]. Preference of an unchanging finite element model for the design area in this method eliminates the necessity for re-meshing [26]. On the basis of homogenization method and fully stressed design concept, ESO method has been presented, and range of application has been extended by subsequent research after then [30].

13.4 Keystones of the Algorithm

Generally, evolutionary algorithms do not have a precise theoretical background, and convergence results have not been proven to be certain. Evolutionary algorithms are believed to mimic natural selection (selection), such as survival of the best and the changes observed in living organisms, and bring the problem to an appropriate solution [11]. Correspondingly, ESO is a heuristic optimization method that works by imitating the development of biological structures in nature. It is observed that naturally formed species, which have fully stressed characteristics, tend to use the material in the most effective way [64]. When inherent changes of natural structures such as shells, bones and trees are observed, it is obviously seen that topology and shape of such structures reach their most effective form after a long process of change and adapt to their environments [18].

ESO is actually a successful design tool performing topology optimization simultaneously with shape and size optimization. The algorithm follows a heuristic process by using the FEM as the analysis engine. Principal logic for optimizing a structure could be represented as iteratively eliminating inefficient elements from the design region defined with loading and boundary conditions [4].

ESO is essentially based on two methods. Michell's truss theory [60], which is one of them, proposes optimal solutions with respect to the amount of material used and the constant stress state of all the members that constitute the structure. Moreover, the other basis, homogenization method that has been presented by Bendsøe and Kikuchi [21] aims to get the optimum structural design by forming microscale voids throughout the optimization process [22]. Dated from the research of Bendsøe and

Kikuchi [21], topological design of structures has steadily improved. With its unique characteristics, specific formulations and solution techniques, topological design can be considered today as a branch of structural and multi-directional optimization [65].

13.5 Basic Principles

In order to start the optimization process, a design domain which covers the area of the expected/desired final design is divided into finite element mesh. Mesh size should be decided by the experience of the designer. By applying the load and boundary conditions on the initial design, a static analysis is carried out to find the stress distribution throughout the whole structure [30]. Based on the acquired results, stress values of the elements are compared with each other and material that is not used efficiently is removed from the structure by applying a rejection criterion. Stress-based early versions of ESO method generally use von Mises stress values of finite elements to guide elimination of elements [66]; however, in the ensuing years, structural optimization with stiffness considerations has been exhaustively studied [67].

Material removal process is accomplished by eliminating the elements from the finite element model one after another. By defining a multiplier or indicator for each element, that element's contribution to the behavior of the entire structure is determined. It is an important step in this method to specify the exact level of contribution for the elements. In a given initial design region, these elements that contribute to the structure at a lower than a specified value are eliminated, taking into account the loads and boundary conditions. Various criteria can also be chosen for the elimination process. These are the stress value determined by the von Mises yield criterion, principal stress, strain energy, displacement, frequency, etc., can be. Thus, it is aimed to maintain the objective function that might be defined in terms of stress distribution, volume, stiffness, frequency, buckling load, moment of inertia, etc [30]. Updated design is re-analyzed after the element removal process. Within the analysis of this new design, new ineffective elements are determined and removed from the structure, and this process continues until the convergence criterion is achieved, that is, the minimum/maximum value of the objective function is maintained.

Updated design is re-analyzed after the element removal process. Within the analysis of this new design, new ineffective elements are determined and removed from the structure, and this process continues until the convergence criterion is achieved, that is, the minimum/maximum value of the objective function is maintained.

Element elimination process among the optimization algorithm can be implemented by completely removing the elements (hard-kill method), as well as reducing the physical properties of the elements such as thickness or their mechanical properties such as modulus of elasticity in order to decrease their contribution to the structural behavior (soft-kill method). However, the second method is generally preferred due to the computational problems to be experienced in finite element analyses in case point contact between elements is encountered. For example, Hinton and Sienz

[68] chose to multiply the modulus of elasticity of the elements to be removed by values such as 10^{-5} or 10^{-6}, while there were many researchers who preferred the same method in the following years.

Optimization algorithm based on the von Mises stress distribution can be summarized systematically as follows [30, 20].

Step 1: Discretize the structure using a fine mesh of finite elements.
Step 2: Analyze the structure under the support and loading conditions.
Step 3: Calculate von Mises stress values for each of the elements and the maximum of the entire structure.
Step 4: Remove a number of elements that have the lowest stress.
Step 5: Repeat steps 2 to 4 until the termination of the process.

Various contributions were made to ESO in the following years, which were initially performed with the element removal algorithm in a way to reduce the volume of the structure by considering von Mises stresses. Through these contributions, structures under dynamic, multi-loading and multi-support conditions can be solved; also, different results can be achieved by adding new elements to regions subject to excessive stress or by increasing element thicknesses, in addition to element removal. Initial and final designs of an optimization process resembling the Michell truss structure obtained by applying topology optimization on a beam, which is also preferred to exhibit the validity of topology optimization results in previous studies, are given in Fig. 13.2. Figures 13.3, 13.4, 13.5 and 13.6 are also provided to demonstrate the efficiency and diversity of the algorithm on various designs.

13.6 Objectives and Constraints

Focal point of most of the optimization approaches in engineering is to determine the maximum stress values for a structure under all loading conditions. This situation generally forms the basis of the design limits and is therefore used to determine the amount of material and weight of the structure [67]. As stated before, main idea of ESO is the expectation of achieving the optimal shape and topology by iteratively eliminating ineffective elements from the structure. Key point of this method is to evaluate the contribution of each element to the structural behavior and to set a guiding criterion to ensure that the least contributing elements are removed [26].

It is possible to achieve many results through optimization methods. However, while applying these methods, some points should be taken into consideration [69], and it should be determined which objective function will be reached with which constraints. Stiffness and strength are generally the major concern for design engineers. Designer often strives to balance the two design goals: increasing stiffness and decreasing maximum stress [70]. Volume or weight, strain energy (stiffness), frequency, buckling load, displacement and stress values of structural elements can be selected as the material and behavior properties to be used for objective function and constraints.

Fig. 13.2 Initial, intermediate and final designs of a beam resembling Michell truss [20]

ESO was firstly created by Xie and Steven [18] by considering the stress constraint for design problems, and then an algorithm with frequency constraint [71] was developed by the same researchers. Various shape and topology optimization algorithms were constructed by Chu et al. using displacement [72] and stiffness [73] constraints. Manickarajah et al. [74], on the other hand, proposed an evolutionary algorithm that could increase the buckling load by resizing the elements [75]. These structural properties can be used in the evolutionary structure optimization algorithm in various combinations, both as an objective function and as a constraint.

Fig. 13.3 Initial, intermediate and final designs of a simple-supported beam [20]

13.7 Optimization Parameters

Before initiating the optimization process, a sufficiently large design region, which could enclose the final design, is defined. Loading and support conditions are applied to the initial design, and stress analysis is performed via finite element method. Based on the obtained results, stress values of elements are compared with each other, and a rejection criterion is applied to eliminate the material that is not used efficiently. After each finite element analysis, considering von Mises stress or other criteria, elements that meet the following condition are extracted from the structure.

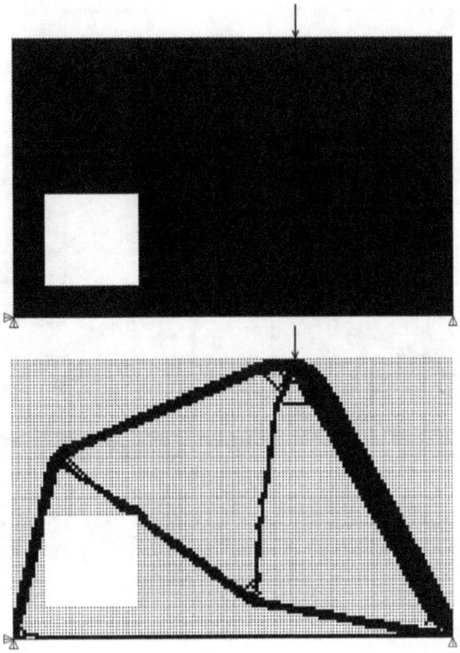

Fig. 13.4 Initial and final designs of a deep beam with web-opening [28]

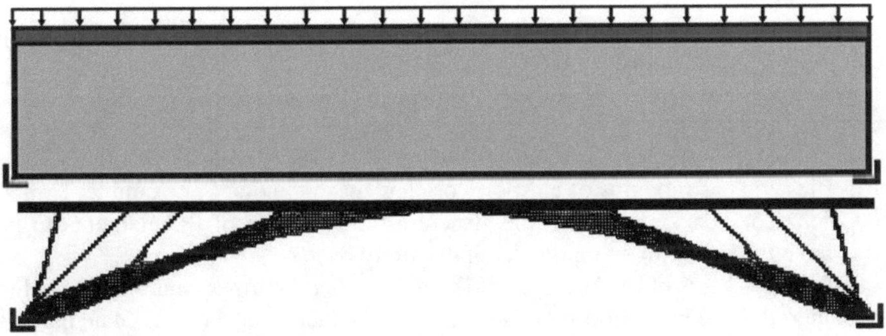

Fig. 13.5 Initial and final designs of a simple bridge structure [17]

$$\sigma_e^{vM} < RR_i \cdot \sigma_{\max}^{vM} \tag{13.1}$$

RR_i in the above inequality represents the current rejection rate (RR). The process is continued with the same RR_i value until a local optimality is achieved, that is, until there are no more elements to be removed within the current iteration. At this stage, an evolutionary rate (*ER*) is defined and added to the rejection rate.

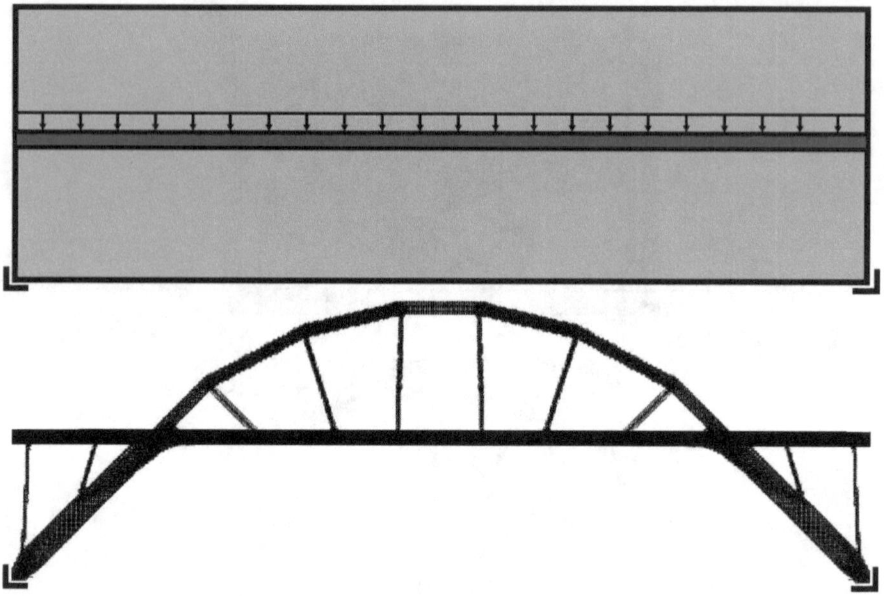

Fig. 13.6 Initial and final designs of a complex bridge structure [17]

$$RR_{i+1} = RR_i + ER \quad i = 0, 1, 2, 3, \ldots \tag{13.2}$$

With this increased rejection rate (RR_{i+1}), finite element analysis and element removal follow each other until a new local optimality [30].

In finite element analysis, presence or absence of an element can be expressed simply by assigning the material property number to 1 or 0 [76]. Moreover, another way of removing elements is to assign very low values to material or dimensional properties such as modulus of elasticity or element thickness. For example, Hinton and Sienz [68] chose to ignore the structural contribution of the elements to be removed by multiplying the modulus of elasticity by 10^{-5} or 10^{-6}.

In the early years of ESO, irregularities in the general stiffness matrix were being encountered that could cause serious failures in displacement and stress calculations as a result of changing modulus of elasticity or thickness values of finite elements. In addition, even if it is assumed that element is removed by changing its modulus of elasticity or thickness, number of equations to be solved in finite analysis remains the same. Therefore, instead of changing these properties, it was preferred to remove the elements to be eliminated from the finite element mesh [27]. However, if the point where computational technology has reached today is observed, it is possible to perform faster and more reliable analyzes than in the past. For this reason, even if it is no longer important which element elimination method is used, such a preference can only be made for extensive problems.

Alongside the main parameters of ESO, constraints and selected mesh size and element type used in finite element analysis have great effects on the optimization process. If the optimization parameters have different values, diverse designs that are not identical in terms of volume, shape and topology can be obtained at the end of the optimization process. In addition, duration of the optimization process, which is highly dependent on these variables, is another issue to consider.

13.7.1 Rejection and Evolutionary Rates

Before initiating the optimization process, two parameters should be defined. First one is the initial rejection rate (RR_0), and second one is the evolutionary rate (*ER*). For many of the problems in previous studies, typically 1% for RR_0 and 1% for *ER* were taken. However, it might be necessary to use lower values for some problems. While designing any new model, optimum values for these parameters could be defined after a few trials. For example, it is understood that lower values should be used for RR_0 or *ER* values if excessive material has been extracted from the structure after a step or when local optimality is achieved [27].

Within the classical approach, if there is no element remaining that satisfies the element removal criterion (Eq. 13.1), it means that local optimality is achieved. When local optimality is achieved, if the optimization process is desired to continue, that is, if the final design has not been obtained yet, optimization cycles continue by increasing *RR* value by *ER* value [77]. If an excessively high value is selected for *ER*, excessive material removal occurs, and the structure could lose its integrity. When this is encountered, it will be advantageous if the algorithm goes one step back and continues the current cycle with an *ER* value that has been reduced to half [18].

13.7.2 Element Removal Ratio

Similar to the move limit or step size in mathematical programming and optimality criteria methods, element removal rate is also an important parameter in ESO [73]. Number of elements to be removed after each analysis is defined by element removal ratio (*ERR*). A detailed comparison study by Chu et al. [26] can be used to examine the effects of the amount of material removed in each cycle on the final structure. Element removal ratio is determined by total number of elements in the current design or the initial design, with typical values considered to be 1 or 2%. However, *ERR* value should be rounded to the nearest integer and should be defined as an even number if symmetry is expected in the optimized structure [27]. Even if more computation time is required, it is thought that a design with higher performance will be achieved as the element removal ratio is reduced. On the other hand, a higher element removal rate will significantly reduce the time required for each cycle to complete [26].

13.7.3 Element Size

In order to achieve accurate results with ESO method, in the first stage, the elements in the design area, that is, in the finite element mesh, should be sized equally. If this requirement is not met, criteria for eliminating elements should be reconstructed in consideration of this situation [11].

Chu et al. [26] stated that element size has a relatively low effect on the shape and topology of the final design. On the other hand, Abolbashari and Keshavarzmanesh [78] stated that the element size is effective on the change of minimum stress and volume values of the structure, and even the shape of the final design is considerably dependent on the element size. Based on the example solutions, the smaller the element used, the more sensitive the reduction in volume will be; by forming a fine finite element mesh, structures with lower maximum stress values and lighter structures can be obtained. Therefore, they stated that the user should choose between less material usage and bigger computation time. However, these comments are of course limited to their example problems. Since all the parameters used in ESO method yield different outcomes in various problems; whenever a new problem is encountered, it should be decided which values are the most appropriate after a few trials.

13.8 Optimality Decision

Increasing importance of the optimization concept in engineering science has brought many different discussions. One of the most important of these is to determine to how much the results achieved are good. Until today, many studies have been conducted to determine the best/suitable/efficient design by creating various algorithms in various optimization methods. There are also different opinions in the studies on evolutionary structural optimization.

During the optimization process, as the finite element analysis is performed one after another and element removal process is repeated, structural volume decreases in each cycle, maximum stress and minimum stress values approach each other, and performance of the new structural design gradually increases. In other words, efficiency of material usage in the structure increases continuously for every new design.

Therefore, in the process of searching for the best design over ESO method, any design in every next iteration could be considered as the best design. However, a certain criteria should be set for this. In other words, purpose of topology optimization should be clearly determined. Especially, one of the main objectives of ESO, "fully stressed design concept", should be taken into consideration and among the designs close to this situation; determination should be performed based on the characteristics of the design such as volume, rigidity, structural integrity and applicability [4].

The concept of "performance index" (*PI*) was first introduced by Querin [79] upon selecting the best design among the designs reached during the ESO process. In the following studies, it is stated that the most accurate choice can be made by only taking into account the relative stiffness (only the displacement of the load point) or the maximum stress values in the structure with only the weight of the structure. However, as mentioned before, since the most appropriate design is not dependent on these terms independently, it has been emphasized that this concept should be further developed, and a generalized method should be used by taking into account other factors that contribute to the behavior of the building.

Even if the methodology of Querin [79] is very limited, subsequent studies on the performance index concept resulted in various formulations. Liang et al. [80] presented a new performance index for the determination of the lightest design with stress constraints for continuous structures (PI_s). Liang et al. [81, 82] considered the displacement of the load point for the termination criterion for the optimization process and determination of the best design (PI_{ds}). Liang et al. [83], who studied the performance index concept on the optimization problems of plates subjected to bending with thickness change, stated that in contrast to plane stress problems, stiffness matrix would lose its linearity in this case and presented a slightly different formula compared to the scaling design approach. According to this formula, performance index is calculated based on the maximum displacement value in the structural member (PI_{dp}) in inverse proportion to the volume. On the other hand, Guan et al. [84], who wanted to develop a performance index on tensile and compression members, took into account principal stresses (PI_t, PI_c). Liang et al [85] and Liang and Steven [86] presented a new performance index (PI_{es}, PI_{ep}) within the context of their researches on linear elastic and continuous structures. Özkal [4] and Özkal and Uysal [22] developed a formula on the basis of the fully stressed design concept, in which volume, stiffness and stress levels within the structure are taken into account, in order to decide which of the designs obtained from the optimization process is the best one and has the highest structural performance (PI_{fsd}).

In order to compare the achieved results within these performance index approaches; PI_s, PI_{ds}, PI_{es} and PI_{fsd} formulations have been applied by Özkal [4] and Özkal and Uysal [22, 29, 87, 88] on optimization problems, and von Mises stress distributions of the designs are presented in Figs. 13.7 and 13.8. PI_{ds}, PI_{es} formulations present same results while PI_s has not provided a meaningful design since this formulation has an increasing characteristic for any optimization problem. In brief, PI_{fsd} formulation (Eq. 13.3) has demonstrated that this approach is capable of selecting the real optimum design among other achieved designs within an optimization process. This formulation evaluates the structural performance of any design regarding stiffness and stress uniformity, and it is used for the determination of the optimum design by selecting the one which has the greatest value.

$$PI_{fsd} = \frac{\left(\sigma_{0,\max}^{vM} / \sigma_{0,avg}^{vM}\right) u_{0,\max} W_0}{\left(\sigma_{i,\max}^{vM} / \sigma_{i,avg}^{vM}\right) u_{i,\max} W_i} \tag{13.3}$$

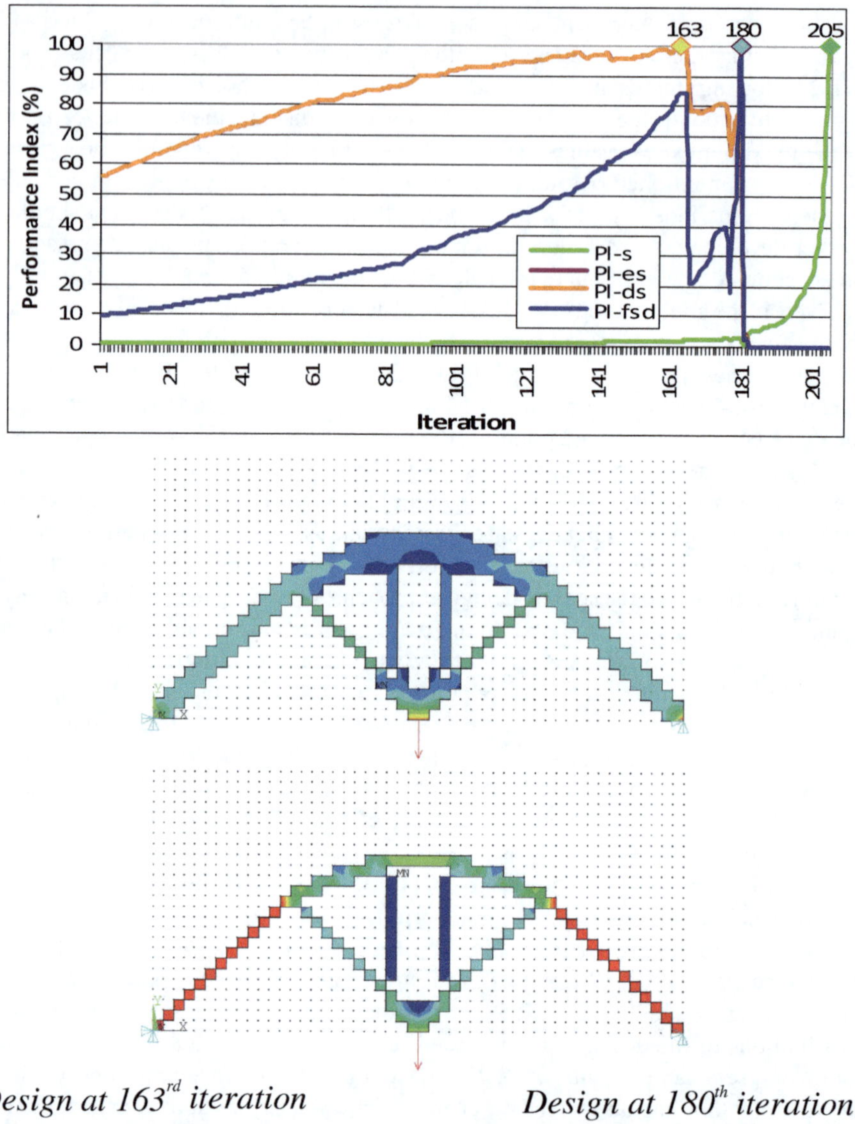

Fig. 13.7 Performance index histories and optimized designs for the Michell truss problem [22]

In this formula, $\sigma_{0,\max}^{vM}$ and $\sigma_{i,\max}^{vM}$ are the maximum von Mises stress values, $\sigma_{0,\mathrm{avg}}^{vM}$ and $\sigma_{i,\mathrm{avg}}^{vM}$ are the average von Mises stress values, $u_{0,\max}$ and $u_{i,\max}$ are the maximum displacement values, W_0 and W_i are the actual weight values of the initial and ith design domains.

However, any performance index is certainly open to improvement or update. Especially, more research and experimentation on the weight of the factors in

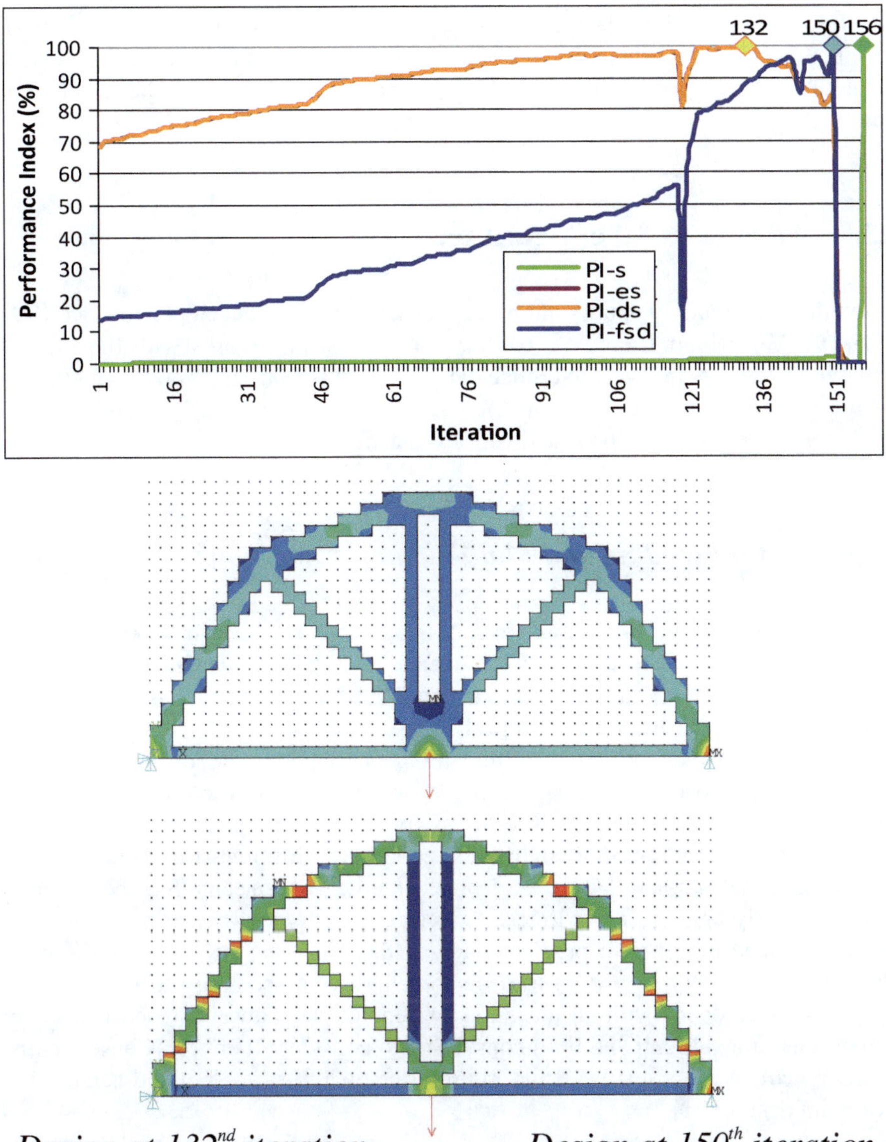

Design at 132nd iteration *Design at 150th iteration*

Fig. 13.8 Performance index histories and optimized designs for a simple-supported beam problem [22] the formula that builds the performance index and inclusion of the concept of producibility with substantial expressions will increase the accuracy of the selection. Ultimately, it is possible to develop new index formulations for the selection of the best design by taking into account many new factors.

13.9 Advances of the Algorithm

Evolutionary structural optimization has been in continuous development since 1992 [30, 49]. Various contributions to ESO, which basically works with the idea of removing elements in a way to reduce the weight by considering von Mises stresses in its initial version, made this method suitable for solving various optimization problems under multiple loads and various constraints.

13.9.1 Multi-loading and Multi-support Conditions

Pioneer studies on multi-loading and multi-support conditions for ESO have been presented by Xie and Steven [89] and Steven et al. [90]. This idea has opened a way for the optimization of structures that are subjected to different loads and different support situations at different times [58].

Designers are often faced with optimization of structures under different loadings. The term "multi-loading" refers to the loads that can affect a structure independently at different times throughout its life. It is expected that the structure stand against various loading situations that might cause completely different structural reactions. These loads are completely different from each other, which can affect the structure simultaneously or at different times. Therefore, it becomes very difficult for even an experienced designer to predict which of these loading situations is the most dangerous. It is very important to design the structure according to the specified criteria in accordance with all these loading conditions. However, even if it is not impossible, it is difficult and time-consuming to consider all effects. Consequently, an intermediate way must be found in order to solve this problem and achieve the optimum design [91].

Thereupon, Xie and Steven [89] developed an optimization approach by taking into account the highest stress state, that is, the most unfavorable situation, on the basis of all loading conditions, in order to ensure extremely efficient material use. According to this approach, elements with the least contribution are eliminated one after the other in all loading situations so that the remaining elements are used relatively efficiently for at least one of the loading situations. In addition to the research of Xie and Steven [89], Young et al. [92] developed a two-way evolutionary topology optimization approach for two and three-dimensional structures in multiple loading situations. In this method, elements are eliminated if they are ineffective in all loading situations, but if they are used effectively even for one of these situations, element

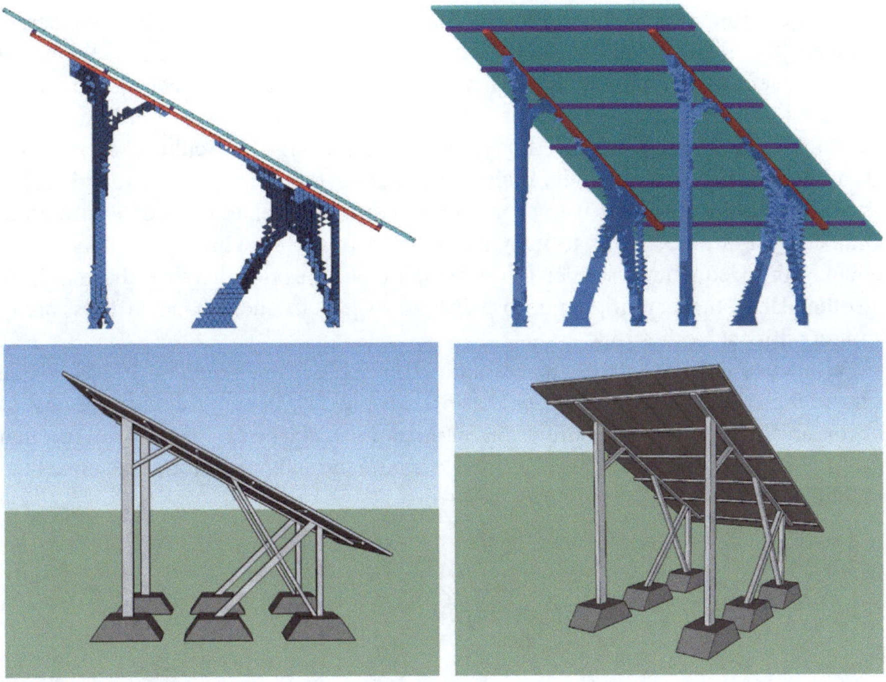

Fig. 13.9 Optimum designs for a solar power panel [93, 94]

reinforcement is made [91]. Recently, Özkal and Küçük [93, 94] demonstrated an optimization approach on the load-bearing parts of solar power panels considering multiple loads such as snow, wind and modal loads (Fig. 13.9). Within the optimization procedure, most ineffective elements for each loading state were eliminated from the finite element mesh, and this process was performed in each iteration.

13.9.2 Multi-criteria Utilization

As mentioned before, ESO was initially performed on a single criterion. However, this situation narrowed the application areas and could not meet all kinds of necessities. Different ideas had to be developed to broaden the horizons of ESO and to increase its usability feature. For this, various other features were added later on the foundation of optimization mechanism [30].

Multi-objective optimization is the method of forming a solution that satisfies a series of conflicting objectives in the best possible way; that is, multi-objective optimization is the generation of a project that obtains the best performance of the structure when several criteria are taken into consideration [95]. In a structural optimization problem, situations that require response to different objective functions might be encountered. Hence, different parameters or criteria should be considered. Since each problem must be solved with a different criterion, reaching the result becomes even more complicated. In many cases, these criteria are in conflict with each other although there is only one possible solution. This dilemma of conflicting goals is encountered in many fields of study: engineering design, agriculture, economics, city planning, etc. For example, main expectation on the design of a load-bearing beam in an aircraft fuselage is to have minimum weight, i.e., to reduce the structural volume. Design objective is to increase stiffness of the structure on the one hand; reducing the frequency in order to decrease the stiffness or increase the mass, on the other. Here, these multiple purpose criteria such as frequency and stiffness are in conflict with each other [58].

Various studies have been carried out on this subject since the late 1970s. While some of these belong to Sawaragi et al. [96], Koski [97] prepared a detailed review of the research on multi-criteria structural optimization in the last 15 years. A review that provides a broad overview of multi-criteria optimization applications was presented by Stadler and Dauer [98].

13.9.3 Bidirectional Optimization

Bidirectional optimization algorithm has been developed to suppress the suspicion of whether there is any material removed inappropriately in ESO. Within this approach, it is also allowed to add material simultaneously with material removal. In other words, material addition is made to the parts of the structure that are subjected to high stress levels. Therefore, evolutionary processes move from the smallest possible structural core toward the optimum design. Final design, which is achieved in this way, has the same shape and topology as the design achieved by element removal mechanism [58].

Conventional and bidirectional algorithms can successfully solve problems using stress, stiffness, frequency or buckling load constraints. Although bidirectional algorithm is computationally more efficient than conventional algorithm for a complex finite element problem, it has a flexible methodology in terms of balancing the computation time and solution efficiency [99].

When compared to initial approach, two superior aspects of bidirectional algorithm stand out. Most importantly, it has a more robust methodology and allows for the retrieval of unnecessarily removed elements. Secondly, it is seen that it is computationally more efficient since it starts with a smaller initial model and therefore uses a smaller finite element model compared to ESO, which initiates the process with a large design area [100].

In addition, another important advantage of bidirectional algorithm is that it assists the design of a self-developing structure in case of change in environmental effects. With the change of any environmental effect, the structure might grow, shrink or completely change its shape and topology. This newer approach made important contributions to the structural design by taking these issues into consideration [101].

13.9.4 Grouping Algorithm

One of the recent contributions to ESO is the configurational optimization called "Group ESO". Within this method, instead of considering each element as a design parameter, element groups or patterns are examined for removal. In this way, design optimization is performed and can be used in the preliminary design stage where the configuration of the structural entities, holes, stiffeners and skin thickness values are not constant [102].

13.9.5 Morphing Algorithm

Similar to shape optimization, a new contribution has been made to the ESO method by Querin [79] termed as "ESO Morphing". Within this method, instead of removing the elements completely as in classical approach, they are removed gradually and selectively. This gradual elimination process can be used on cross-sectional area of beams; thickness, modulus of elasticity or unit weight of plates; and modulus of elasticity or unit weight for bricks [58].

13.9.6 Combination with Strut-and-Tie Method

If each element in a structure reaches its maximum allowable stress limit for at least one of the loading conditions, it is defined as a fully stressed structure [103]. Relatively less research has been done on the fully stressed design of frame structures compared to more complex design approaches such as optimization algorithms using mathematical programming and other methods [87, 104].

Strut-and-tie model (STM) technique is an appropriate design system for structural elements that require exceptional detailing for all support and loading conditions, such as beam-column connections, deep beams or short columns. Since discontinuity regions do not show linear elastic behavior, many difficulties arise in achieving realistic designs from traditional techniques. Therefore, strut-and-tie modeling has been widely accepted as one of the most realistic design methodologies because it is a much simpler model [105].

Inclined tensile stresses in reinforced concrete beams under shear effect and inclined cracking behavior resulting from this should be formulated with the help of a "mechanical-mathematical" model. The best model to formulate a concrete beam with shear reinforcements in this way is the beam-truss model. These analogy models are effective and realistic methods as they consider behavior of the structure after cracking. Beam-truss, which initially has high degree of indeterminacy, is made isostatic with some assumptions based on the plasticity theory. I n this case, it is quite easy to solve this isostatic beam with the help of equilibrium equations [106].

Strut-and-tie modeling method is an analysis and design tool in which internal stresses are distributed to the structure over a truss system. In this method, compression and tension bars connected at nodes can be thought of as members of a fictional truss. This internal truss plays an important role in the distribution of stresses caused by bending moment and shear [107]. Sudden changes in the cross-sectional dimensions of reinforced concrete members, individual loads or support reactions cause discontinuities due to the complexity in the flow of internal forces in these regions. These regions are called "D-regions" (discontinuous, irregular, complex regions) [108]. B-regions are the areas where the beam theory (Bernoulli Hypothesis) could be applied, that is, internal stresses are uniformly distributed throughout the cross section of the members. Although the analysis of the B-regions and determination of the reinforcement layout can be performed by traditional methods, a suitable strut-and-tie model can be designed for the whole of the structural member even if both B and D-regions are present in this member [109].

However, it is possible to encounter many strut-and-tie models in the design process carried out with this method, and it is not certain whether the most suitable model has been selected [110]. Some researchers have pioneered the use of structural optimization techniques to facilitate the assessment of internal stress trajectories and to automate the creation of strut and tie patterns. These studies show that strut-and-tie modeling method is an important assistant for integrating topology optimization into the civil engineering field. After preliminary studies, Özkal [4] proposed an integrated design approach to determine the reinforcement layout in concrete members by topology optimization and strut-and-tie modeling methods, respectively [88]. Demonstrations in Fig. 13.10 provide a reinforcement layout design over this integrated design approach.

13.9.7 Combination with Other Metaheuristic Algorithms

Heuristics and metaheuristics algorithms are the major components of stochastic methods. These algorithms are preferred to expedite the mechanism of optimization in order to achieve high-quality solutions in difficult conditions. Discrimination between the two algorithms is not obvious [111, 112]. Heuristics algorithms aim the solution of a specific problem while application of the generated approach is not compatible on a different problem even it has a similar characteristic [113], while metaheuristic algorithms could be placed higher than heuristic algorithms in hierarchical system by appealing a generalized mechanism [111]. In other words, a heuristic algorithm has a problem-dependent mechanism aiming to find an optimal solution to a specific problem without knowing the level of the optimality, while metaheuristics algorithms could be applied on a wide range of problems without any critical limitation.

In case conceptual definition should be provided alongside technical explanation of these algorithms, heuristic means "to find" or "to discover by trial and error" while meta-means "beyond" or "higher level". Despite the popularity of metaheuristics, in

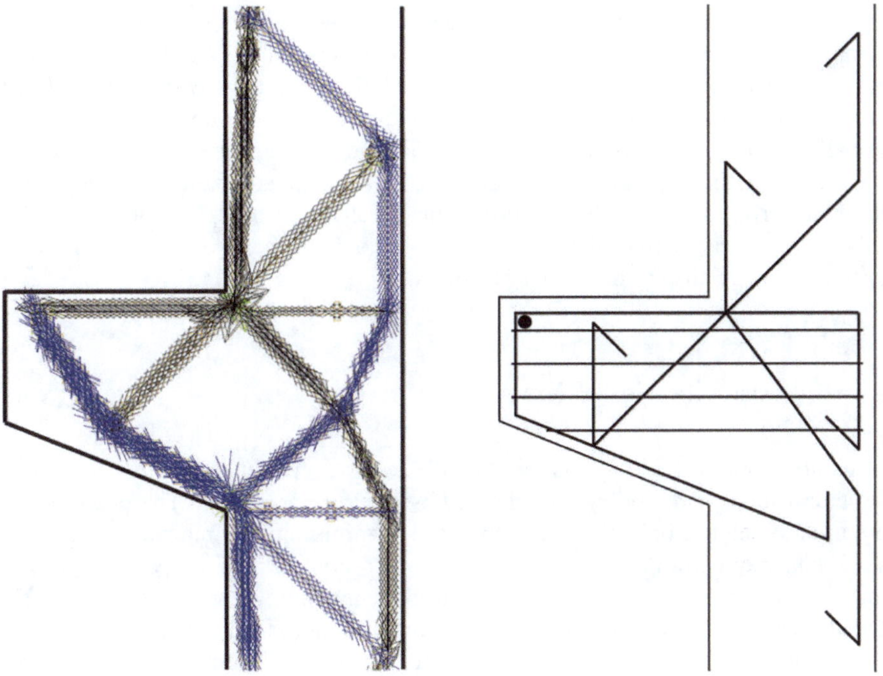

Fig. 13.10 Optimal topology and reinforcement details for a concrete corbel [88]

fact, a general agreement upon the definition of heuristics and metaheuristics in the literature does not exist. It is highly possible to come across recent studies that use "heuristics" and "metaheuristics" terms used a reciprocal manner [111]. The most interesting and most widely used metaheuristic algorithms are swarm-intelligence algorithms which are based on a collective intelligence of colonies of ants, termites, bees, flock of birds, etc. [114].

It is a necessity to highlight that ESO, which has a very different theoretical basis, is not only an intuitive method, but also offers much convenience in applying to engineering design problems. In this respect, ESO has great potential to be a generalized and convenient tool for design engineers [11]. Nevertheless, heuristic-metaheuristic characteristics of ESO have not been adequately scrutinized up to the present. In order to response to this ambiguity, it should be emphasized ESO algorithm is actually qualified and should be defined as metaheuristic since this method is obviously inspired by nature and has not a problem-dependent mechanism.

Recent studies on combining ESO or generalized topology optimization with other metaheuristic algorithms have generally been oriented toward improving the accuracy or speed of topology optimization. As a general brief, it has been observed that heuristic algorithms such as artificial bee colony, modified ant colony, particle swarm optimization, harmony search, binary bat, big bang-big crunch and genetic algorithms have been applied topology optimization of structural beam members

[115–124]. Modified approaches have been suggested for topology optimization to be applied on nonlinear systems using various constraints [125–128]. Topology optimization has also been performed on truss and frame systems with many different heuristic algorithms such as Jaya algorithm and genetic algorithm. [129–139]. Additionally, some of the recent metaheuristic algorithms might be considered for the integration with topology optimization regarding their compatible characteristics on engineering design problems. For instance, chaos game optimization [3, 140], tribe-interior search algorithm [141, 142] and quantum behaved developed swarm optimizer especially for real-size structural design problems [143].

13.10 Superiorities of the Algorithm

Structural optimization process traditionally follows a parametric path in which the structural shape and topology are defined by a set of parameters. These parameters are used to achieve objective functions such as minimum volume without ignoring some constraints. In the last decades, these operations, which are performed using mathematical programming, have only reached challenging and limited results [27]. Due to the fact that mathematical programming is inefficient and often inadequate in design optimization problems, optimality criteria and optimal design theories have been developed to overcome these difficulties [18]. Compared to mathematical programming algorithms, optimality criteria methods are more effective in large optimization problems. However, these methods need to be generalized in order to solve various type of problems [11]. Since a slightly changed version of the optimality criteria has been included in the ESO, so it is superior to mathematical programming methods even in this respect.

As stated before, Michell [60], who studied isostatic systems under particular loading and support conditions, pioneered topology optimization studies. Analytical solutions, called Michell trusses, have an infinite number of members with various sizes. In Michell trusses, each bar is subjected to constant strain or stress. It has also been proven analytically that Michell trusses cannot have more elasticity than any truss system using the same amount of material. However, since Michell trusses have infinite number of structural elements, they are quite unsuitable for engineering applications. If the solutions obtained from ESO are compared with analytical methods such as Michell truss systems, it will be seen that the results are quite successful [11].

Unlike traditional shape optimization, topology optimization method, which forms the basis of ESO, does not aim to achieve optimum shape by referencing the initial shape. In addition, it has a remarkable feature because it provides more advantages compared to the size optimization method and provides significant increases in structural performance. Topology optimization not only considerably increases structural performance, but also offers options for detailed size and shape optimization [144].

When considered in terms of engineering, it is seen that ESO has many attractive characteristics. Since it is used via finite element method, its application is very simple, and it requires relatively little computation time compared to other methods [11]. One of the main advantages of this method is that it is very easy to interact with finite element analysis and can be designed considering various criteria [66]. In addition, compared to other finite element-based structural optimization methods, although final design is very different compared to initial design, ESO method is one step ahead because it does not require re-meshing within the iterative process [27].

ESO is an easy and straightforward method to implement. It uses standard finite elements to define the initial design region. Achieved optimum design is also based on the layout of the elements in the finite element mesh that was created at the beginning. Design cycle applied in this method consists of analysis, calculation of sensitivity numbers and lastly, element elimination. It can be easily applied to any FEA software. Even if the use of constant FEA model in defining the design region, such as the homogenization method, causes the formation of design surfaces, it covers this defect by eliminating continuous re-meshing action. Results of all other structural optimization methods, which are much more mathematically complex, can be achieved by ESO method [26]. Comprehensive mechanism of ESO proves that it can overcome any structural optimization problem in the field of applied engineering [145].

13.11 Conclusions

Within the optimization research field, it is often stated that a completely new method has been developed. In fact, most of these algorithms have a similar approach to an existing method presented years ago, which could also be valid for evolutionary structural optimization. However, this certainly does not decrease the significance of ESO or other optimization methods. Furthermore, using a comprehensive topology optimization method that is not limited by complex optimization algorithms and theories is of great importance for the designer.

Implementation of ESO method is actually very simple, and following contributions to the algorithm have transformed this method into a versatile design tool. Results obtained with ESO offer truss-like fully stressed designs with maximum stiffness depending on the structural volume as expected. Generally, such structures are analogous to minimal weight truss systems. In case the design problem is restricted by any set of criteria considering structural behavior, design parameters of ESO could be rearranged so that these constraints are adequately satisfied. One of the obvious advantages of the algorithm compared to other classical methods is that it does not involve complex mathematical operations and programming and does not even require much FEA knowledge, while the other one is that relatively small amount of computation time is required.

ESO, which is generally defined as being intuitive, also has a distinct theoretical and analytical basis. Therefore, it could be summarized that ESO is a standard mathematical programming method that minimizes/maximizes an objective function. Although ESO does not have evidence to demonstrate that shape and topology of the optimum design have mathematically minimum or maximum values; it could be suggested that this algorithm operates within a metaheuristic mechanism, considering that ESO is obviously inspired by nature and has not a problem-dependent mechanism.

References

1. Lloyd P (2000) Storytelling and the development of discourse in the engineering design process. Des Stud 21(4):357–373
2. Skepper N, Straker L, Pollock C (2000) A case study of the use of ergonomics information in a heavy engineering design process. Int J Ind Ergon 26(3):425–435
3. Talatahari S, Azizi M (2020) Chaos game optimization: a novel metaheuristic algorithm. Artif Intell Rev
4. Özkal FM (2012) Design and experimental verification of the optimum reinforcement layout for reinforced concrete members with the aid of topology optimization and strut-and-tie modeling methods. Ph.D. thesis, Atatürk University, Turkey
5. Tang D, Zheng L, Li Z, Li D, Zhang S (2000) Re-engineering of the design process for concurrent engineering. Comput Ind Eng 38(4):479–491
6. Eversheim W, Bochtler W, Gräßler R, Kölscheid W (1997) Simultaneous engineering approach to an integrated design and process planning. Eur J Oper Res 100(2):327–337
7. Yoely YM, Hanniel I, Amir O (2020) Structural optimization with explicit geometric constraints using a B-spline representation. Mech Based Des Struct Mach
8. Talaslıoğlu T (2004) Evrimsel gelişim algoritmaları ile yapı elemanları tasarım optimizasyonu. Ph.D. thesis, Çukurova University, Turkey
9. Pourazady M, Fu Z (1996) An integrated approach to structural shape optimization. Comput Struct 60(2):279–289
10. El Hami A, Kadry S (2016) Global optimization method for design problems. Eng Rev 36(2):149–155
11. Tanskanen P (2002) The evolutionary structural optimization method: theoretical aspects. Comput Methods Appl Mech Eng 191(47–48):5485–5498
12. Wang L, Zhang H, Zhu M, Chen YF (2020) A new evolutionary structural optimization method and application for aided design to reinforced concrete components. Struct Multidisc Optim 62:2599–2613
13. Chen L, Xu L, Wang K, Li B, Hong J (2020) Parametric structural optimization of 2D complex shape based on isogeometric analysis. Comput Model Eng Sci 124(1):203–225
14. Klarbring A (2008) An introduction to structural optimization. Springer, Netherlands, Berlin Heidelberg
15. Sokolowski J, Zolesio JP (2003) Introduction to shape optimization. Society for Industrial and Applied Mathematics (SIAM), Philadelphia, PA
16. Wall WA, Frenzel MA, Cyron C (2008) Isogeometric structural shape optimization. Comput Methods Appl Mech Eng 197(33–40):2976–2988
17. Özkal FM, Uysal H (2010) Examination of a novel approach for the best topology selection on bridge type structures. In: 9th international congress on advances in civil engineering, Trabzon, Turkey
18. Xie YM, Steven GP (1993) A simple evolutionary procedure for structural optimization. Comput Struct 49(5):885–896
19. Li Q, Steven GP, Xie YM (1999) Displacement minimization of thermoelastic structures by evolutionary thickness design. Comput Methods Appl Mech Eng 179(3–4):361–378

20. Özkal FM, Uysal H (2009) General aspects of evolutionary structural optimization: a review. Pamukkale Univ J Eng Sci 15(3):383–393
21. Bendsøe MP, Kikuchi N (1988) Generating optimal topologies in structural design using a homogenization method. Comput Methods Appl Mech Eng 71(2):197–224
22. Özkal FM, Uysal H (2010) A new performance index formulation aiming to attain fully stressed designs for topology optimization problems. Sci Res Essays 5(15):2027–2036
23. Maier G (1973) Limit design in the absence of a given layout: a finite element, zero-one programming problem. In: Gallagher RH, Zienkiewicz OC (eds) Optimum structural design: theory and applications. Wiley, London, pp 223–239
24. Rodriguez-Velazquez J, Seireg AA (1985) Optimizing the shapes of structures via a rule-based computer program. Comput Mech Eng 4:20–28
25. Atrek E (1989) SHAPE: a program for shape optimization of continuum structures. In: Brebbia CA, Hernandez S (eds) Computer aided optimum design of structures: applications. Computation Mechanics Publications, Southampton, pp 135–144
26. Chu DN, Xie YM, Hira A, Steven GP (1997) On various aspects of evolutionary structural optimization for problems with stiffness constraints. Finite Elem Anal Des 24(4):197–212
27. Xie YM, Steven GP (1997) Evolutionary structural optimization. Springer, Berlin
28. Özkal FM, Uysal H (2008) Determination of the best topology for deep beams with web opening by the evolutionary method. In: 8th international congress on advances in civil engineering, Gazi Mağusa, North Cyprus
29. Özkal FM, Uysal H (2016) A computational and experimental study for the optimum reinforcement layout design of an RC frame. Eng Comput 33(2):507–527
30. Özkal FM (2006) Optimum design of beams by evolutionary structural optimization. M.Sc. thesis, Atatürk University, Turkey
31. Zheng Y, Wang Y, Lu X, Liao Z, Qu J (2020) Evolutionary topology optimization for mechanical metamaterials with auxetic property. Int J Mech Sci 179:105638
32. Radman A, Huang XD, Xie YM (2013) Topological optimization for the design of microstructures of isotropic cellular materials. Eng Optim 45(11):1331–1348
33. Liu YS, Wang XJ, Wang L, Liu DL (2019) A modified leaky ReLU scheme (MLRS) for topology optimization with multiple materials. Appl Math Comput 352:188–204
34. Bakhtiari-Shahri M, Moeenfard H (2018) Topology optimization of fundamental compliant mechanisms using a novel asymmetric beam flexure. Int J Mech Sci 135:383–397
35. Eschenauer HA, Olhoff N (2001) Topology optimization of continuum structures: a review. Appl Mech Rev 54(4):331–390
36. Chang KH, Tang PS (2001) Integration of design and manufacturing for structural shape optimization. Adv Eng Softw 32(7):555–567
37. Bletzinger KU, Ramm E (2001) Structural optimization and form finding of light weight structures. Comput Struct 79(22–25):2053–2062
38. Azad SK, Bybordiani M, Azad SK, Jawad FKJ (2018) Simultaneous size and geometry optimization of steel trusses under dynamic excitations. Struct Multidisc Optim 58:2545–2563
39. Sigmund O, Maute K (2013) Topology optimization approaches. Struct Multidiscip Optim 48:1031–1055
40. Wu Y, Li E, He ZC, Lin XY, Jiang HX (2020) Robust concurrent topology optimization of structure and its composite material considering uncertainty with imprecise probability. Comput Methods Appl Mech Eng 364:112927
41. Schmit LA (1960) Structural design by systematic synthesis. In: Proceedings of the 2nd ASCE conference on electronic computation. ASCE, Pittsburgh, pp 105–122
42. Chen A, Cai K, Zhao ZL, Zhou Y, Xia L, Xie YM (2020) Controlling the maximum first principal stress in topology optimization. Struct Multidisc Optim
43. Prager W, Shield RT (1967) A general theory of optimal plastic design. J Appl Mech 34(1):184–186
44. Prager W, Rozvany GIN (1977) Optimization of structural geometry. In: Bednarek AR, Cesari L (eds) Dynamical systems proceedings of a University of Florida international symposium. Academic Press, New York, pp 265–293

45. Bendsøe MP (1989) Optimal shape design as a material distribution problem. Struct Optim 1(4):193–202
46. Moreira JBD, de Souza Lisboa E, Rodrigues GC, Link FB, Casas WJP (2020) Multiscale topology optimization for frequency domain response with bi-material interpolation schemes. Optim Eng
47. Uysal H (2002) Duyarlılık analizi kullanılarak kabuk yapıların şekil optimizasyonu. Ph.D. thesis, Atatürk University, Turkey
48. Tang W, Tong L, Gu Y (2005) Improved genetic algorithm for design optimization of truss structures with sizing, shape and topology variables. Int J Numer Methods Eng 62(13):1737–1762
49. Querin OM, Steven GP, Xie YM (2000) Evolutionary structural optimisation using an additive algorithm. Finite Elem Anal Des 34(3–4):291–308
50. Haftka RT, Gürdal Z (1992) Elements of structural optimization. Kluwer Academic Publishers, Netherlands
51. Uysal H, Gul R, Uzman U (2007) Optimum shape design of shell structures. Eng Struct 29(1):80–87
52. Li A, Liu C, Feng S (2018) Topology and thickness optimization of an indenter under stress and stiffness constraints. J Mech Sci Technol 32:211–222
53. Shobeiri V (2016) Topology optimization using bi-direction at evolutionary structural optimization based on the element-free Galerkin method. Optim Eng 48:380–396
54. Lee SJ, Hinton E (2000) Dangers inherited in shells optimized with linear assumptions. Comput Struct 78(1):473–486
55. Wang YJ, Wang ZP, Xia ZH, Poh LH (2018) Structural design optimization using isogeometric analysis: a comprehensive review. Comput Model Eng Sci 117(3):455–507
56. Grierson DE, Khajehpour S (2002) Conceptual design optimization of engineering structures. In: Burns SA (ed) Recent advances in optimal structural design. American Society of Civil Engineers, Reston, pp 81–95
57. Kim H, Querin OM, Steven GP (2002) On the development of structural optimisation and its relevance in engineering design. Des Stud 23(1):85–102
58. Proos K (2002) Evolutionary structural optimisation as a robust and reliable design tool. Ph.D. thesis, The University of Sydney, Sydney, Australia
59. Barr AS, Sarin SC, Bishara AG (1989) Procedure for structural optimization. ACI Struct J 86(5):524–531
60. Michell AGM (1904) The limits of economy of material in frame-structures. Philos Mag 8(47):589–597
61. Wang SY, Wang MY, Tai K (2006) An enhanced genetic algorithm for structural topology optimization. Int J Numer Methods Eng 65(1):18–44
62. Rozvany GIN (1998) Exact analytical solutions for some popular benchmark problems in topology optimization. Struct Optim 15(1):42–48
63. Zhou K, Li J (2004) The exact weight of discretized Michell trusses for a central point load. Struct Multidiscip Optim 28(1):69–72
64. Das R, Jones R, Xie YM (2005) Design of structures for optimal static strength using ESO. Eng Fail Anal 12(1):61–80
65. Fuchs MB, Shemesh NNY (2004) Density-based topological design of structures subjected to water pressure using a parametric loading surface. Struct Multidiscip Optim 28(1):11–19
66. Steven GP, Li Q, Querin O (2001) Some thoughts on the physics and mechanics of the evolutionary structural optimization process. In: 3rd ASMO UK-ISSMO conference on engineering design optimization, Harrogate, North Yorkshire, UK
67. Li Q, Steven GP, Xie YM (2000) Evolutionary structural optimization for stress minimization problems by discrete thickness design. Comput Struct 78(6):769–780
68. Hinton E, Sienz J (1995) Fully stressed topological design of structures using an evolutionary procedure. Eng Comput 12(3):229–244
69. Pedersen P (2000) On optimal shapes in materials and structures. Struct Multidiscip Optim 19(3):169–182

70. Steven GP, Li Q, Xie YM (2002) Multicriteria optimization that minimizes maximum stress and maximizes stiffness. Comput Struct 80(27–30):2433–2448
71. Xie YM, Steven GP (1996) Evolutionary structural optimization for dynamic problems. Comput Struct 58(6):1067–1073
72. Chu DN, Xie YM, Hira A, Steven GP (1995) An evolutionary procedure for structural optimization with displacement constraints. In: Proceedings of the 5th East Asia-Pacific conference on structural engineering and construction, Building for the 21st Century, vol 2, pp 1091–1096
73. Chu DN, Xie YM, Hira A, Steven GP (1996) Evolutionary structural optimization for problems with stiffness constraints. Finite Elem Anal Des 21(4):239–251
74. Manickarajah D, Xie YM, Steven GP (1995) A simple method for the optimization of columns, frames and plates against buckling. In: Kitipornchai S, Hancock GJ, Bradford MA (eds) Structural stability and design. AA Balkema Publishers, Rotterdam, p 180
75. Chu DN, Xie YM, Steven GP (1998) An evolutionary structural optimization method for sizing problems with discrete design variables. Comput Struct 68(4):419–431
76. Li Q, Steven GP, Querin OM, Xie YM (1999) Evolutionary shape optimization for stress minimization. Mech Res Commun 26(6):657–664
77. Kim H, Querin OM, Steven GP, Xie YM (2003) Improving efficiency of evolutionary structural optimization by implementing fixed grid mesh. Struct Multidiscip Optim 24(6):441–448
78. Abolbashari MH, Keshavarzmanesh S (2006) On various aspects of application of the evolutionary structural optimization method for 2D and 3D continuum structures. Finite Elem Anal Des 42(6):478–491
79. Querin OM (1997) Evolutionary structural optimisation: stress based formulation and implementation. Ph.D. thesis, University of Sydney, Sydney, Australia
80. Liang QQ, Xie YM, Steven GP (1999) Optimal selection of topologies for the minimum-weight design of continuum structures with stress constraints. Proc Inst Mech Eng Part C J Mech Eng Sci 213(8):755–762
81. Liang QQ, Xie YM, Steven GP (2000) Topology optimization of strut-and-tie models in reinforced concrete structures using an evolutionary procedure. ACI Struct J 97(2):322–332
82. Liang QQ, Xie YM, Steven GP (2000) Optimal topology selection of continuum structures with displacement constraints. Comput Struct 77(6):635–644
83. Liang QQ, Xie YM, Steven GP (2001) A performance index for topology and shape optimization of plate bending problems with displacement constraints. Struct Multidiscip Optim 21(5):393–399
84. Guan H, Chen YJ, Loo YC (2001) Topology optimization of bridge type structures with stress and displacement constraints. Int J Comput Eng Sci 2(2):199–221
85. Liang QQ, Xie YM, Steven GP (2000) Optimal topology design of bracing systems for multistory steel frames. J Struct Eng 126(7):823–829
86. Liang QQ, Steven GP (2002) A performance-based optimization method for topology design of continuum structures with mean compliance constraints. Comput Methods Appl Mech Eng 191(13–14):1471–1489
87. Özkal FM, Uysal H (2012) A fully stressed design method to determine the optimum strut-and-tie model for beam-column connections. Int J Comput Methods 9(3):1250035
88. Özkal FM, Uysal H (2017) Reinforcement detailing of a corbel via an integrated strut-and-tie modeling approach. Comput Concr 15(5):589–597
89. Xie YM, Steven GP (1994) Optimal design of multiple load case structures using an evolutionary procedure. Eng Comput 11:295–302
90. Steven GP, Querin OM, Xie YM (1995) Multiple constraint environments for evolutionary structural optimization. In: Olhoff N, Rozvany GIN (eds) Proceedings of the first world congress on structural and multidisciplinary optimization. Goslar, Germany, pp 213–218
91. Li W, Li Q, Steven GP, Xie YM (2005) An evolutionary shape optimization for elastic contact problems subject to multiple load cases. Comput Methods Appl Mech Eng 194(30–33):3394–3415

92. Young V, Querin OM, Steven GP, Xie YM (1999) 3D and multiple load case bi-directional evolutionary structural optimization (BESO). Struct Optim 18(2–3):183–192
93. Özkal FM, Küçük M (2017) Güneş enerjisi panellerinin taşıyıcı sistem tasarımı için yapısal eniyileme yaklaşımı. In: 7th international symposium on steel structures, Gaziantep, Turkey, pp 270–280
94. Özkal FM, Küçük M (2018) A structural optimization approach for the load-bearing system design of solar panels. Eur Counc Civ Eng: E J 17:18–24
95. Simonetti HL, Almeida VS, das Neves FA, Greco M (2019) Multi-objective topology optimization using the boundary element method. Struct 19:84–95
96. Sawaragi Y, Nakayama H, Tanino T (1985) Theory of multiobjective optimization. Academic Press Inc, Orlando
97. Koski J (1993) Multicriteria optimization in structural design: state of the art. Adv Des Autom 1:621–629
98. Stadler W, Dauer J (1993) Multicriteria optimization in engineering: a tutorial and survey. In: Kamat M (ed) Structural optimization: status and promise. AIAA Inc, Washington
99. Yang XY, Xie YM, Steven GP (2005) Evolutionary methods for topology optimisation of continuous structures with design dependent loads. Comput Struct 83(12–13):956–963
100. Yang XY, Xie YM, Liu JS, Parks GT, Clarkson PJ (2003) Perimeter control in the bidirectional evolutionary optimization method. Struct Multidiscip Optim 24(6):430–440
101. Querin OM, Young V, Steven GP, Xie YM (2000) Computational efficiency and validation of bi-directional evolutionary structural optimisation. Comput Methods App Mech Eng 189(2):559–573
102. Lencus A, Querin OM, Steven GP, Xie YM (1999) Modifications to the evolutionary structural optimisation (ESO) method to support configurational optimisation. In: 3rd world congress of structural and multidisciplinary optimization, New York, USA
103. Mueller KM, Burns SA (2001) Fully stressed frame structures unobtainable by conventional design methodology. Int J Numer Methods Eng 52(12):1397–1409
104. Liu M, Burns SA (2003) Multiple fully stressed designs of steel frame structures with semi-rigid connections. Int J Numer Methods Eng 58(6):821–838
105. Bakir PG, Boduroglu HM (2006) Nonlinear analysis of beam–column joints using softened truss model. Mech Res Commun 33(2):134–147
106. MacGregor JG (1997) Reinforced concrete: mechanics and design. Prentice-Hall International Inc, New Jersey
107. Ahyat AB (2005) Analysis and design of hammerhead bridge pier using a strut and tie method. M.Sc. thesis, Malaysia Technology University, Malaysia
108. Cook WD, Mitchell D (1988) Studies of disturbed regions near discontinuities in reinforced concrete members. ACI Struct J 85(2):206–216
109. Altan M, Taşkın B (2006) A comparative study of the strut-and-tie model for shear critical structures. In: Proceedings of the 5th GAP engineering congress, vol 2. Şanlıurfa, Turkey, pp 889–896
110. Matraji M (1997) Design of reinforced concrete members by the strut-and-tie method. M.Sc. thesis, Middle East Technical University, Ankara, Turkey
111. Stojanović I, Brajević I, Stanimirović PS, Kazakovtsev LA, Zdravev Z (2017) Application of heuristic and metaheuristic algorithms in solving constrained Weber problem with feasible region bounded by arcs. Math Prob Eng 2017:8306732
112. Gandomi AH, Yang XS, Talatahari S, Alavi AH (2013) Metaheuristic algorithms in modeling and optimization. In: Gandomi AH, Yang XS, Talatahari S, Alavi AH (eds) Metaheuristic applications in structures and infrastructures. Elsevier, Massachusetts, pp 1–24
113. Martí R, Reinelt G (2011) The linear ordering problem: exact and heuristic methods in combinatorial optimization, vol 175. Springer, New York
114. Yang X, Chien SF, Ting TO (2014) Computational intelligence and metaheuristic algorithms with applications. Sci World J 2014:425853
115. Park JY, Han SY (2013) Application of artificial bee colony algorithm to topology optimization for dynamic stiffness problems. Comput Math Appl 66(10):1879–1891

116. Yoo KS, Han SY (2013) A modified ant colony optimization algorithm for dyna topology optimization. Comput Struct 123:68–78
117. Park JY, Han SY (2015) Topology optimization for nonlinear structural problems based on artificial bee colony algorithm. Int J Precis Eng Manuf 16(1):91–97
118. Alberdi R, Murren P, Khandelwal K (2015) Connection topology optimization of steel moment frames using metaheuristic algorithms. Eng Struct 100:276–292
119. Kim YH, Han SY (2017) Topological shape optimization scheme based on the artificial bee colony algorithm. Int J Precis Eng Manuf 18(10):1393–1401
120. Lee SM, Han SY (2017) Topology optimization based on the harmony search method. J Mech Sci Technol 31:2875–2882
121. Millan C, Millan E (2018) Topology optimization of two-dimensional beams: a comparative study. Contemp Eng Sci 11:5075–5080
122. Di Cesare N, Domaszewski M (2019) A new hybrid topology optimization method based on I-PR-PSO and ESO. Application to continuum structural mechanics. Comput Struct 212:311–326
123. Ahn HK, Han DS, Han SY (2019) A modified big bang–big crunch algorithm for structural topology optimization. Int J Precis Eng Manuf 20:2193–2203
124. Jaafer AA, Al-Bazoon M, Dawood AO (2020) Structural topology design optimization using the binary bat algorithm. Appl Sci 10(4):1481
125. Gao X, Ma H (2015) Topology optimization of continuum structures under buckling constraints. Comput Struct 157:142–152
126. Lee K, Ahn K, Yoo J (2016) A novel P-norm correction method for lightweight topology optimization under maximum stress constraints. Comput Struct 171:18–30
127. Jankovics D, Gohari H, Tayefeh M, Barari A (2018) Developing topology optimization with additive manufacturing constraints in ANSYS®. IFAC-PapersOnLine 51(11):1359–1364
128. Abdi M, Ashcroft I, Wildman R (2018) Topology optimization of geometrically nonlinear structures using an evolutionary optimization method. Eng Optim 50(11):1850–1870
129. Šešok D, Belevičius R (2007) Use of genetic algorithms in topology optimization of truss structures. Mechanika 64(2):34–39
130. Kociecki M, Adeli H (2014) Two-phase genetic algorithm for topology optimization of free-form steel space-frame roof structures with complex curvatures. Eng Appl Artif Intell 32:218–227
131. Kaveh A, Mahdavi VR (2015) Colliding bodies optimization for size and topology optimization of truss structures. Struct Eng Mech 53(5):847–865
132. Mortazavi A, Toğan V (2016) Simultaneous size, shape, and topology optimization of truss structures using integrated particle swarm optimizer. Struct Multidiscip Optim 54(4):715–736
133. Maheri MR, Askarian M, Shojaee S (2016) Size and topology optimization of trusses using hybrid genetic-particle swarm algorithms. Irani J Sci Technol Trans Civ Eng 40(3):179–193
134. Assimi H, Jamali A, Nariman-Zadeh N (2017) Sizing and topology optimization of truss structures using genetic programming. Swarm Evol Comput 37:90–103
135. Gholizadeh S, Ebadijalal M (2018) Performance based discrete topology optimization of steel braced frames by a new metaheuristic. Adv Eng Software 123:77–92
136. Mortazavi A, Toğan V, Daloğlu A, Nuhoğlu A (2018) Comparison of two metaheuristic algorithms on sizing and topology optimization of trusses and mathematical functions. Gazi Univ J Sci 31(2):416–435
137. Degertekin SO, Lamberti L, Ugur IB (2018) Sizing, layout and topology design optimization of truss structures using the Jaya algorithm. Appl Soft Comp 70:903–928
138. Tejani GG, Savsani VJ, Bureerat S, Patel VK, Savsani P (2019) Topology optimization of truss subjected to static and dynamic constraints by integrating simulated annealing into passing vehicle search algorithms. Eng Comput 35(2):499–517
139. Mortazavi A (2020) A new fuzzy strategy for size and topology optimization of truss structures. App Soft Comput 93:106412
140. Talatahari S, Azizi M (2020) Optimization of constrained mathematical and engineering design problems using chaos game optimization. Comput Ind Eng 145:106560

141. Talatahari S, Motamedi P, Farahmand Azar B, Azizi M (2019) Tribe–charged system search for parameter configuration of nonlinear systems with large search domains. Eng Optim
142. Talatahari S, Azizi M (2020) Optimum design of building structures using tribe-interior search algorithm. Struct 28:1616–1633
143. Talatahari S, Azizi M (2020) Optimal design of real-size building structures using quantum-behaved developed swarm optimizer. Struct Des Tall Special Build 29(11):e1747
144. Guan H, Chen YJ, Loo YC, Xie YM, Steven GP (2003) Bridge topology optimisation with stress, displacement and frequency constraints. Comput Struct 3:131–145
145. Rong JH, Xie YM, Yang XY, Liang QQ (2000) Topology optimization of structures under dynamic response constraints. J Sound Vib 234(2):177–189

Chapter 14
An Extensive Review of Charged System Search Algorithm for Engineering Optimization Applications

Siamak Talatahari and Mahdi Azizi

14.1 Introduction

All of the design problems in nature such as engineering problems can be determined as optimization problem that require efficient optimization methods and techniques to be dealt with. Nowadays, engineering problems have become more complex, and the classic mathematical methods such as methods based on the gradient of the objective function are incapable of producing acceptable results in a reasonably practical time. In recent decades, introducing new methods based on dealing with deficiencies of classical methods have been concerned. Advances in engineering technology have required new methods to provide greater accuracy, efficiency and speed in solving engineering problems. In addition, avoiding local optima and dealing with problems having convexity and smoothness in search space have been of great concern. These concerns have led the researchers to present a new methodology in solving optimization problems called "metaheuristics." The "metaheuristic," term firstly proposed by Glover [1] in 1986, is combined of a suffix and a word that have their origins in the Greek words. The word "heuristic" comes from the old Greek word "heuriskein" which denotes on discovering new strategies (rules) to deal with problems and the "meta" means "upper level methodology." The metaheuristics are some kinds of solution methods that combine the improvement in higher-level strategies and search processes to conduct an optimization process to perform a powerful search in search space with ability of escaping the local optima. As discussed by Sörensen [2], the

S. Talatahari (✉) · M. Azizi
Department of Civil Engineering, University of Tabriz, Tabriz, Iran
e-mail: siamak.talat@gmail.com

M. Azizi
e-mail: mehdi.azizi875@gmail.com

S. Talatahari
Engineering Faculty, Near East University, North Cyprus Mersin 10, Nicosia, Turkey

S. Carbas et al. (eds.), *Nature-Inspired Metaheuristic Algorithms for Engineering Optimization Applications*, Springer Tracts in Nature-Inspired Computing,
https://doi.org/10.1007/978-981-33-6773-9_14

history of considering metaheuristic as methods for considering real-life problems can be discussed in five distinct periods. The first time period is called the "pre-theoretical" period (until 1940) that both heuristics and metaheuristics had been used for solving some problems but there have not been any formal discussion. In the "early" period as second period, which is from 1940 to 1980, researchers have conducted some studies on heuristics that was the first formal discussion in this field. In the third period called "method-centric" which is from 1980 to 2000, the field of metaheuristics had been developed and various methods had been proposed. In the "framework-centric" period (2000 until now), the intuition grows and the metaheuristics not as methods but as frameworks have been more successfully described. In the last period as the "scientific" or "future" period, the design of different metaheuristics turns into a matter of science rather than art.

Based on the improvement of different metaheuristic algorithms, three classifications can be made in terms of their inspiration. The first category is termed as "evolutionary algorithms" consisting of the genetic algorithm (GA) [3], evolution strategies (ES) [4], and the differential evolution (DE) [5] that have been proposed by consideration of biological reproduction and evolution. The swarm intelligence-based metaheuristic algorithms are the second category that has been proposed by consideration of cooperative behavior for decentralized and self-organized artificial and natural systems. The ant colony optimization (ACO) [6], particle swarm optimization (PSO) [7], and firefly algorithm (FA) [8] are some of the fully known algorithms in this group. The third group consists of optimization algorithms that are motivated by physical laws. The simulated annealing (SA) [9], gravitational search algorithm (GSA) [10], and the harmony search (HS) [11] are some the well-known algorithms in this area. In addition to these algorithms, some of the other algorithms were also developed regarding the life style of humans and animals such as imperialistic competitive algorithm (ICA) [12], bat algorithm (BA) [13], and cuckoo search algorithm (CSA) [14]. In addition to these algorithms, some other efforts in improving or hybridizing standard metaheuristic algorithms have also been conducted [15–23].

In recent years, Kaveh and Talatahari [24] have proposed a novel metaheuristic algorithm as charges system search (CSS) for providing proper solutions to optimization problems. The main theory of this algorithm is derived of some mechanics and physics principles. It mimics the Newtonian laws of mechanics and the governing Coulomb law from electrostatics to formulate a search technique. In the CSS algorithm as a multi-agent intelligent approach, each search agent is considered as a charged particle (CP) and CPs can impress others based on their separation distances and fitness (objective) values. As a fact in electrical physics, an electric field is generated by each electric charge which exerts a special force toward other (electrically) charged objects; therefore, an electrical force is exerted on each CP from other CPs. The interaction between the CPs is based on the qualities of the movement that is obtained by utilizing Newtonian mechanics laws and the quantities of resultant force that is obtained by utilization of the electrostatics laws. The CSS is considered as a powerful metaheuristic algorithm that contains both exploration and exploitation strategies. Some advantages of this algorithm are its simplicity in implementation, flexibility in dealing with complex optimization problems, adaptively with computational efforts, and efficiency in engineering application. As a metaheuristic algorithm, the CSS contains the essential operators of evolutionary algorithms such as crossover and mutation, alongside the fast convergence rate similar to the swarm intelligent based techniques. The CSS has been utilized for a wide range of optimization problems as a successful search technique. Some of these applications were conducted mostly in the structural design optimization while some other applications were in optimization of engineering frameworks, control strategies, efficient networks and economic systems. In addition, in spite of the capability of the CSS in solving difficult problems, many developments have been done in order to calculate better results with lower computational time. These developments have been considered as improving the existing algorithms or hybridizing with other metaheuristics to accomplish better results for objective functions.

The main aim and contribution of this book chapter is to provide an extensive review on related applications of the CSS and its developments in different optimization area as well as bringing out future challenges and possibilities. In this regard, the published papers by various publishers such as Science Direct, Springer, Taylor & Francis, IEEE Explore, and some others. As mentioned by Google Scholar, the main reference of the CSS [Ref. 24] and other papers that had been discussed about CSS and its applications, modifications and hybridizations, have been totally cited for 2389 times. Regardless of the fact that some of these citations are about literature review but most of them are directly about application, improvement, and hybridization of the CSS. In Fig. 14.1, the cumulative sum of citations per year corresponding to CSS is depicted.

This book chapter is organized in sections and subsections as follows. The general formulation of the CSS algorithm is provided in Sect. 14.2 while the reviews of CSS algorithm in relation to its applications, modifications, and hybridizations are given in Sects. 14.3, 14.4 and 14.5 respectively. The multi-objective approaches for the CSS alongside the final discussions and conclusions are outlined in Sects. 14.6 and 14.7.

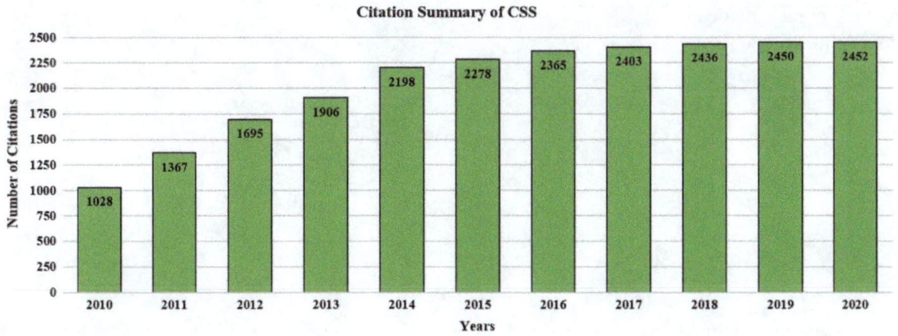

Fig. 14.1 Publication summary for CSS

14.2 General Formulation of CSS

14.2.1 Inspiration

As mentioned before, some basic principles of physics alongside the important aspects of mechanics are utilized in formulation of the CSS optimization algorithm. In physics, Coulomb's law denotes that the electric force generated between two specific charged particles that are stationary, which is conversely proportional to the separation distance square directed with the line joining two particles. This force is proportional to the direct product of the two particles' charges while it is repulsive if the charges have the same sign and attractive if the charges are of opposite sign. By this law and its relations, the total magnitude of the electric (Coulomb) force is generated between charges of these two points and is formulated as follows [24]:

$$F_{ij} = k_e \frac{q_i q_j}{r_{ij}^2} \tag{14.1}$$

where q_i and q_j are the two point charges, k_e is the Coulomb constant, and r_{ij} is the total distance between these two charges.

An sphere is considered as an insulating solid with radius of a with uniform volume density which carries a total amount of positive charge q_i, Gauss's law provides the magnitude of the required electric field (E_{ij}) at a point outside and inside of a charge as displayed in Fig. 14.2.

The main topic in Newtonian mechanics is the movement of objects. In this topic, the moving object is presented as a particle regardless of its size having infinitesimal size (as a point-like round mass). The displacement of a particle in space (Δr) is

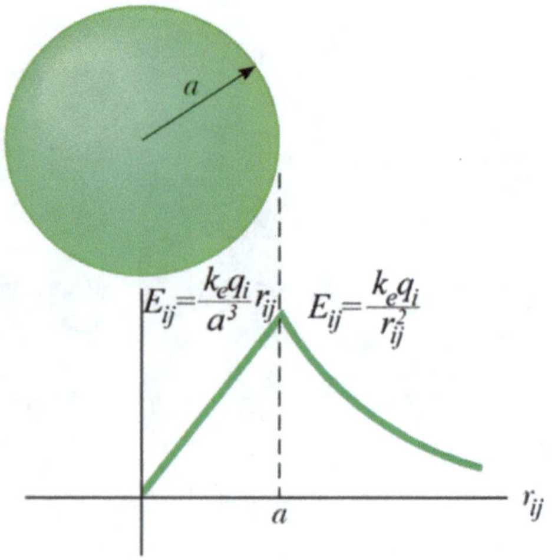

Fig. 14.2 Magnitude of the required electric field at a specific point outside and inside of a charge [24]

defined as it moves out of an initial position r_{old} to a final destination r_{new} and is as follows [24]:

$$\Delta r = r_{\text{new}} - r_{\text{old}} \tag{14.2}$$

Any object's displacement as a function of time (Δt) in terms of the acceleration of particles (a), and the velocity of particles (v) is approximated as [24]:

$$r_{\text{new}} = \frac{1}{2}a \cdot \Delta t + v_{\text{old}} \cdot \Delta t + r_{\text{old}} \tag{14.3}$$

As second law of Newton, any object's acceleration is conversely proportional to its mass (m) and is straightly proportional to the acting net force on the particle. In this regard, the displacement of an object is presented as [24]:

$$r_{\text{new}} = \frac{1}{2}\frac{F}{m} \cdot \Delta t + v_{\text{old}} \cdot \Delta t + r_{\text{old}} \tag{14.4}$$

The electrical and mechanical laws presented in this section are considered as main characteristics of the CSS algorithm. In the CSS algorithm, each solution nominee (X_i) consists of some decision variables (i.e., $X_i = \{x_{i,j}\}$) that is determined as a charged particle (CP). Each CP is concerned by the electrical fields of some other CPs by exerting electrical forces. By using the electrostatics and the mechanic laws of Newton, the total quantity of this force and the specific quality of the movement are determined. It seems that a CP with proper position has to exert a total force with greater extent than the bad CP, so the charge's amount will be determined regarding the value of the objective (fitness) function [24].

14.2.2 Mathematical Model

In the CSS, a number of CPs with special magnitudes of charge (q_i) that creates a specific electrical field nearby its space is considered. Regarding the quality of its fitness, the magnitude of the CP's charge is defined as follows [24]:

$$q_i = \frac{\text{fit}(i) - \text{fitworst}}{\text{fitbest} - \text{fitworst}}, \quad i = 1, 2, \ldots, N. \tag{14.5}$$

where fitbest and fitworst are the best fitness and the worst fitness between all CPs, fit(i) denotes on the fitness value obtained by objective function for the agent i, and N is the required number of CPs. r_{ij} is the separation distance within two specific CPs and is formulated as follows [24]:

$$r_{ij} = \frac{\|X_i - X_j\|}{\|X_i + X_j/2 - X_{\text{best}} + \varepsilon\|}, \tag{14.6}$$

where X_i and X_j are the position vectors of the ith and jth CPs, X_{best} is the best current position vector of the CPs, and ε is a predefined small positive value for avoiding singularity.

By utilizing a random process, the initial position vectors of CPs are specified in the predefined search space and the initial velocity vectors of CPs are supposed to be zero. In the CSS, all CPs with better behavior can seduce CPs with bad behavior and only some bad search agents can seduce good search agents [24]. The specific value of the required resultant electrical force which acts on a CP is specified as follows where the resultant force acts on the jth CP, are presented in Fig. 14.3.

$$F_j = q_j \sum_{i,i \neq j} \left(\frac{q_i}{a^3} r_{ij}.i_1 + \frac{q_i}{r_{ij}^2}.i_2 \right) p_{ij} (X_i - X_j), \quad \begin{cases} j = 1, 2, \ldots, N, \\ i_1 = 1, i_2 = 0 \Leftrightarrow r_{ij} < a, \\ i_1 = 1, i_2 = 0 \Leftrightarrow r_{ij} \geq a, \end{cases} \tag{14.7}$$

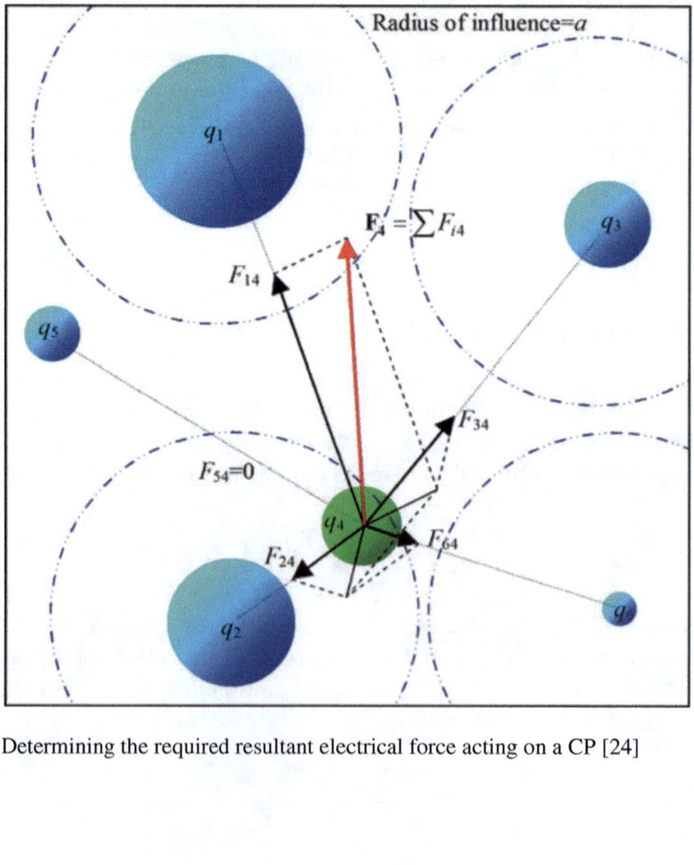

Fig. 14.3 Determining the required resultant electrical force acting on a CP [24]

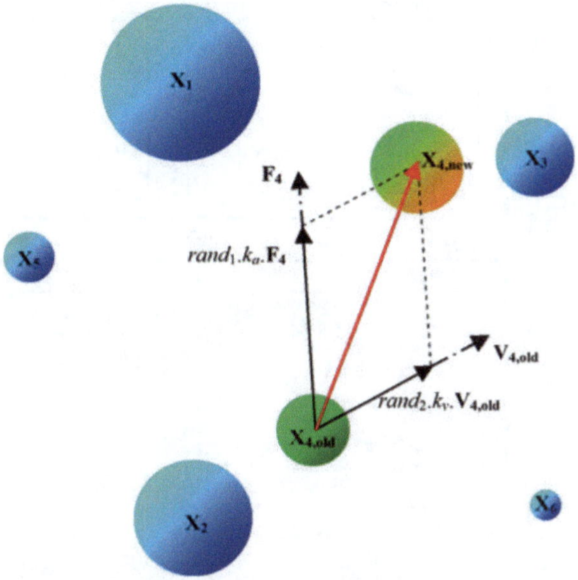

Fig. 14.4 Schematic view for a CP moving to a new position [24]

For each CP, the new position and also velocity vectors are determined as follows [24]:

$$X_{j,\text{new}} = \text{rand}_{j1} \cdot k_a \cdot F_j + rand_{j2} \cdot k_v \cdot V_{j,\text{old}} + X_{j,\text{old}} \tag{14.8}$$

$$V_{j,\text{new}} = X_{j,\text{new}} - X_{j,\text{old}}, \tag{14.9}$$

where k_a and k_v represent the acceleration and velocity coefficient. rand_{j1} and rand_{j2} are two randomly generated numbers that are distributed uniformly in the specific range of (0, 1). The movement of a CP to the new position is shown in Fig. 14.4.

14.2.3 Implementation of the CSS

The implementation of the CSS algorithm in order to use as a programming code is simple. This implementation can be presented in two main steps each one having multiple sub steps that translate the CSS theory into programming code. The first main steps are the "initialization" phase that has three sub steps. In the first sub-step, the initialize specifications of the optimization problem and algorithm parameters are initialized while the initial position of charged particles based on random values for their positions and associated velocities are determined. In the second sub-step, the values of the objectives functions for the CPs are established while the fitness values of each CP are compared with other CPs and are arranged in increasing order.

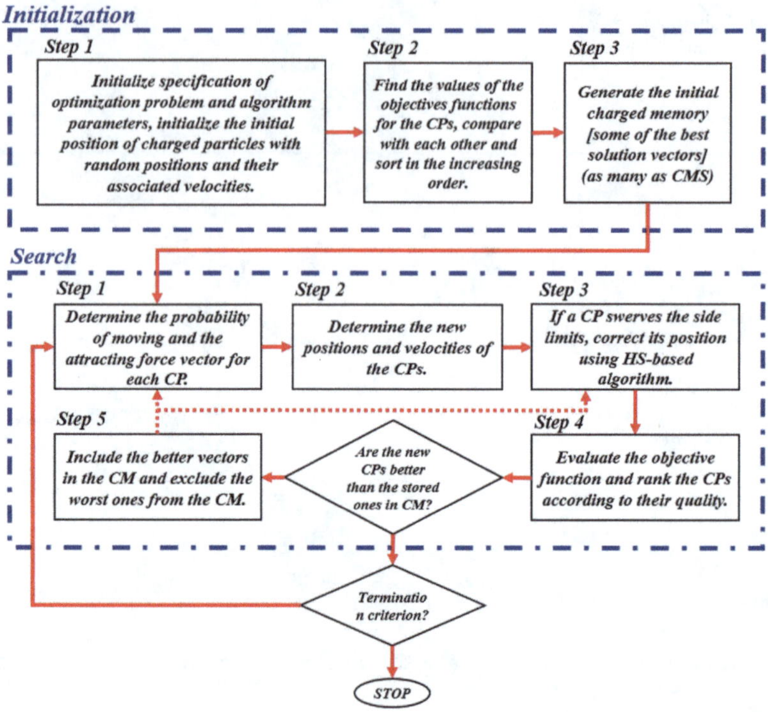

Fig. 14.5 Flowchart of the CSS algorithm [24]

The third sub-step, the initial charged memory (some of the best solution vectors as many as CMS) is generated.

The second main steps are the "search" phase that has five sub-steps. In the first sub-step, the probability of moving and attracting force vector for each CP is determined. In the second one, the new positions and velocities of the CPs are defined. Controlling over the position limits of the CPs is conducted in terms of an HS-based control method in the third sub-step. In the fourth sub-step, evaluation of the objective function is concerned alongside the ranking of the CPs according to their qualities. In the last sub-step, the better vectors are saved in the CP while the worst ones are expelled. The flowchart of the CSS is demonstrated in Fig. 14.5.

14.3 Applications of CSS

Several applications of CSS in popular benchmark and real-world optimization problems have been conducted while in most of them; its performance was compared to other metaheuristic algorithms. Most of these applications are about optimization of engineering structures with emphasis on topology, layout and sizing optimization however, many other optimization problems with different kinds of objective functions were solved by this algorithm. Some of the important applications of the CSS are presented in the next subsections.

14.3.1 Applications to Structural Engineering Design

Kaveh and Talatahari [25] firstly utilized the CSS in optimization of truss and frame structures. They conducted a thorough comparing study of the outcomes of the CSS algorithm with other optimization algorithms and proved that the CSS provides very robust results. Kaveh and Talatahari [26] discussed the optimization of cross-sectional properties in transverse and longitudinal beams of grillage systems and studied the ability of the CSS in dealing with large-scale engineering problems. They formulated the optimization problem based on the optimum selections of W-sections for the grillage system members while the code of practice is selected as the LRFD-AISC and the allowable stress constraints alongside the deflection limitations were considered as the constraints. They compared the results of the selected algorithm with the outcomes the HS and GA and proved the efficiency of the CSS approach in balancing between the exploitation and the exploration. Kaveh and Talatahari [27] developed a new approach for optimization of dome structures with capability of determining automatically the number of elements and the element configurations during the optimization process. They used CSS as the optimization algorithm and unlike the previous studies that concern only geometry or size optimization of the domes by defining the cross-sectional areas of the elements and the number of rings; they utilized the automatically selected number of elements as the design variables. The results prove that the CSS algorithm is a robust method that can effectively be used in the practical optimal topology configuration of the domes. Kaveh and Talatahari [28] proposed a discrete version of the CSS in order to deal with the optimum design of frame structures. They compared the performance of presented method to other classical and new metaheuristic algorithms while design examples are selected as three different frame structures to evaluate the effectiveness of this new method. Kaveh et al. [29] utilized CSS for optimum design of composite channels. They presented four different models while in the first one; an identical coefficient of roughness is computed without any section division. The channel section is separated into multi-segmental areas by drawing vertical lines and contemplating a constant velocity for each segment in the second model. In the third model, the maximum velocity of each segment is controlled in order to safely convey the required discharge. The fourth model divided the channel into horizontal slices. The results of the CSS are equated to the results of ACO and the GA method to highlight its superiority for determining the optimum design of composite open channel structures. Kaveh and Behnam [30] discussed the optimum configuration of reinforced concrete cantilever retaining walls considering the static loads using the CSS. The optimization process is based on the materials cost function used in retaining walls and their constructions. Four design examples are considered, and the reliability of the CSS algorithm is compared to HS algorithm. Kaveh and Behnam [31] utilized CSS for optimum design of multi-story three dimensional reinforced concrete structures. In this study, the structural members are subjected to axial loads and biaxial moments while the structural analysis is performed by the standard stiffness methodology. They also considered the second-order effects, and the end moments of the columns are magnified when requested. Kaveh and Nasrollahi [32] discussed the performance-based optimal design of steel building frames considering seismic behavior with CSS algorithm. In this study, two numerical examples are selected and based on concept of semirigid connection, a pushover analysis method is considered as design and analysis techniques. The proved that by utilizing the CSS algorithm, a substantial enhancement in structural total weight related to the classical design methodologies is performed. They also compared the capabilities of the CSS in dealing with seismic design of structures with other metaheuristics such as ACO and GA. Kaveh and Shokohi [33] discussed the optimal design of castellated beam compared to some steel structural design codes and specifications. Design problems are considered as the castellated beams with circular and hexagonal openings are considered as design problems. They concluded that castellated beams with hexagonal openings have less cost in comparison with the cellular beams. Comparing to other metaheuristics, the CSS algorithm is faster in finding the optimum solution. Kooshkbaghi et al. [34] utilized the CSS for optimal topological configuration of latticed single-layer domes which are geometrically nonlinear. Tavakkoli et al. [35] presented the shape optimization of structures based on isogeometric analysis (IA) and sequential quadratic programming (SQP) methods utilizing the CSS algorithm. For estimation and interposition of the displacement field alongside the modeling geometry of the structure, they used non-uniform rational B-Spline (NURBS) basis functions. Sharafi et al. [36] presented a preliminary layout design of orthogonal buildings using CSS algorithm. They intended to determine the optimum forms for conceptual design in terms of cost and plan regularity. A numerical bi-objective problem of rectilinear building frames is selected as design example, and the results of the optimization problem with CSS are compared with other metaheuristic as ACO algorithm. Sharafi et al. [37] discussed the free shape optimizations of thin-walled steel sections, represented by some popular graph theory-based methods using CSS algorithm. The main purpose of this research study is to achieve shapes of minimum mass and/or maximum strength for thin-walled steel sections that satisfy design constraints, which results in a general formulation for a bi-objective combinatorial optimization problem. They utilized a numerical example involving the shape optimization of thin-walled open and closed steel sections is presented to represent the robustness of the method. They concluded that the combination of the graph theory methods and the CSS algorithm could offer an effective method for optimal shape configuration of thin-walled steel sections with multiple conflicting objectives o f mass minimization and strength maximization, as well as accounting for the strength and geometric constraints. Uz et al. [38] utilized CSS to formulate an automated technique for cost optimization of geometric configuration design of reinforced concrete beams with multi-spans excited by dynamic loading. They proposed the constrained optimization problem in structures based on the dynamic and static equilibrium, structural action effects and architectural dimensions. The results demonstrate that the utilized method using the new cost function for optimization concludes some satisfactory outcomes and is able to result in over 7% cost saving.

14.3.2 Applications on Control Systems

Precup et al. [39] utilized the CSS algorithm for the optimum tuning of proportional–integral (PI) controllers while the focused is on a category of second-order processes with variable parameters and an integral component. They selected the integral of squared control error extended by the integral of squared output sensitivity functions as the objective function. The results of the CSS algorithm were compared to the results of other metaheuristics including the PSO and GSA while the performance of the CSS is proved to be good. Talatahari et al. [40] presented a new inspiring identification concept for extremely nonlinear systems with hysteretic behavior determined by utilization of different mechanical models with CSS. They formulated the optimization problem as the parameter configuration of Bouc–Wen damping models for fluid dampers of type MR. Precup et al. [41] utilized CSS for the optimum tuning of the Takagi–Sugeno PI fuzzy controllers (TS-PI-FC). The sum of squared output sensitivity function plus the absolute control error is selected as the objective functions and leads to an optimum control system of type fuzzy with decreased sensitivity. Kaveh et al. [42] utilized CSS to optimize the parameters of tuned mass dampers (TMD) that result in utmost reduction in the structural response to earthquake loading. They utilized a ten-story shear building as numerical example and compared the capability of the CSS in reducing the response of the structure with other published papers. They concluded that the performance of CSS is reasonable compared to other methods. Shahrouzi et al. [43] discussed the utilization of seismic excitation with multiple support as the source of dynamic loading for optimal tuning of TMD systems in controlling resultant vertical response of simply supported steel bridges. They utilized CSS as optimization algorithm while the supremacy of many TMD over one TMD is investigated after unifying their parameters via optimization process. They compared the optimized parameters of the TMD alongside the optimized responses of the bridges obtained by CSS with GA approach and proved the efficiency of the CSS.

14.3.3 Applications on Damage Detection

Kaveh and Maniat [44] discussed the problem of damage detection in engineering structures with vibration data utilizing CSS optimization algorithm. They used the mode shapes and natural frequencies to form the needed objective function in order to detect the position and range of multi-damage in a wide range of structures such as beams, frames, and trusses. They concluded that the proposed methodology can certainly specify different damage configuration using incomplete noisy and data measurements. Tabrizian et al. [45] discussed the damage assessment and damage detection methods regarding the variation in dynamic components of the complex structural systems. Saberi and Kaveh [46] proposed a new two-stage algorithm for damage assessment of large structures. In this study, a modified residual force method is used to locate the damage regions in structures, especially barrel vaults, and then the CSS algorithm is used to quantify the amount of damage. They utilized a modified vector for enhancing efficiency of the modal residual force method and suppressing effect of noise while results prove that severities of damaged elements are assessed by utilizing CSS as optimization algorithm. Safari and Gholizad [47] developed a method to specify and evaluate structural damages by considering modifications in mode shapes. They applied the mechanism of using two-dimensional continuous wavelet transform for damage localization and utilized the finite element model updating technique as an inverse optimization problem by applying the CSS to assess the damage in each element sited in the first stage. They studied a series of numerical examples with different damage scenarios have been carried out in the double-layer space structures while the results confirm the certainty and relevance of introduced method with CSS.

14.3.4 Applications on Robotics and Power Systems

Precup et al. [48] proposed an optimal path planning technique in the basis of CSS algorithm to deal with mobile robots with multiple configuration upon holonomic wheeled patterns. They formulated the optimum design problem in order to reduce the weighted summation of four fitness values as specific targets of path plannings for each robot. They validated the new path planning approach with different algorithms and concluded that the new techniques have provided challenging results. Kanagaraj and Ponnambalam [49] utilized the CSS to deal with robotic drill path problem involved in printed circuit board (PCB) manufacturing industries. They compared the performance of CSS with four other case studies from the literature and concluded that the CSS is capable of effectively finding the optimum path for the process of PCB holes drilling with logical computational time. Özyön et al. [50] utilized the CSS algorithm in order to optimize the emission constrained economic power dispatch problem. They formulated the optimization problem for solving the test system with 30-bus 6-generator. They compared the performance of the selected algorithm with other metaheuristics. Lenin et al. [51] discussed solving different power dispatch problems in the power systems of reactors. They selected the loss maximization and minimization of stability margin in voltage as the objective functions and the CSS as the optimization algorithm. They compared the performance of the selected algorithm for optimization of the 30-bus 6-generator test system. Wu et al. [52] proposed a novel wind power forecasting approach based on stratification and presented a hybrid forecasting model at different stratifications by utilizing

CSS algorithm. They utilized the theory of optimal stratification in order to apply the concept of segmentation to forecast short-term wind power outputs. The results proved the capability of the presented forecasting model for exact wind power forecasting. Zhang et al. [53] formulated an optimization problem to minimize the loss estimation in the transmission line of power dispatch systems in order to satisfy all the constraints of the system with CSS. They compared the results of the CSS with those of PSO and concluded that the utilized optimization technique is possible for power dispatch problems in reactive, decreasing the active power loss in the transmission line and keeping the load bus voltage within a reasonable range.

14.3.5 Applications on Other Optimization Problems

Pereira et al. [54] improved the training phase of a neural network-based classifier using CSS algorithm. They carried out some experiments in the field of non-technical loss in power distribution systems in records obtained from an electrical power company in Brazil. They compared the results of CSS with several others optimization methods for training neural networks and concluded that the CSS results are robust. Sheikholeslami et al. [55] utilized the CSS for optimization of the water distribution networks. They used benchmark instances, Hanoi network, double Hanoi network, and New York City tunnel problem to assess the performance of CSS. They compared the results of the CSS with GA, HS, PSO, and ACO. They concluded that the computational cost is reduced in CSS and this algorithm is able to specify the optimal outcomes with a minimum number of system analyses. Faraji et al. [56] presented a robust technique for modeling and decision making with respect to the uncertainties involved in climate changes. They utilized the non-probabilistic regret approach and selected the CSS algorithm as the optimization algorithm. Arzani et al. [57] presented an adaptive node rearrangement in the basis of the estimated error in various domains by discrete least-squares meshless (DLSM) method for dealing with some problems in the fracture mechanics. They utilized CSS as a tool for repositioning process or adaptive rearrangement. They evaluated the Efficiency and effectiveness of the adaptive rearrangement method by some two-dimensional benchmark crack examples with feasible analytical solution around crack tips. Alami et al. [58] combined the CSS algorithm and a semi-distributed hydrological model to obtain cost-effective combination of different best management practices. For evaluating the effectiveness and applicability of the coupled model, they conducted a study for watershed upstream of the Sofichai in Alavian Reservoir from the northwestern part of Iran to match four reduction levels of nitrate nitrogen, sediment, and phosphate phosphorous loads at the watershed outlet. They compared the performance of the proposed method by other metaheuristics such as PSO, GA, and CBO. Huang et al. [59] proposed an innovative technique to provide a solution to the capacity allocation problem in dealing with standalone micro-grid system. They combined the Monte Carlo simulation and CSS algorithm, and they compared the evaluation of the new method by other nature-inspired approaches such as PSO.

14.4 Modifications of CSS

Several modifications for the standard version of the CSS have been proposed with the purpose of improving the performance and potential of the algorithms in dealing with different optimization or optimum design problems. Some of the important modified approaches alongside the applications of them are presented here.

Kaveh and Talatahari [60] performed the first development in the standard version of the CSS called "enhanced CSS" (ECSS) with emphasis on the conception of fields of forces. In the standard CSS, the calculations of the forces, updating the new locations of agents, and updating the CM are all achieved after dispersing all CPs to the new positions. Therefore, this action corresponds to discrete time methodology. On the contrary, in the enhanced CSS, continuous time changing and all of the updating processes including fitness values are conducted after creating each solution (CP). In Fig. 14.6, the pseudo-code of the ECSS algorithm is provided.

Kaveh and Zolghadr [61] utilized this development in shape and also size optimization of multiple truss structures with some frequency constraints. Talatahari et al. [62] studied the optimum design of industrial tunnel sections with this development. Talatahari and Sheikholeslami [63] utilized this development in optimum design of reinforced and gravity retaining walls.

Talatahari et al. [64] combined the standard CSS with chaos to solve mathematical global optimization problems. They introduced chaos into the CSS for the purpose of increasing its global search mobility for an enhancement in global optimization. The proposed method utilizes different chaotic systems in substituting some random numbers for different parameters of the CSS. They developed nine chaos-based CSS (CCSS) methods and evaluated the performance of each one in identifying the most powerful and robust CCSS. They compared the results of these methods with the standard CSS, and results prove the suitability and superiority of the selected methods for the mathematical benchmark optimization problems. They proposed nine different approaches based on multiple chaotic maps (cm) for utilizing chaos theory in CSS formulation. The tent map, logistic map, sinusoidal iterator, circle map, gauss map, sinus map, Henon map, Ikeda map, Liebovtech map, and Zaslavskii map

Fig. 14.6 Pseudo-code of the ECSS algorithm [60]

```
Initialize random array for CPs
Assume that the initial velocities of the CPs are zero
while iteration<max iteration
    for each CP
        Calculate the charge magnitude
        Calculate the resultant force
        Calculate the new position and velocity
        Control the limiting states of position
        Evaluate fitness value (objective function)
        Update the Best Solution so far
        Update the CM
    end
end
```

are utilized for comparing studies. The first approach denotes that the initialization of CPs should be configured chaotically by utilization of iterative chaotic maps.

In the second approach, two kinds of interactive forces including the attracting or repelling are evaluated by utilization of different chaotic maps. This approach is summarized as follows:

$$\mathrm{ar}_{ij} = \begin{cases} +1 & k_t < \mathrm{cm}_{ij} \\ -1 & k_t > \mathrm{cm}_{ij} \end{cases} \tag{14.10}$$

where cm_{ij} is the chaotic map variable.

In the third approach, the probability of approaching CPs to each other is dealt with based on the chaos maps and is formulized as follows:

$$p_{ij} = \begin{cases} 1 \ \frac{\mathrm{fit(i)-fitworst}}{\mathrm{fitbest-fitworst}} > \mathrm{cm}_{ij} \quad \vee \quad \mathrm{fit}(i) > \mathrm{fit}(j), \\ 0 \ \mathrm{else} \end{cases} \tag{14.11}$$

In the fourth one, the second and third approaches are combined while a kind of the forces is determined by using Eq. (14.10), and p_{ij} value is adjusted by the selected chaotic map. In the fifth one, the coefficient of the force term in Eq. (14.8) is altered by the considered chaotic maps and update equation for position is changed by the following:

$$X_{j,\mathrm{new}} = \mathrm{cm}_{j1} \cdot F_j + \mathrm{rand}_{j2} \cdot k_v \cdot V_{j,\mathrm{old}} + X_{j,\mathrm{old}}, \tag{14.12}$$

where cm_{j1} is a variable for chaos based on the selected map.

In the sixth one, the coefficient of the velocity term in Eq. (14.8) is updated as follows:

$$X_{j,\mathrm{new}} = \mathrm{rand}_{j1} \cdot k_a \cdot F_j + \mathrm{cm}_{j2} \cdot V_{j,\mathrm{old}} + X_{j,\mathrm{old}} \tag{14.13}$$

where cm_{j2} is a variable for chaos based on the selected map.

In the seventh approach, the coefficients of the both force and velocity terms in Eq. (14.8) are changed as follows:

$$X_{j,\mathrm{new}} = \mathrm{cm}_{j1} \cdot F_j + \mathrm{cm}_{j2} \cdot V_{j,\mathrm{old}} + X_{j,\mathrm{old}}, \tag{14.14}$$

In the eighth approach, the fourth and seventh ones are combined while in the ninth one, the first and eighth approaches are combined.

Talatahari et al. [65] presented the complete methodology of the CCSS and its application in engineering design optimization. Nouhi et al. [66] utilized this development for solving practical optimization problems. Kumar and Sahoo [67] utilized this development for data clustering and Nouhi et al. [68] discussed constraint optimization problems with this development.

Talatahari et al. [40] combined the standard CSS with sub-optimization mechanism (SOM) to enhance the total effectiveness and adaptability of this algorithm and mentioned the new method as adaptive CSS (ACSS).

Kaveh and Zolghadr [69] presented the improved CSS (ICSS) and utilized it in order to deal with the inverse problem of damage assessment in beams and frames. The improvements allow the standard CSS to change the number of variables (number of damaged members) dynamically as the optimization procedure proceeds. In the standard CSS, the number of variables is kept unchanged during the optimization procedure while in structural damage identification problems, the number and location of the damaged problems are not previously known. However, it is demonstrated that the total number of damaged elements is usually far smaller than the total number of considered structural members. In order to address the abovementioned problem, they modified the standard CSS to provide a solution to the optimum design problems while the number of variables is not previously known. In other words, the number of variables (the number of damaged members) is treated as a variable and the algorithm tries to optimize its value. A brief summary on the formulation of the ICSS is presented as follows.

In ICSS, the number of design variables (or specially the total number of damaged members) is treated as a variable and the algorithm tries to optimize its value. Each solution candidate is a $(2n + 1)$-dimensional vector, n being the largest possible number of damaged elements. The first variable determines the number of damaged elements. The value of this variable is shown by n_d which may be different for different CPs. The next n_d entries in the solution vector represent the indices of the damaged members. This part of the solution vector is called the index part. And finally, the next n_d entries represent the percentage of damage in the damaged members. This part of the solution vector is called the percentage part. The rest of the variables $(2n + 1, (2\text{nd} + 1))$ will be filled by zeros to keep the dimensions of all the solution vectors equal. In the next iterations, once the values of the first variable of all CPs (n_d) are established, all the redundant variables in the percentage part (n, nd) should be set to zero. Occurrence of these redundant variables is inevitable due to the movements of the charged particles within the search space. In fact, nothing guarantees the number of nonzero elements in the percentage part of a solution vector to be equal to n_d.

Saberi and Kaveh [70] utilized this development for structural damage identification. Kaveh and Zolghadr [71] discussed the damage assessment of truss structures with changes in mode shapes and natural frequencies using the ICSS. Kaveh and Zolghadr [72] also utilized this development to deal with the problem of determining as many global optimum solutions as possible in a single run for damage detection in truss structures.

Kaveh et al. [73] proposed an improved edition of the CSS algorithm that is termed magnetic CSS (MCSS). In this method, magnetic forces as well as electrical forces are concerned, using the Biot–Savart law. Kaveh and Zolghadr [74] utilized this modification method in structural design optimization. Kaveh et al. [75] studied the construction site layout planning optimization by MCSS. Talatahari and Kaveh [76] utilized the MCSS for optimal design of multiple truss structures with large scales. In recent years, some improvements in the MCSS method (called IMCSS) are conducted to enhance the performance of harmony search scheme in the algorithm and achieve more optimal results. In this approaches, the most important parameters in convergence aspects of the standard CSS algorithm are changed. Kaveh et al. [77, 78] utilized IMCSS in optimum design configuration of different single-layer and double-layer frames in barrel vault. Kaveh et al. [79] discussed the optimum design of different truss structures with discrete and continuous variables with IMCSS. Kaveh [80] presented the shape and size simultaneous optimization of single-layer barrel vaults by the utilization of the IMCSS.

Lenin [81] presented the adaptive CSS algorithm (ACA) as a multi-agent technique in which search agents have effects on each other by consideration of separation distances and fitness values. They have tested the new method in some standard bus systems and real power loss systems.

Moez et al. [82] proposed scaled CSS (SCSS) with the purpose of solving the infeasible moves of CPs. The proposed method contains a mapping function and if any of the entries of a design variable exceeds the limits, the whole vector will be adjusted by scaling and fitting the entries. They utilized the new development in two benchmark structural optimization problem for evaluating the performance of this method.

Zakian and Kaveh [83] proposed an adaptive CSS algorithm (ACSS) for providing solution to the well-known economic dispatch optimum problems. In order to present the new development, they utilized two effective strategies. The first strategy is an improved initialization based on opposite based learning and sub-spacing techniques. The second strategy is for updating process of the algorithm and is based on the Levy flight random walk for enriching. They concluded that the ACSS has better behavior regarding optimization problems such as finding optimized fuel costs compared to the CSS and other optimization algorithms.

By comparison with the CSS that utilizes one search space for initialization, the proposed ACSS utilizes a fourfold initialization space instead of onefold. In this method, three new spaces are joined to the standard CSS. In the first initialization process, the ACSS is initialized like the CSS. In the second one, the concept of the opposition-based learning (OBL) is utilized to initialize. The third and the fourth spaces are the initializations from lower bound and upper bound subdomains. Accordingly, these four spaces are presented as initialization part of the ACSS as follows:

$$X_{i,j}^{\text{initial 1}} = X_{i,\min} + \text{rand} \cdot \left(X_{i,\max} - X_{i,\min}\right), \quad i = 1, 2, \ldots, nv \tag{14.15}$$

$$X_{i,j}^{\text{initial 2}} = X_{i,\max} + X_{i,\min} - X_{i,j}^{\text{initial 1}}, \quad i = 1, 2, \ldots, nv \tag{14.16}$$

$$X_{i,j}^{\text{initial3}} = X_{i,\min} + \text{rand} \cdot \left(\frac{X_{i,\max} + X_{i,\min}}{2} - X_{i,\min}\right), \quad i = 1, 2, \ldots, nv \tag{14.17}$$

$$X_{i,j}^{\text{initial4}} = \frac{X_{i,\max} + X_{i,\min}}{2} + \text{rand} \cdot \left(X_{i,\max} - \frac{X_{i,\max} + X_{i,\min}}{2} \right),$$
$$i = 1, 2, \ldots, nv \tag{14.18}$$

$X_{i,\max}$ and $X_{i,\min}$ are upper bounds and lower bounds of ith design variables, respectively, while nv is the number of variables. The flowchart of the ACSS is shown in Fig. 14.7.

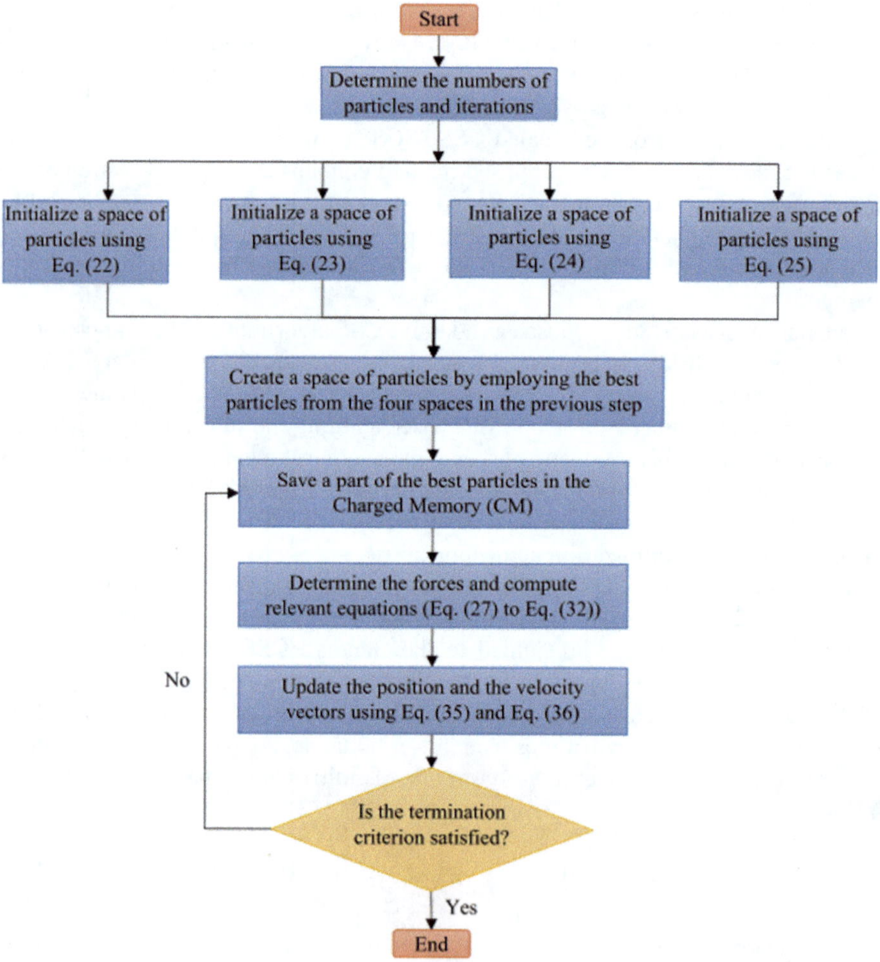

Fig. 14.7 Flowchart of the ACSS [83]

14.5 Hybridizations of CSS

Kaveh and Talatahari [84] proposed a new hybrid method by inserting some positive characteristics of the PSO algorithm to the CSS in order to deal with constrained engineering optimization problems. Kaveh et al. [85] utilized the PSO-CSS hybrid method in optimal design of laminated composite structures. Talatahari et al. [86] discussed the optimization of the hydrologic models for instance linear Muskingum model including a minimum number (one or two) of model parameters with PSO-CSS hybrid approach. Kaveh and Nasrollahi [87] utilized PSO-CSS hybrid approach in optimization of some benchmark examples which are optimized by many other methods and are suitable for comparison. They proved that the new method has better performance and higher convergence rate for the problem studied. Talatahari and Jahani [88] discussed the optimum design of single-layer barrel vault structures with PSO-CSS approach.

In the PSO-CSS hybrid approach, the superiority of the PSO including the utilization of global best optimum solution and the local best solutions is implemented to the standard CSS algorithm. In this approach, the updating process of the CM is defined as follows:

$$\mathrm{CM}_{i,\mathrm{new}} = \begin{cases} \mathrm{CM}_{i,\mathrm{old}} & \mathrm{fit}\left(X_{i,\mathrm{new}}\right) \geq \mathrm{fit}\left(\mathrm{CM}_{i,\mathrm{old}}\right) \\ X_{i,\mathrm{new}} & \mathrm{fit}\left(X_{i,\mathrm{new}}\right) \leq \mathrm{fit}\left(\mathrm{CM}_{i,\mathrm{old}}\right) \end{cases} \tag{14.19}$$

Considering the new CM mentioned before, the generated electric forces by agents are modified as follows:

$$\begin{aligned} F_j = & \sum_{i \in S_1} \left(\frac{q_i}{a^3} r_{ij} \cdot i_1 + \frac{q_i}{r_{ij}^2} \cdot i_2 \right) a r_{ij} p_{ij} \left(CM_{i,\mathrm{old}} - X_j \right) \\ & + \sum_{i \in S_2} \left(\frac{q_i}{a^3} r_{ij} \cdot i_1 + \frac{q_i}{r_{ij}^2} \cdot i_2 \right) a r_{ij} p_{ij} \left(X_i - X_j \right), \end{aligned} \tag{14.20}$$

$$S_1 = \{ t_1, t_2, \ldots, t_n | q(t) > q(j), j = 1, 2, \ldots, N, j \neq i, g \} \tag{14.21}$$

$$S_2 = S - S_1 \tag{14.22}$$

where the g as a subtitle presents the number of so far stored elite position among all of the CPs.

Talatahari et al. [89] developed a new hybrid method regarding the coalition of the MCSS algorithm, Big Bang–Big Crunch (BBBC) algorithm, and artificial neural network methods called quick hybrid CSS (QHCSS). They utilized the hybrid method in finding the optimal shapes of double curvature arch dams with emphasis fluid structure interaction excited by earthquake loads. The new method is implemented for optimization of arch dams as follows:

1. Initialize a number of random designs of arch dams based on their geometric parameters and consider them as initial CPs.
2. Calculate principal stresses of each design (CP) using exact finite element analysis.
3. Calculate the maximum value of Willam–Warnke failure criterion among a time history results (called here MFC) for each dam.
4. Consider the geometric parameters of generated dams as the inputs and their corresponding MFC as the targets of backpropagation neural network (BPNN).
5. Train BPNN.
6. Define the CM.
7. Determine both the electric and magnetic forces on CPs.
8. Perform solution construction.
9. Do position updating process.
10. Employ the trained BPNN for predicting the MFC of new CPs.
11. Evaluate the objective and constraint function by BPNN method.
12. Update CM.
13. Control termination criterion, if satisfied, stop otherwise go to the step (7).

Aryan and Alizadeh [90] proposed a new hybrid method called Bayesian CSS (BCSS) for engineering optimization applications. The normal search process of CSS has been utilized in order to include an additional part dedicated to investigating the most important regions of the global searching space regarding to the existing knowledge around the searching landscapes. They have utilized the concept of decreasing the influential particles by utilizing an easy selection mechanism. They proved that the proposed method improved the exploitation rate of the standard CSS. Prasad and Kumar [91] proposed a new hybrid method based on the state estimation utilized in radial distribution networks implemented in the fuzzy logic-based frameworks and the CSS algorithm. Wu et al. [92] presented a novel hybrid method that incorporates modified priority list (MPL) with CSS, called MPL-CSS. They utilized the new hybrid method in solving the operational optimum design problems including the crucial power systems, named as unit commitment (UC) scheduling. They proved that the cos and execution time have been optimized by the hybrid method successfully. Talatahari et al. [93] presented a hybrid metaheuristic by utilization of the CSS and BBBC methods. They applied the new method in optimization of the failure cost optimization of different arch dams.

14.6 Multi-Objective CSS Approaches

In most of the research papers, the CSS has been utilized for the single objective optimizing problems while some multi-objective configurations of this algorithm are developed. Kaveh and Massoudi [94] presented a new optimization method for the solution of multi-objective optimization problems with the CSS algorithm.

They utilized the new method for optimum design configuration of different structures. They also evaluated the efficiency of the proposed method by solving several numerical examples. They compared the optimized results with other multi-objective approaches. Chu and Tsai [95] developed the non-dominated sorting CSS (NDSCSS) for solving the multi-objective optimum design of distribution systems. Sharbatdar et al. [96] discussed the concept of conditional network to find a proper way to illustrate these kinds of networks and their related time–cost configuration. Talatahari et al. [97] developed a special technique in which Pareto dominance is integrated into the CSS algorithm in order to allow this algorithm to deal with problems with some multi-objective functions. They implemented the proposed method on some test problems and proved that the new method outperforms the other algorithms in terms of maximum spread, generational distance, spacing, coverage of two set and hyper-volume Indicator.

14.7 Conclusion

In this book chapter, an extensive review on CSS as one of the recently proposed metaheuristic algorithms is provided based on the published papers on different journals. The applications of this algorithm alongside the modified and hybrid approaches based on the standard CSS in different optimization fields are discussed properly. Based on the number of published papers regarding the CSS and its developed and hybridized approaches as optimization algorithms denotes that the utilization rate of this algorithm in different fields has been increased despite the fact that many different optimization algorithms have been proposed in recent decade. According to the reviewed articles, it can be noted that there is still much more to do in upgrading the performance of CSS and its modified or hybridized versions. Additional studies based on the adaptation of this algorithm to other domains, self-adaptation of algorithm parameters, and some theoretical studies are required to be investigated in the nearest future. It can be seen from the reviewed papers that there are still many interesting research directions ahead that can be achieved by utilization of the CSS algorithm. In addition, the applicability of CSS and its improved versions can be evaluated by considering some real-size complex problems including optimum design of building structures which can be investigated in the future.

References

1. Glover F (1986) Future paths for integer programming and links to artificial intelligence. Comput Oper Res 13(5):533–549
2. Sörensen K, Sevaux M, Glover F (2018) A history of metaheuristics. Handb Heuristics 1–8
3. Holland JH (1992) Adaptation in natural and artificial systems: an introductory analysis with applications to biology, control, and artificial intelligence. MIT Press

4. Beyer HG, Schwefel HP (2002) Evolution strategies—a comprehensive introduction. Nat Comput 1(1):3–52
5. Storn R, Price K (1997) Differential evolution—a simple and efficient heuristic for global optimization over continuous spaces. J Global Optim 11(4):341–359
6. Dorigo M, Maniezzo V, Colorni A (1996) Ant system: optimization by a colony of cooperating agents. IEEE Trans Syst Man Cybern B Cybern 26(1):29–41
7. Eberhart R, Kennedy J (1995) A new optimizer using particle swarm theory. In: MHS'95. Proceedings of the sixth international symposium on micro machine and human science. IEEE, pp 39–43
8. Yang XS (2010) Nature-inspired metaheuristic algorithms. Luniver Press
9. Kirkpatrick S, Gelatt CD, Vecchi MP (1983) Optimization by simulated annealing. science. 220(4598):671–680
10. Rashedi E, Nezamabadi-Pour H, Saryazdi S (2009) GSA: a gravitational search algorithm. Inf Sci 179(13):2232–2248
11. Geem ZW, Kim JH, Loganathan GV (2001) A new heuristic optimization algorithm: harmony search. Simulation 76(2):60–68
12. Atashpaz-Gargari E, Lucas C (2007) Imperialist competitive algorithm: an algorithm for optimization inspired by imperialistic competition. In: 2007 IEEE congress on evolutionary computation. IEEE, pp 4661–4667
13. Yang XS (2010) A new metaheuristic bat-inspired algorithm. In: Nature inspired cooperative strategies for optimization (NICSO 2010) 2010. Springer, Berlin, Heidelberg, pp 65–74
14. Yang XS, Deb S (2009) Cuckoo search via Lévy flights. In: 2009 World Congress on Nature & Biologically Inspired Computing (NaBIC). IEEE, pp 210–214
15. Talatahari S, Azizi M (2020) Chaos Game Optimization: a novel metaheuristic algorithm. Artif Intell Rev 22:1–88. https://doi.org/10.1007/s10462-020-09867-w
16. Talatahari S, Azizi M (2020) Optimization of Constrained Mathematical and Engineering Design Problems Using Chaos Game Optimization. Comput Ind Eng 20:106560. https://doi.org/10.1016/j.cie.2020.106560
17. Talatahari S, Azizi M (2020) Optimal design of real-size building structures using quantum-behaved developed swarm optimizer. Struct Des Tall Spec Build 15:e1747. https://doi.org/10.1002/tal.1747
18. Talatahari S, Motamedi P, Farahmand Azar B, Azizi M (2019) Tribe-charged system search for parameter configuration of non-linear systems with large search domains. Eng Optim. https://doi.org/10.1080/0305215X.2019.1696786
19. Azizi M, Ejlali RG, Ghasemi SA, Talatahari S (2019) Upgraded Whale Optimization Algorithm for fuzzy logic based vibration control of nonlinear steel structure. Eng Struct 1(192):53–70. https://doi.org/10.1016/j.engstruct.2019.05.007
20. Azizi M, Ghasemi SA, Ejlali RG, Talatahari S (2020) Optimum design of fuzzy controller using hybrid ant lion optimizer and Jaya algorithm. Artif Intell Rev 53(3):1553–1584. https://doi.org/10.1007/s10462-019-09713-8
21. Azizi M, Ghasemi SA, Ejlali RG, Talatahari S (2019) Optimal tuning of fuzzy parameters for structural motion control using multiverse optimizer. Struct Des Tall Spec Build 28(13). https://doi.org/10.1002/tal.1652
22. Talatahari S, Azizi M (2020) Optimum design of building structures using tribe-interior search algorithm. Structures 28:1616–1633
23. Azizi M, Ghasemi SA, Ejlali RG, Talatahari S (2021) Optimization of fuzzy controller for nonlinear buildings with improved charged system search. Struct Eng Mech 76(6):781
24. Kaveh A, Talatahari S (2010) A novel heuristic optimization method: charged system search. Acta Mech 213(3–4):267–289
25. Kaveh A, Talatahari S (2010) Optimal design of skeletal structures via the charged system search algorithm. Struct Multidisciplinary Optim. 41(6):893–911
26. Kaveh A, Talatahari S (2010) Charged system search for optimum grillage system design using the LRFD-AISC code. J Constr Steel Res 66(6):767–771

27. Kaveh A, Talatahari S (2011) Geometry and topology optimization of geodesic domes using charged system search. Struct Multidisciplinary Optimization 43(2):215–229
28. Kaveh A, Talatahari S (2012) Charged system search for optimal design of frame structures. Appl Soft Comput 12(1):382–393
29. Kaveh A, Talatahari S, Azar BF (2012) Optimum design of composite open channels using charged system search algorithm. Iranian J Sci Technol Trans Civl Eng 36(C1):67
30. Kaveh A, Behnam AF (2013) Charged system search algorithm for the optimum cost design of reinforced concrete cantilever retaining walls. Arab J Sci Eng 38(3):563–570
31. Kaveh A, Behnam AF (2013) Design optimization of reinforced concrete 3D structures considering frequency constraints via a charged system search. Sci Iranica 20(3):387–396
32. Kaveh A, Nasrollahi A (2014) Performance-based seismic design of steel frames utilizing charged system search optimization. Appl Soft Comput 22:213–221
33. Kaveh A, Shokohi F (2014) Cost optimization of castellated beams using charged system search algorithm. Iran J Sci Technol Trans Civ Eng 38(C1+):235
34. Kooshkbaghi M, Zarfam P, Fereydoni O (2015) Optimal geometry design of single layer lamella domes using charged system search algorithm. J Appl Environ Biol Sci 5(12S):846–854
35. Tavakkoli SM, Mashmoul SS, Khadem Hosseini O (2016) Application of isogeometric analysis and charged system search algorithm in structural shape optimization. Modares Mech Eng 16(5):251–260
36. Sharafi P, Askarian M, Uz ME, Abaci H (2017) Form finding for rectilinear orthogonal buildings through charged system search algorithm. Int J Optim Civ Eng 7(1):129–142
37. Sharafi P, Mortazavi M, Askarian M, Uz ME, Zhang C, Zhang J (2017) Thin walled steel sections'free shape optimization using charged system search algorithm. Iran Univ Sci Technol 7(4):515–526
38. Uz ME, Sharafi P, Askarian M, Fu W, Zhang C (2018) Automated layout design of multi-span reinforced concrete beams using charged system search algorithm. Eng Comput 35(3):1402–1413
39. Precup RE, David RC, Petriu EM, Preitl S, Radac MB (2012) Charged system search algorithms for optimal tuning of PI controllers. IFAC Proc 45(4):115–120
40. Talatahari S, Kaveh A, Rahbari NM (2012) Parameter identification of Bouc-Wen model for MR fluid dampers using adaptive charged system search optimization. J Mech Sci Technol 26(8):2523–2534
41. Precup RE, David RC, Petriu EM, Preitl S, Rădac MB (2014) Novel adaptive charged system search algorithm for optimal tuning of fuzzy controllers. Expert Syst Appl 41(4):1168–1175
42. Kaveh A, Mohammadi S, Hosseini OK, Keyhani A, Kalatjari VR. Optimum parameters of tuned mass dampers for seismic applications using charged system search. Iranian Journal of Science and Technology. Transactions of Civil Engineering. 2015; 39(C1):21
43. Shahrouzi M, Nouri G, Salehi N (2017) Optimal seismic control of steel bridges by single and multiple tuned mass dampers using charged system search. Int J Civ Eng 15(2):309–318
44. Kaveh A, Maniat M (2014) Damage detection in skeletal structures based on charged system search optimization using incomplete modal data. Int J Civ Eng IUST 12(2):291–298
45. Tabrizian Z, Ghodrati Amiri G, Hossein Ali Beigy M (2014) Charged system search algorithm utilized for structural damage detection. Shock Vib. http://dx.doi.org/10.1155/2014/194753
46. Saberi M, Kaveh A (2015) Damage detection of space structures using charged system search algorithm and residual force method. Iran J Sci Technol Trans Civ Eng 39:215–229
47. Safari H, Gholizad A (2018) Application of finite element model updating for damage assessment of space structure using charged system search algorithm. Iran Univ Sci Technol 8(2):275–293
48. Precup RE, Petriu EM, Fedorovici LO, Radac MB, Dragan F (2014) Multi-robot charged system search-based optimal path planning in static environments. In: 2014 IEEE international symposium on intelligent control (ISIC). IEEE, pp 1912–1917
49. Kanagaraj G, Ponnambalam SG, Loo CK (2015) Charged system search algorithm for robotic drill path optimization. In: 2015 international conference on advanced mechatronic systems (ICAMechS). IEEE, pp 125–130

50. Özyön S, Temurtaş H, Durmuş B, Kuvat G (2012) Charged system search algorithm for emission constrained economic power dispatch problem. Energy 46(1):420–430
51. Lenin K, Reddy BR, Kalavathi MS (2014) A new charged system search for solving optimal reactive power dispatch problem. Int J Comput 14(1):22–40
52. Wu YK, Su PE, Hong JS (2015) Stratification-based wind power forecasting in a high penetration wind power system using a hybrid model with charged system search algorithm. In: 2015 IEEE industry applications society annual meeting. IEEE, pp 1–9
53. Zhang YM, Chen CR, Lee CY (2018) Solution of the optimal reactive power dispatch for power systems by using novel charged system search algorithm. In: 2018 7th international symposium on next generation electronics (ISNE). IEEE, pp 1–4
54. Pereira LA, Afonso LC, Papa JP, Vale ZA, Ramos CC, Gastaldello DS, Souza AN (2013) Multilayer perceptron neural networks training through charged system search and its application for non-technical losses detection. In: 2013 IEEE PES conference on innovative smart grid technologies (ISGT Latin America) 2013. IEEE, pp 1–6
55. Sheikholeslami R, Kaveh A, Tahershamsi A, Talatahari S (2014) Application of charged system search algorithm to water distribution networks optimization. Iran Univ Sci Technol 4(1):41–58
56. Faraji E, Afshar A, Rasekh A (2015) Regret-based TMDL optimization under climate change with charged system search algorithm. In: World Environmental and Water Resources Congress, pp 2449–2458
57. Arzani H, Kaveh A, Kaveh A, Taheri Taromsari M (2017) Optimum two-dimensional crack modeling in discrete least-squares meshless method by charged system search algorithm. Sci Iran 24(1):143–152
58. Alami MT, Abbasi H, Niksokhan MH, Zarghami M (2018) Charged system search for optimum design of cost-effective structural best management practices for improving water quality. Iran Univ Sci Technol 8(2):295–309
59. Huang CM, Huang YC, Huang KY (2018) Capacity optimization of a stand-alone microgrid system using charged system search algorithm. In: 2018 IEEE international conference on industrial electronics for sustainable energy systems (IESES). IEEE, pp 345–350
60. Kaveh A, Talatahari S (2011) An enhanced charged system search for configuration optimization using the concept of fields of forces. Struct Multidisciplinary Optim 43(3):339–351
61. Kaveh A, Zolghadr A (2011) Shape and size optimization of truss structures with frequency constraints using enhanced charged system search algorithm. Asian J Civ Eng (Build Hous) 12(4):487–509
62. Talatahari S, Veladi H, Nouhi B (2014) Enhanced charged system search for optimum design of industrial tunnel sections. Int J Optim Civil Eng 4(3):309–319
63. Talatahari S, Sheikholeslami R (2014) Optimum design of gravity and reinforced retaining walls using enhanced charged system search algorithm. KSCE J Civ Eng 18(5):1464–1469
64. Talatahari S, Kaveh A, Sheikholeslami R (2011) An efficient charged system search using chaos for global optimization problems. Int J Optim Civil Eng 1(2):305–325
65. Talatahari S, Kaveh A, Sheikholeslami R (2012) Engineering design optimization using chaotic enhanced charged system search algorithms. Acta Mech 223(10):2269–2285
66. Nouhi B, Talatahari S, Kheiri H (2013) Chaos embedded charged system search for practical optimization problems. Int J Optimiz Civ Eng 3(1):23–36
67. Kumar Y, Sahoo G (2014) A chaotic charged system search approach for data clustering. Informatica 38(3)
68. Nouhi B, Talatahari S, Kheiri H, Cattani C Chaotic charged system search with a feasible-based method for constraint optimization problems. Math Prob Eng. http://dx.doi.org/10.1155/2013/391765
69. Kaveh A, Zolghadr A (2012) An improved charged system search for structural damage identification in beams and frames using changes in natural frequencies. Int J Optim Civil Eng 2(3):321–339
70. Saberia M, Kavehb A (2014) Structural damage identification using enhanced charged system search algorithm. Sci Iranica 21(6):1793–1802

71. Kaveh A, Zolghadr A (2015) An improved CSS for damage detection of truss structures using changes in natural frequencies and mode shapes. Adv Eng Softw 80:93–100
72. Kaveh A, Zolghadr A (2012) Damage detection of truss structures using an improved charged system search algorithm. In: Proceedings of the eleventh international conference on computational structures technology, Civil-Comp Press, Stirlingshire, UK, Paper, vol 82
73. Kaveh A, Share MA, Moslehi M (2013) Magnetic charged system search: a new meta-heuristic algorithm for optimization. Acta Mech 224(1):85–107
74. Kaveh A, Zolghadr A (2014) Magnetic charged system search for structural optimization. Periodica Polytech Civ Eng 58(3):203–216
75. Kaveh A, Khanzadi M, Moghaddam MR, Rezazadeh M (2018) Charged system search and magnetic charged system search algorithms for construction site layout planning optimization. Periodica Polytech Civ Eng 62(4):841–850
76. Talatahari S, Kaveh A (2015) Magnetic charged system search algorithm for optimum design of large-scale truss structures. Iran J Struct Eng 1(1)
77. Kaveh A, Mirzaei B, Jafarvand A (2013) Optimal design of single-layer barrel vault frames using improved magnetic charged system search. Int J Optim Civil Eng 3(4):575–600
78. Kaveh A, Mirzaei B, Jafarvand A (2014) Optimal design of double layer barrel vaults using improved magnetic charged system search. Asian J Civ Eng (Build Hous) 15(1):135–154
79. Kaveh A, Mirzaei B, Jafarvand A (2015) An improved magnetic charged system search for optimization of truss structures with continuous and discrete variables. Appl Soft Comput 28:400–410
80. Kaveh A (2017) Simultaneous shape–size optimization of single-layer barrel vaults using an improved magnetic charged system search algorithm. In: Applications of metaheuristic optimization algorithms in civil engineering. Springer, Cham, pp 117–146
81. Lenin K (2017) Reduction of active power loss by adaptive charged system search algorithm. Int J Res Granthaalayah 5(10):35–45
82. Moez H, Taghizadieh N, Kaveh A (2017) Scaled charged system search algorithm for optimum design of steel frames. J Civ Environ Eng 47(1):81–90
83. Zakian P, Kaveh A (2018) Economic dispatch of power systems using an adaptive charged system search algorithm. Appl Soft Comput 73:607–622
84. Kaveh A, Talatahari S (2011) Hybrid charged system search and particle swarm optimization for engineering design problems. Eng Comput 28(4):423–440
85. Kaveh A, Hashemi SB, Sheikholeslami R (2013) Optimal design of laminated composite structures via hybrid charged system search and particle swarm optimization. Asian J Civ Eng (Build Hous) 14(4):587–604
86. Talatahari S, Sheikholeslami R, Farahmand Azar B, Daneshpajouh H (2013) Optimal parameter estimation for Muskingum model using a CSS-PSO method. Adv Mech Eng 5:480954
87. Kaveh AA, Nasrollahi A (2014) Charged system search and particle swarm optimization hybridized for optimal design of engineering structures. Sci Iranica Trans A Civ Eng 21(2):295
88. Talatahari S, Jahani Y (2015) Hybrid charged system search-particle swarm optimization for design of single-layer barrel vault structures. Asian J Civ Eng (BHRC) 16(4):515–533
89. Talatahari S, Aalami MT, Parsiavash R (2016) Optimum design of double curvature arch dams using a quick hybrid charged system search algorithm. Iran Univ Sci Technol 6(2):227–243
90. Aryan M, Alizadeh BA (2016) Bayesian charged system search: a hybrid method for multi-modal optimization problems. In: 2016 IEEE congress on evolutionary computation (CEC) 2016. IEEE, pp 1501–1508
91. Prasad S, Kumar DV (2017) Hybrid fuzzy charged system search algorithm based state estimation in distribution networks. Int J Eng Sci Technol 20(3):922–933
92. Wu YK, Huang CC, Lin CL, Chang SM (2017) A hybrid Unit commitment approach incorporating modified priority list with charged system search methods. Smart Grid Renew Energy 8(06):178
93. Talatahari S, Aalami MT, Parsiavash R (2018) Risk-based arch dam optimization using hybrid charged system search. ASCE-ASME J Risk Uncertainty Eng Syst Part A Civ Eng 4(2):04018008

94. Kaveh A, Massoudi MS (2014) Multi-objective optimization of structures using charged system search. Sci Iranica. Trans A Civ Eng 21(6):1845
95. Chu CC, Tsai MS (2016) Multiple Objectives Reconfiguration in Distribution System Using Non-Dominated Sorting Charged System Search. In: International conference on industrial, engineering and other applications of applied intelligent systems. Springer, Cham, pp 944–955
96. Sharbatdar MK, Talatahari S, Mousavi MR (2016) The optimization of the time-cost tradeoff problem in projects with conditional activities using of the multi-objective charged system search algorithm (SMOCSS). J Hist Cult Art Res 5(3):178–194
97. Ranjbar A, Talatahari S, Hakimpour F (2019) The application of multi-objective charged system search algorithm for optimization problems. Sci Iranica 26(3):1249–1265

Part II
Electrical and Electronics, Computer, and Communication Engineering

Chapter 15
Artificial Bee Colony Algorithm and Its Application to Content Filtering in Digital Communication

Bilge Kagan Dedeturk, Bahriye Akay, and Dervis Karaboga

15.1 Introduction

The nature-inspired algorithms deal with developing models that simulate various natural phenomena such as natural selection, genetic operations simulated annealing or collective behavior of some species. These models generally establish a basis for optimization algorithms used for solving hard problems due to dimensionality, multi-objectives, constraints, nonlinearity or non-separability. One of the concepts in the nature-inspired paradigms is swarm intelligence which refers to intelligent behavior patterns of collaborating creatures. Self-organization and division of labor are two fundamental characteristics of the swarm intelligence. By means of the division of labor, agents are distributed to various roles crucial for the survival of the colony, and they are able to adjust their roles depending on the environmental conditions. The self-organization means that each agent adapts its actions without a central instruction and supervision. Rather, they use local interactions among the other agents and with the environment. Positive feedback, negative feedback, multiple interactions and fluctuations reveal the self-organization. The positive feedback means that agents repeat the profitable patterns to promote colony benefits. The negative feedback is a kind of abandoning some patterns to avoid getting stuck to the same situation,

B. K. Dedeturk
Graduate School of Natural and Applied Sciences,
Erciyes University, Melikgazi, Kayseri 38039, Turkey
e-mail: kagandedeturk@gmail.com

B. Akay (✉) · D. Karaboga
Department of Computer Engineering, Erciyes University, Melikgazi, Kayseri 38039, Turkey
e-mail: bahriye@erciyes.edu.tr

D. Karaboga
e-mail: karaboga@erciyes.edu.tr

S. Carbas et al. (eds.), *Nature-Inspired Metaheuristic Algorithms for Engineering Optimization Applications*, Springer Tracts in Nature-Inspired Computing,
https://doi.org/10.1007/978-981-33-6773-9_15

and the fluctuation explores new patters to obtain diversification in the colony. As a component of self-organization, the multiple interaction is the way of building a global information based on the local information collected from the other agents about their states.

In nature, we can observe the swarm intelligence in the vital activities of some species, including ants, termites, bees, bacteria, fireflies, birds and fishes. As a research field, the swarm intelligence generally copes with creating algorithms by mimicking the collective and intelligent behavioral patterns of colonies or swarms. The researchers working in this field have proposed several novel algorithms and problem-solving techniques, including ant colony optimization [3], particle swarm optimization [7], firefly algorithm [18] and bacterial foraging algorithm [9].

A bee colony exhibits many intelligent behavior patterns, including division of labor in nest building, mating to generate an eusocial colony, nest site selection based on various coalitions of scouts by attracting others and foraging [5]. The foraging is the most life-sustaining activity in a hive performed by forager bees grouped as employed bees, onlooker bees and scout bees depending on how they search for food sources. Artificial bee colony (ABC) algorithm proposed by Karaboga [4] is a swarm intelligence-based algorithm simulating the foraging behavior of a honey bee colony. It has been successfully applied to solve various types of design problems [1, 5, 6].

This chapter is devoted to the ABC algorithm and its analogy to the foraging behavior of real honey bee colony. Once a broad explanation of its phases is provided, an application of the algorithm in the field of digital communication is presented.

15.2 Foraging in a Real Honey Bee Colony

Honeybees (Apis mellifera) are eusocial insects living as colonies that consist of one queen, from none to approximately 1000 haploid drones and about 60,000 female worker bees. There are many tasks in a honeybee colony such as nursing, feeding larvae, ventilating the hive, defending the nest and foraging. Each task is performed by a specialized group of bees classified b y the division of labor in a colony. Among these tasks, foraging is the most crucial task to the survival and reproduction of the colony. Animals have to select the optimal decision among the alternatives with different profitabilities to be the fittest individual and not to be eliminated by natural selection. Similarly, honeybees have to choose the most profitable forage sites offering the maximum energy and minimum foraging cost [12]. The honeybee colony has ability to adapt the foraging task according to changing resources in the hive, colony state and environmental conditions. To select food source sites, the forager bees act collectively or individually based on their own experience or social information, which yields a global behavior pattern in whole colony [13].

There are various types of foragers according to their tendency to leave the nest to search for new food sources [11]. Foragers that explore the environment compulsively or based on an external sign are referred to ‘scouts’. When a scout discovers a rich food source, then it is referred to ‘employed bee’. The employed bees perform repeated vis-

Fig. 15.1 Onlooker bees watching the waggle dance of an employed bee

its to the sources memorized for exploiting their nectar. When employed bees return hive to unload the nectar into honeycomb cells, they communicate to the other bees in the hive to recruit them to the flower patches they discovered by waggle dancing or other dancing types on the dance floor. The communication transmits the information such as the direction, location, distance and profitability of forage sites discovered and yields to social or collective intelligence which minimizes the energy and time consumption to find new sources and maximizes the quality nectar amount unloaded to the hive. Bees that watch the dances of employed bees and fly to the sources recruited by the employed bees are called onlooker bees (Fig. 15.1). In addition to waggle dance, the tremble dance, the stop signal can also influence the bees in making a foraging decision. The number of bees attracted and recruited to a specific site by dancing is proportional to the profitability of that site [14]. Another decision the foragers make is whether they should repeat the visit to the site in their memories (exploitation) or abandon the source to find out a new one (exploration). There is a trade-off between the exploration and exploitation behavior patterns, and they should be performed in a balanced and adaptive manner for energy minimization. When a food source is exhausted or if there is more profitable source than the source in the bee's memory, the current source is discarded and new source is attempted to be exploited.

15.3 Artificial Bee Colony Algorithm

Artificial bee colony (ABC) [4] simulates the optimal foraging theory which states that the fittest species foraging the most profitable sources to obtain the maximum energy are not eliminated by natural selection. By an analogy with the foraging food source sites (environment), a problem search space can be swept to find out global maximum/minimum decision vector optimizing a fitness function. In other words, the aim is to find the fittest solution in the search space. In the algorithm, food sources' positions are represented by solution vectors encoded by representation form suitable for the problem. Since there are three types of food source searches in the real foraging behavior conducted by scout bees, employed bees and onlooker bees, the ABC algorithm has three phases performing the search in the search space: scout bee, employed bee and onlooker bee phases. Main steps of the ABC algorithm are given in algorithm 1.

```
Data: Determine the values of the control parameters
begin
    Initialization;
    while Termination Criteria is not Satisfied do
        Employed Bee Phase;
        Onlooker Bee Phase;
        Memorize the best solution found so far;
        Scout bee phase;
    end
end
```

Algorithm 1: Main steps of ABC algorithm

As shown in Fig. 15.2 and in algorithm 1, the foraging cycle starts with the flight of scout bees to discover food sources because the colony has no information or experience at the beginning (Edge 1 in Fig. 15.2). The initialization phase performs exploration pattern. When the scout bees discover a food source and evaluate its quality, they start to work as an employed bee (Edge 2 in Fig. 15.2). An employed bee memorizes the location of the food source and performs repeated visits to the vicinity of the sources to exploit its nectar (Edge 3 and 4). In terms of the algorithmic concept, a local search is conducted around the present solution. When the employed bee

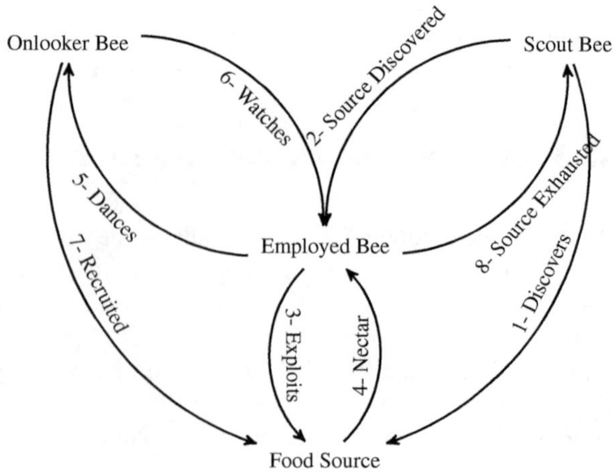

Fig. 15.2 Foraging cycle in ABC algorithm

returns to the hive to unload the nectar, she performs a dance (Edge 5) to communicate and recruit the onlooker bees waiting in the hive (Edge 6). A high-quality source is likely to be selected by an onlooker bee, and more bees are recruited to better sources (Edge 7). The way of food source selection by an onlooker bee is a stochastic selection proportional to food source quality. Once an onlooker bee has selected a food source to fly, she works like an employed bee in terms of exploiting the source. In consequence of exploitations, the nectar of the source decreases gradually and is finally exhausted. Therefore, this source is abandoned, and its employed bee converts into a scout bee to discover a new rich source instead of the exhausted one (Edge 8). Each step is described in the subsections.

15.3.1 Initialization

The first step in the initialization phase (Algorithm 2) is assigning values for the control parameters, including the number of food sources (SN), abandonment criterion (limit parameter) that is used to determine the exhausted food sources, and the maximum number of cycles (MCN) or maximum evaluation number to terminate the algorithm or another termination criterion can be used.

```
Data: Set values for the control parameters
SN: The number of food sources,
MCN: Termination criteria, the maximum number of cycles,
limit: Abandonment criteria, the maximum number of exploitations for a
solution
begin
    //Initialization;
    for i = 1 to SN do
        x_i ← a random food source location by Eq. 15.1;
        f_i = f(x_i);
        c_i = 0;
    end
end
```

Algorithm 2: Initialization in the ABC algorithm

Each food source position is represented by D-dimensional decision vector as shown in Fig. 15.3, and an initial population is generated randomly in the search space by Eq. 15.1:

$$x_i^j = x_{\min}^j + \text{rand}[0, 1](x_{\max}^j - x_{\min}^j) \tag{15.1}$$

where $i = 1, \ldots, \text{SN}, j = 1, \ldots, D$, $x_{\max}^j$ and $x_{\min}^j$ are upper and lower bounds of jth parameter, respectively.

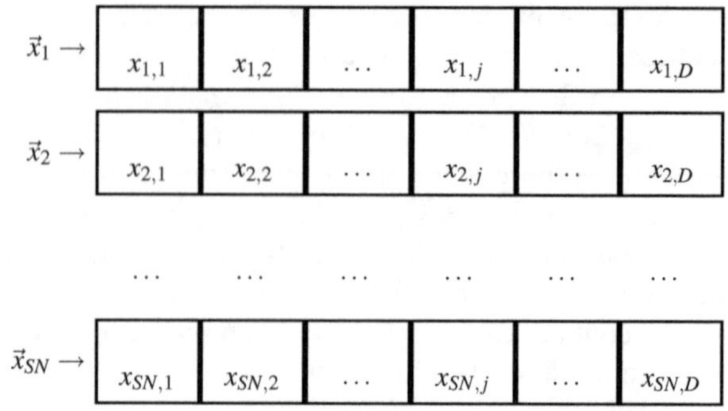

Fig. 15.3 Solution representation in the ABC algorithm

The solutions are evaluated by using the cost function to be optimized and assigned a fitness value. Each source, x_i, is assigned a counter (c_i) for keeping the number of exploitations. In the initialization phase, these counters are set to zero because new solutions have not been exploited yet.

15.3.2 Employed Bee Phase

begin
  `//Employed Bees' Phase;`
  for $i = 1$ *to SN* **do**
    $\upsilon_i \longleftarrow$ a neighbor food source location generated by Eq. 15.2;
    if $f(\upsilon_i) < f_i$ **then**
      $x_i = \upsilon_i$;
      $f_i = f(\upsilon_i)$;
      $c_i = 0$;
    else
      $c_i + +$;
    end
  end
  $p \longleftarrow$ assign probability values by Eq. (15.3);
end

Algorithm 3: The employed bee phase of the ABC algorithm

In the employed bee phase (Algorithm 3), generated food source positions are memorized by the employed bees, and they search the vicinity of these sources by using Eq. 15.2.

$$\upsilon_{ij} = x_{ij} + \phi_{ij}(x_{ij} - x_{kj}) \tag{15.2}$$

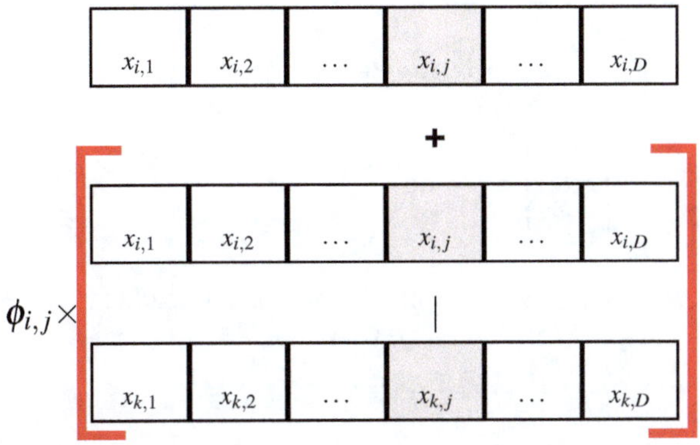

Fig. 15.4 Local search in the employed and onlooker bee phases, $j \in [1, D]$ is a randomly selected dimension and $k \in [1, \text{SN}]$ is a randomly selected neighbor, $\phi_{i,j}$ is a random number within $[-1, 1]$

where $(k \neq i) \in \{1, 2, \ldots, \text{SN}\}$ is a randomly chosen neighbor index that is different from i and $j \in \{1, 2, \ldots, D\}$ is a random dimension index. To generate a new solution by a slight modification on the current solution, basic ABC algorithm modifies the value in one dimension of the decision vector. $\phi_{ij} \in [-1, 1]$ is a random number from uniform distribution. This operation is shown in Fig. 15.4.

New solution, υ_i, is replaced with current solution, x_i, in the employed bee's memory and its counter is reset provided that υ_i is better in terms of fitness value. Otherwise, the current solution remains in the population, and its counter is incremented by 1. Employed bees' communication to recruit onlooker bees to higher quality sources is simulated by assigning probability values, p_i, to each source by Eq. 15.3. Since the frequency and duration of the dance are related to the source quality, the probability value is associated with the fitness of the source. The communication emerging as the information sharing among the bees is multiple interactions property in the ABC algorithm.

$$p_i = \frac{\text{fitness}_i}{\sum_{n=1}^{\text{SN}} \text{fitness}_n} \tag{15.3}$$

15.3.3 *Onlooker Bee Phase*

While each employed bee performs visit to the source in her memory, an onlooker bee performs stochastic selection to determine food source to exploit. In basic ABC, a roulette wheel like selection is applied to recruit onlooker bees to more profitable sources. Each solution takes turn, and its probability value is compared to a random value within [0, 1] (Algorithm 4). If the probability value of the solution is higher than the random value, an onlooker bee is recruited to this source. Better sources are more likely to be selected, which is positive feedback property in the ABC algorithm. Once a solution is selected by an onlooker bee, she searches the vicinity of the selected solution by Eq. 15.2, and the better one remains in the population. The counters are updated according to the solutions selected.

```
begin
    //Onlooker Bees' Phase;
    i = 0 ;
    t = 0 ;
    while t < SN do
        if rand(0, 1) < p_i then
            t = t + 1;
            υ_i ⟵ a neighbor food source location generated by Eq. 15.2;
            if f(υ_i) < f_i then
                x_i = υ_i ;
                f_i = f(υ_i);
                c_i = 0;
            else
                c_i ++;
            end
        end
        i = (i + 1) mod (SN − 1);
    end
end
```

Algorithm 4: The onlooker bee phase of the ABC algorithm

15.3.4 Scout Bee Phase

In the scout bee phase (Algorithm 5), exhausted sources are determined by checking the counters, c_i. If the value of any counter passes the limit value, the associated solution is considered to be exhausted. Therefore, this solution is discarded, and a new solution is generated by Eq. 15.1 by a scout bee. As it is a new solution, its counter is reset to 0. In the basic ABC algorithm, only one scout is allowed in each cycle, while in some other variants of ABC algorithm, more than one may occur. This abandonment is negative feedback property, and generating random solution to be substituted with the exhausted source is fluctuation property in the ABC algorithm.

As seen from the phases of the ABC algorithm, there is no supervision to assign tasks to the bees. But, there is self-organization (positive feedback, negative feedback, multiple interactions and fluctuations) based on the local interaction among the bees. These local interactions and self-organization lead to collective intelligence having a global behavior pattern.

```
begin
    //Scout bee phase;
    si = {i : c_i = max(c⃗)};
    if c_si > limit then
        x_si ⟵ a random food source location by Eq. 15.1;
        c_si = 0;
    end
end
```

Algorithm 5: The Scout Bee Phase of the ABC algorithm

15.4 How the ABC Algorithm Evolves Food Sources

To see the distribution of the food sources in the search space and how they evolve, we can examine the behavior of solution population for Rastrigin benchmark function given by Eq. 15.4.

$$f(\mathbf{x}) = A * D + \sum_{i=1}^{n} \left[x_i^2 - A\cos(2\pi x_i)\right] \tag{15.4}$$

where A constant is 10 and D represents the dimension. Surface plot and contour plots are given in Eq. 15.5a and 15.5b, respectively. As seen from these figures, the function is a multimodal function with many local minima. In our case, D is 2, SN is 50, MSN is 200, limit value is 20, and the search range is between $[-5.12, 5.12]$. The value of the global minimum is 0, and it is located at (0, 0).

After setting the control parameters, the food sources are randomly generated and shown in Fig. 15.6a. These initial food sources are memorized by the employed bees. The blue marks in the figure are food sources exploited by the employed. The green marks are scout bees generated for the exhausted sources. The purple mark shows the best value found so far. The red mark shows the global minima of the problem. In the 10th cycle given in Fig. 15.6a, the bees fly to the better sources close to the global minima as a result of positive effect. In the 20th cycle shown in Fig. 15.6c, we see that some bees are located on the food sources which are the local minima of the problem seen in the contour plot (Fig. 15.5b), and the diversity among the solutions is low. In this cycle, because a solution cannot be improved sufficiently and finally exhausted, one scout is generated instead of that source, which is negative feedback property of ABC. This new random solution yields fluctuation and brings diversity. The new solution is green mark in red circle. Evolution of the solutions in the next cycles is shown in Fig. 15.6d–h.

As seen from the convergence graph in Fig. 15.7, the best solution is found approximately in 10 cycles. The best fitness shows the fitness value of the best solution found so far. Average fitness value shows the mean of fitness of the solutions in the population. The iterations in which a scout occurred are marked with red marker on the

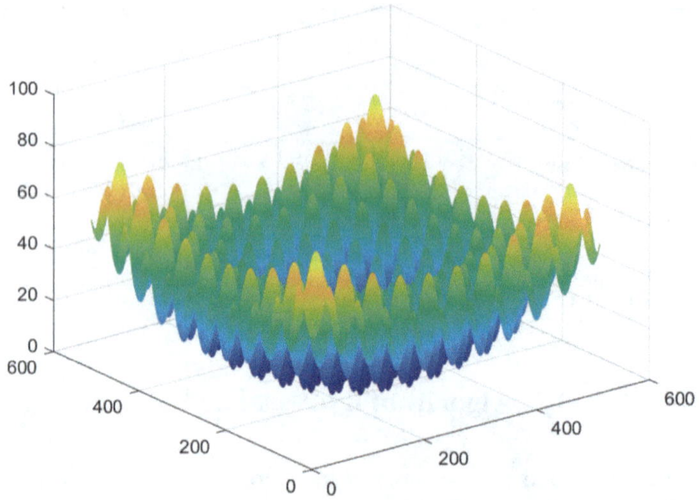

(a) Rastrigin Surface Plot

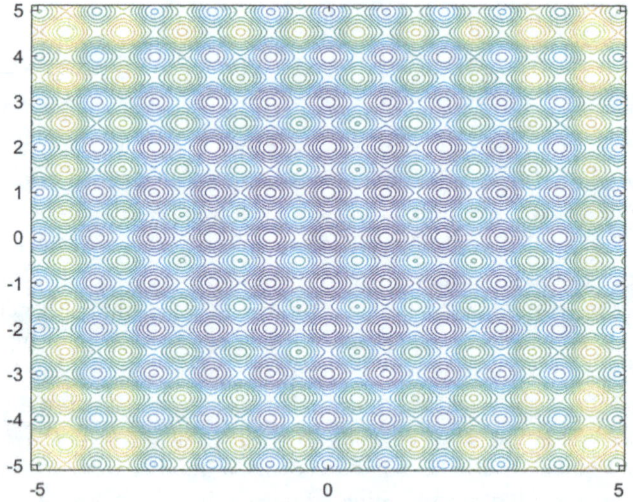

(b) Rastrigin Contour Plot

Fig. 15.5 Two-dimensional Rastrigin surface and contour plot

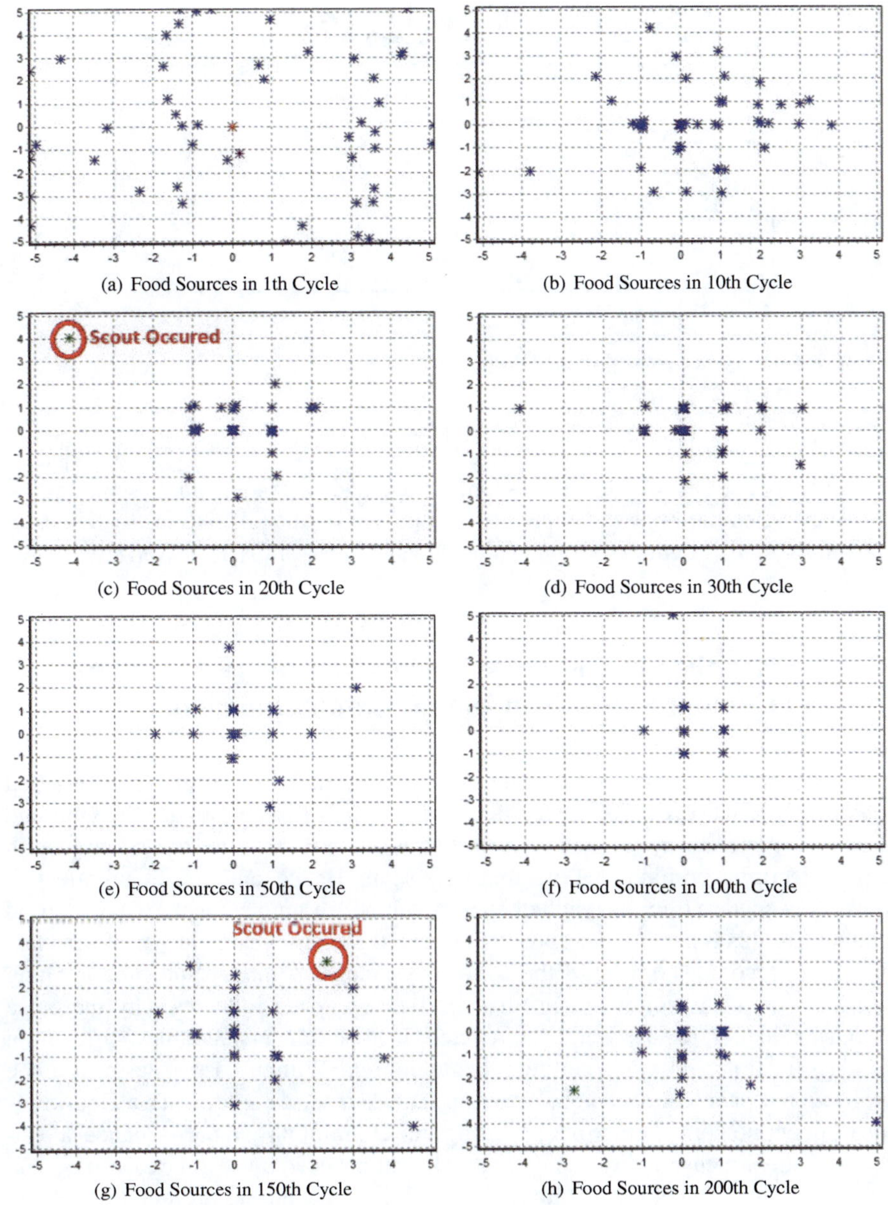

(a) Food Sources in 1th Cycle

(b) Food Sources in 10th Cycle

(c) Food Sources in 20th Cycle

(d) Food Sources in 30th Cycle

(e) Food Sources in 50th Cycle

(f) Food Sources in 100th Cycle

(g) Food Sources in 150th Cycle

(h) Food Sources in 200th Cycle

Fig. 15.6 Evolution of the food sources in some cycles

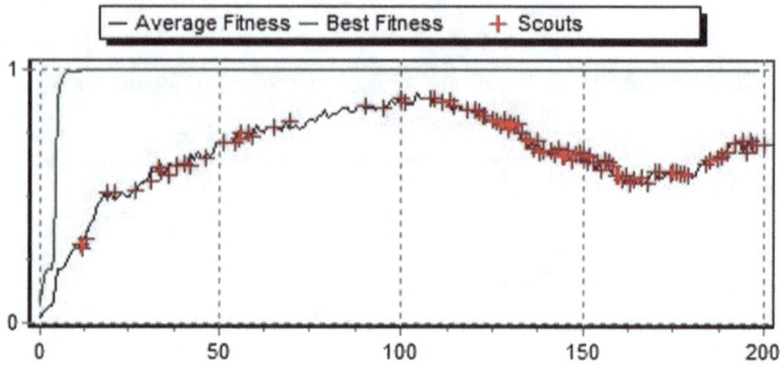

Fig. 15.7 Convergence graph produced during the search

average fitness. Because new discovered solutions do not need to be high quality and diversity increases in the population, the average fitness falls in these iterations. However, when the fitness of individuals improves by the positive feedback effect, it again rises. This demo program can be downloaded from ABC's Web site.[1]

15.5 An Application of the Artificial Bee Colony Algorithm to Content Filtering in Digital Communication

One of the challenging problems in the communication field is receiving unsolicited content in digital mediums, such as emails, which is time wasting, annoying and a security vulnerability for users. According to the reports [10, 15–17], more than half of electronic communication emails are spam. Hence, automated and effective tools are needed to filter unsolicited ones. Machine learning algorithms can be used to automate this issue. Without pre-definitions, they can dynamically perform classification process according to the content of emails. Although machine learning classifiers are easy to implement, their performances are influenced by the curse of dimensionality, feature weights, predetermined parameters, and they suffer from misclassification rates when getting stuck to minima. Without loss of easiness in the application of machine learning algorithms, nature-inspired optimization algorithms can be integrated into them to avoid getting stuck to minima and performance dependence on the dimensionality and dataset. Algorithms need powerful exploration and efficient exploitation ability for this purpose. In this chapter, we present the problem description for filtering unsolicited mails and aim to overcome the drawbacks of logistic regression classifier using ABC algorithm.

[1] http://abc.erciyes.edu.tr.

15.5.1 Problem Description

In spam filtering, first, some preprocessing techniques, such as tokenization, pruning, removing stop words, stemming and feature extraction, are applied to express the email with a convenient representation. Describing an informative representation is important because it directly affects the classification rate. The second step is feature selection to reduce the dimensionality of the feature space. In the third step, a classifier assigns an email, e_i, to spam or legitimate class, $\text{class}_j \in C = \{\text{class}_{\text{spam}}, \text{class}_{\text{legitimate}}\}$. The aim is to maximize the classification rate on nonlinearly separated, high-dimensional, balanced or imbalanced, in any language, personalized or non-personalized datasets without sensitivity to parameters of the system. A robust classifier can be defined by using a training algorithm in the training phase.

In the preprocessing step, non-informative expressions are removed from the content. First, html tags are eliminated from the content, and all the letters are converted to lowercase. Then, the strings are divided to substrings (tokenization), and stop words and punctuation (except for exclamation mark) in these substrings are eliminated. The bag of words (BOW) model is established, which includes all unique terms in the email set, $\text{BOW} = \{\text{term}_1, \text{term}_2, \ldots, \text{term}_{|B|}\}$ where $|B|$ is the number of unique terms. The frequency of each term is different in each mail. Hence, each email is expressed as a weight vector $\vec{w}_{e_i} = \{w_1, w_2, \ldots, w_{|B|}\}$ indicating the term frequency inverse document frequency (tf-idf) (Eq. 15.5) values which show how significant a word in a mail.

$$\text{tf-idf}(\text{term}_j, e_i) =< w_j, e_i >= \text{tf}(\text{term}_j, e_i) \times \log \frac{|E|}{|E_{\text{term}_j}|}, \tag{15.5}$$

where $\text{tf}(\text{term}_j, e_i)$ refers to the total number of occurrences of term_j in mail e_i, $|E|$ is the total number of mails, and $|E_{\text{term}_j}|$ indicates the total number of emails containing the term term_j. As the content of a mail includes considerable number of words, dimensionality of the vector space increases, and this makes the computation very expensive. Therefore, the most informative terms are selected to reduce the dimensionality. Feature selection methods discover a set of significant terms that constitutes the feature set ($S = \text{term}_1, \ldots, \text{term}_n$). Each email can be defined by a feature vector $\vec{f} = [f_1, f_2, \ldots, f_n]$, where f_i is the value of term i, and n is the size of feature vector. A subset of the terms is selected to be used in training of text classification. Feature selection enhances the classification success by eliminating non-discriminative terms that exist almost all messages. The reduced feature vector is passed to a supervised classifier to map the mails to one of the classes. The supervised classifiers are given a training set $T = (\vec{f}, \text{Label})$ where $\text{Label} \rightarrow \text{class}_j \in C = \{\text{class}_{\text{spam}}, \text{class}_{\text{legitimate}}\}$ hand-crafted by a human. In this study, we have chosen a supervised logistic regression (LR) classifier due to its simplicity and speed.

15.5.2 Logistic Regression

Given a training set $\{(\vec{f}_1, y_1), \ldots, (\vec{f}_m, y_m)\}$, $\vec{f}_i \in R^n$ and $y_i \in \{0, 1\}$, $1 \le i \le m$, LR assigns a class to $\vec{f}_i$ by Eq. (15.6):

$$y_i' = \begin{cases} 0, \ \mathrm{pr}_i < 0.5 \\ 1, \ \mathrm{pr}_i \ge 0.5 \end{cases} \tag{15.6}$$

where pr_i is calculated by Eq. 15.7 using sigmoid function:

$$\mathrm{pr}_i = \frac{1}{1 + e^{-\vec{w}\vec{f}_i}}. \tag{15.7}$$

A cost function (such as the cross-entropy) can be defined by Eq. 15.8.

$$J(\vec{w}) = -\sum_{i=1}^{m} y_i \log(\mathrm{pr}_i) + (1 - y_i)\log(1 - \mathrm{pr}_i) + \frac{\lambda}{2}\sum_{j=1}^{n} w_j^2, \tag{15.8}$$

where λ is a regularization parameter. The aim is to adjust the weights ($\vec{w}$) that minimize the cost function, which can be formulated by Eq. 15.9:

$$\arg \min_{w} J(\vec{w}) \tag{15.9}$$

15.5.3 ABC-Based LR Classifier

As stated in the problem statement section and Eq. 15.9, the aim is to find the weights ($\vec{w}$) that minimize the cost function. In the ABC algorithm, the weights and bias are encoded in the food source locations, and the sum squared error at the output is considered as the cost function. Instead of the neighbor solution production (Eq. 15.2) in the basic ABC algorithm, a modified operator defined by Eq. 15.10 is used to find a new solution in the vicinity of the current solution.

$$v_{ij} = \begin{cases} w_{ij} + \varphi_{ij} \times (w_{ij} - w_{kj}), & R_{ij} < \mathrm{MR} \\ w_{ij}, & \text{otherwise} \end{cases} \tag{15.10}$$

where MR $\in [0, 1]$ is a control parameter that controls the number of dimensions to be changed, and $R_{ij} \in [0, 1]$ is drawn randomly from a uniform distribution. Since more than one dimension of the current solution might change, the convergence ability of the algorithm is improved by Eq. 15.10.

A recent study [2] has shown that classification performance of the ABC-LR outperforms the performance of logistic regression, Naive Bayes, support vector machine classifiers and some state-of-the-art methods. Authors have concluded that

ABC-LR can handle highly nonlinear and high-dimensional data due to its efficient local and global search capabilities and have proved the efficiency of the ABC-LR on a wide range of spam datasets.

15.5.4 Feature Representation and Selection

One of the feature selection methods used in the experiments is a tf-idf-based feature selection approach. Once the tf-idf vectors are generated, they are normalized by the Euclidean norm (Eq. 15.11). At the length of document gets longer, a term will appear in that document more likely, which causes the term seem more significant than it is. The goal of normalization is to avoid this bias especially in lengthy documents. The total weight of each term is calculated by summing the associated normalized tf-idf weights of that term in all documents. Then, the weights are arranged in descending order and the top n terms with the highest total weight are chosen to generate the feature set. Then, each document is expressed by a normalized vector of tf-idf values.

$$v^i_{\text{norm}} = \frac{v^i}{v^i_1 + v^i_2 + \cdots + v^i_n} \quad (i = 1, \ldots, N) \tag{15.11}$$

The second feature representation method used in the experiments is mutual information that measures the amount of information in the presence or absence of a word to perform the classification correctly. If a word is seen frequently in one class but not in the other class, the mutual information value of this word is expected to be high. If a word is used frequently or rarely in both classes, the mutual knowledge value of this word is expected to be low. The following equation is used to calculate the mutual information values of words.

$$mi(T; S) = \sum_{\substack{t \in \{0,1\}, \\ s \in \{\text{spam}, \text{normal}\}}} P(T = t, S = s) \times \log \frac{P(T = t, S = s)}{P(T = t), P(S = s)} \tag{15.12}$$

where S represents the class (spam or normal) and takes the value 1 if the mail belongs to the s class, otherwise it takes the 0. T represents the term, if the mail contains the term t, it takes the value 1, and if it does not, it takes the value 0.

Terms in the feature vector created using the mutual information method are expected to represent a class well. So if a term appears frequently in spam classes, but rarely or not at all in regular classes, this term is a good representative for the spam class. Likewise, if a term appears frequently in regular classes, but is not seen in spam classes, or very little, this term is a good representative for the normal class. Terms with these characteristics are expected to have high mutual information values. The mutual information method enables the classification process to be done successfully with a much smaller number of terms compared to all terms. The better the representation ability of the terms that make up the attribute vector, the better the classifier success is obtained. In other words, the selection of the terms that make up the feature vector directly contributes to the success of the classifier.

15.5.5 Experimental Settings

In the experiments, to validate the efficiency of the algorithms, the dataset presented in [8] was used. The dataset consists of a total of 760 emails, of which 405 are spam and 355 are normal emails. In the experiments, two different feature selection techniques, which are tf-idf and mutual information representation-based selection, were employed.

The performance of the proposed ABC-based classifier was compared to those of gradient descent-based (LR), differential evolution-based (DE-LR) and particle swarm optimization-based (PSO-LR) LR classifiers. LR, ABC-LR, DE-LR and PSO-LR were trained with five different feature vector sizes $\{50, 100, 150, 200, 300\}$ and a maximum iteration number of 1000. Modification rate (MR) and limit values in ABC-LR were set to 0.1 and 100, respectively. Hyper-parameters of PSO-LR - inertia weight, acceleration coefficients c_1 and c_2 were set to 0.6, 2 and 2, respectively. Crossover probability (CR) and differential weight (F) in DE-LR were set to 0.8 and 0.5, respectively. In LR classification model, learning rate α and regularization parameter λ were set to 0.001 and 0. Hyper-parameters of the methods, achieving best classification performance for each method, were determined by using exhaustive grid search approach. Lower bound x_j^{min} and upper bound x_j^{max} for each parameter learned by ABC-LR, PSO-LR and DE-LR were set to -64 and 64. After determining the best hyper-parameters, a total of five experiments were performed for each classification model and each feature selection approach. Each experiment was repeated 20 times to prevent the randomized data effect on classification performance and tenfold cross-validation technique is used to wipe out the influence splitting scheme or the order of the data.

15.5.6 Results

Tables 15.1 and 15.2 show some statistical measures of classification accuracies, including the best, median, mean, worst and standard deviation values gathered through 20 runs of each experiment. The average elapsed training time in seconds is also reported in the tables. Results in Table 15.1 were obtained using tf-idf-based feature selection approach while results in Table 15.2 were obtained using mutual information. Experimental result shows that ABC-LR classification model has better classification performance than LR, PSO-LR and DE-LR in terms of classification accuracy. Compared to the other classification models, ABC-LR has smaller standard deviation values, especially in tf-idf-based feature selection approach. In terms of running time, LR is advantageous. Also, the empirical results demonstrate that ABC-

Table 15.1 Using tf-idf as feature selection approach, classification performances of LR, ABC-LR, DE-LR and PSO-LR with respect to best, worst, median and mean accuracies, standard deviation values and average training time in seconds

Method	FVS	Best (%)	Worst (%)	Median (%)	Mean (%)	Std	Time
LR	50	**92.67**	90.67	92.07	91.93	0.57	**0.061**
	100	93.07	91.60	92.33	92.26	0.47	**0.080**
	150	94.13	91.60	92.60	92.58	0.63	**0.122**
	200	93.87	92.13	93.33	93.16	0.48	**0.136**
	300	94.27	91.73	93.47	93.43	0.63	**0.154**
ABC-LR	50	92.40	**92.40**	**92.40**	**92.40**	**0.00**	3.76
	100	**93.60**	**93.07**	**93.20**	**93.28**	**0.13**	4.71
	150	**94.40**	**93.47**	**94.00**	**93.96**	**0.18**	5.19
	200	**95.47**	**94.80**	**95.07**	**95.07**	**0.18**	5.65
	300	**96.13**	**95.07**	**95.47**	**95.53**	**0.25**	6.67
DE-LR	50	92.40	91.07	92.00	91.89	0.47	12.64
	100	93.33	91.07	92.07	92.04	0.58	66.92
	150	93.60	92.13	93.27	93.04	0.53	93.12
	200	94.53	93.07	93.80	93.77	0.52	117.54
	300	94.80	92.67	94.07	93.91	0.68	165.57
PSO-LR	50	92.00	90.80	91.27	91.31	0.41	10.78
	100	90.13	87.87	89.47	89.33	0.81	61.01
	150	91.33	88.13	89.40	89.51	1.09	88.70
	200	91.07	88.00	89.07	89.24	1.01	114.13
	300	91.07	87.20	88.87	89.01	1.16	161.63

The best results are given in bold face. FVS: size of feature vector

LR is more robust and reliable classification model because the difference between best and worst classification accuracy is low compared to LR, PSO-LR and DE-LR. ABC-LR attained the best overall results, with success rate of 96.13%.

15.6 Conclusion

This chapter is devoted to one of the algorithms simulating swarm intelligence, artificial bee colony (ABC) algorithm which mimics the foraging behavior of a honeybee colony. The self-organization properties and division of labor in a honey bee colony are described, and an analogy is established between these features and optimization algorithm. The phases of the ABC algorithm are presented by a broad explanation, and an application of the ABC algorithm is presented for filtering unsolicited content in digital communication. Two different feature selection techniques, including tf-idf and mutual information-based methods, were employed, and the ABC-based LR

Table 15.2 Using mutual information as feature selection approach, classification performances of LR, ABC-LR, DE-LR and PSO-LR with respect to best, worst, median and mean accuracies, standard deviation values and average training time in seconds

Method	FVS	Best (%)	Worst (%)	Median (%)	Mean (%)	Std	Time
LR	50	92.27	90.00	91.27	91.29	0.67	**0.102**
	100	93.33	90.53	91.33	91.53	0.80	**0.115**
	150	92.93	90.80	91.67	91.69	0.65	**0.126**
	200	92.53	89.73	91.13	91.13	0.78	**0.139**
	300	93.07	89.33	91.27	91.17	0.83	**0.120**
ABC-LR	50	**93.60**	**92.80**	**93.20**	**93.17**	**0.32**	4.91
	100	**94.67**	**92.93**	**93.67**	**93.67**	0.49	6.57
	150	94.27	92.93	93.40	93.44	0.41	7.64
	200	**94.40**	**93.73**	**94.13**	**94.12**	**0.22**	8.63
	300	**95.47**	**94.00**	**94.80**	**94.89**	0.47	10.61
DE-LR	50	93.07	92.00	92.67	92.60	0.40	11.94
	100	93.60	92.00	92.80	92.76	**0.46**	62.61
	150	**94.53**	**93.07**	**94.07**	**93.87**	0.55	91.95
	200	93.60	92.13	93.13	92.97	0.59	115.62
	300	94.40	93.20	93.73	93.75	**0.40**	163.03
PSO-LR	50	92.53	91.20	91.40	91.67	0.46	10.76
	100	93.20	90.27	92.00	91.89	0.80	66.78
	150	92.27	91.07	91.87	91.79	**0.36**	92.96
	200	92.00	90.27	91.73	91.41	0.68	120.37
	300	93.20	90.27	91.53	91.55	0.80	176.85

The best results are given in bold face. FVS: size of feature vector

classifier was compared to LR variants using gradient descent, DE and PSO algorithms. It was shown that ABC-LR obtained the best overall results, with success rate of 96.13%, and tf-idf-based feature representation yields better results compared to mutual information-based representation. Therefore, ABC-LR might be a good method for filtering purpose in digital communication.

Acknowledgements This research was partially supported by the Scientific and Technological Research Council of Turkey (TUBITAK) under project number 116E947.

References

1. Akay B, Karaboga D (2015) A survey on the applications of artificial bee colony in signal, image, and video processing. Signal Image Video Process 9(4):967–990
2. Dedeturk B, Akay B (2020) Spam filtering using a logistic regression model trained by an artificial bee colony algorithm. Appl Soft Comput 91:106229

3. Dorigo M, Maniezzo V, Colorni A (1991) Positive feedback as a search strategy. Technical Report 91-016, Politecnico di Milano, Italy
4. Karaboga D (2005) An idea based on honey bee swarm for numerical optimization. Technical Report TR06, Erciyes University, Engineering Faculty, Computer Engineering Department
5. Karaboga D, Akay B (2009) A survey: algorithms simulating bee swarm intelligence. Artif Intell Rev 31(1–4):61–85
6. Karaboga D, Gorkemli B, Ozturk C, Karaboga N (2012) A comprehensive survey: artificial bee colony (ABC) algorithm and applications. Artif Intell Rev 1–37
7. Kennedy J, Eberhart R (1995) Particle swarm optimization. In: IEEE international conference on neural networks, Piscataway, NJ, pp 1942–1948
8. Ozgur L, Gungor T, Gurgen F (2004) Adaptive anti-spam filtering for agglutinative languages: a special case for Turkish. Pattern Recogn Lett 25(16):1819–1831
9. Passino KM (2002) Biomimicry of bacterial foraging for distributed optimization and control. IEEE Control Syst Mag 22(3):52–67
10. Radicati Group (2019) Email statistics report, 2019–2023. In: Executive summary
11. Richter MR, Keramaty JM (2003) Chapter 11–honey bee foraging behavior. In: Ploger BJ, Yasukawa K (eds) Exploring animal behavior in laboratory and field. Academic, San Diego, pp 133–145
12. Richter MR, Keramaty JM (2003) Honey bee foraging behavior. In: Exploring animal behavior in laboratory and field. Elsevier, pp 133–145
13. Rivera MD, Donaldson-Matasci M, Dornhaus A (2015) Quitting time: when do honey bee foragers decide to stop foraging on natural resources? Front Ecol Evol 3:50
14. Seeley TD (1994) Honey bee foragers as sensory units of their colonies. Behav Ecol Sociobiol 34(1):51–62
15. Statista (2018) Tech. rep. global e-mail spam rate from 2012 to 2018
16. Symantec (2018) Internet security threat report (ISTR)
17. Vergelis TMS, Demidova N, Scherbakova T (2018) Kaspersky lab. report: spam and phishing in q3 2018
18. Yang X-S (2010) Nature-inspired metaheuristic algorithms. Luniver Press

Chapter 16
Multi-objective Design of Multilayer Microwave Dielectric Filters Using Artificial Bee Colony Algorithm

Abdurrahim Toktas

16.1 Introduction

Metaheuristic optimization algorithms which have been developed by inspiring natural phenomenon have been achieved amazing results in optimum engineering designs in the last decades. They derive their achievements from the perfection in nature. Therefore, the more similar the algorithm is modeled to the processing of the nature, the more successful it is. Almost all-natural phenomenon has attempted to model for an optimization algorithm [1]. In general, the algorithms are built on controllable stochastic computations that iteratively try to improve the candidate solutions. The most prominent ones are herein: Genetic algorithm (GA) [2] and differential evolution (DE) [3] were developed by inspiring the mutation phenomenon of living organism. Particle swarm optimization (PSO) was simulated the flock behavior organisms such as birds and fishes [4]. Ant colony optimization (ACO) was mimicked the communication and direction finding paradigm of the biological ants [5]. Harmony search (HS) was inspired by the improvisation process of jazz musicians [6]. Cuckoo search (CS) was modeled the obligate brood parasitism of some cuckoo species by laying their eggs in the nests of other species [7]. Bat algorithm (BA) was simulated the echolocation behavior of the natural bats [8]. Artificial bee colony (ABC) was emerged by making analogy of nectar foraging behavior of the honey bees [9]. ABC algorithm and its variants have been successfully applied to a variety of the engineering problems [10–14]. Although several multi-objective versions of ABC have been proposed [11, 12, 15–17], its success on multi-objective electromagnetic (EM) computational expensive problems has been still remaining a curiosity. When metaheuristic optimization algorithms are first developed, they are

A. Toktas (✉)
Department of Electrical and Electronics Engineering, Faculty of Engineering, Karamanoglu Mehmetbey University, Karaman 70200, Turkey
e-mail: atoktas@kmu.edu.tr

S. Carbas et al. (eds.), *Nature-Inspired Metaheuristic Algorithms for Engineering Optimization Applications*, Springer Tracts in Nature-Inspired Computing,
https://doi.org/10.1007/978-981-33-6773-9_16

emerged for single objective optimizations. Additional techniques, i.e., Pareto optimality, are integrated to them for adapting multi-objective capability, in general [18–20]. Thanks to Pareto optimality, it is possible to refine optimal solutions according to all objective vectors and herewith to determine a global optimal solution by taking account of the trade-offs among the objective vectors.

Microwave dielectric filters (MDFs) are interested components for microwave and wireless applications such as communication, antenna, radar and biomedical systems [21]. MDFs are used to reflect, transmit or absorb EM fields at the desired frequency band of the field. The most commonly designed MDFs are low-pass (LP), high-pass (HP), and band-pass (BP) filters. It is desired that the incident EM wave passes through the layers of the MDF at the pass band and reflects from the MDF at the stop band [22]. It is inconvenient to obtain near ideal filter characteristic using single layer MDF design. On the other hand, by using multilayer MDF (MMDF) design composed of different dielectric layers, it is more likely to achieve rigorous and sharp filter characteristic. The design of MMDF requires a multi-objective optimization strategy in which the reflected EM wave must be synchronously minimized at the pass band and maximized at the stop band. A computational EM model is also required for this multi-objective optimization scheme, taking into account the incident wave angle and polarization at all layer interfaces. The design parameters of the MMDF regarding both the thickness and material types of the layers should be simultaneously determined by searching within an existing material database with frequency-dependent complex permittivity through a powerful multi-objective optimization procedure as well as considering the trade-off between the objective functions.

Few studies have attempted to design MMDFs in recent years. In [23], a procedure for designing MMDFs was presented a binary-coded GA. Several five-layer MMDFs were designed between 8 GHz and 18 GHz for single objective through non-defined dielectric materials. In other words, dielectric constants for layer materials were searched continuously between 1.03 and 10.0. The two objectives were combined by weighting each of them. In [24], an evolutionary programming algorithm was used to design MMDFs between 24 and 36 GHz for single objective as similar to the procedure in [23]. Dielectric constants were Rogers Corporation with values ranging from 1.01 to 10.2. In [25], a few MMDFs were designed using multi-objective GA between 24 and 36 GHz by selecting material types from a material database including dielectric constants. They are all designed at incident angle 45°. In real applications, material types are usually with frequency dependent complex permittivity and permeability. However, material database with only real permittivity (dielectric constant) was employed.

In this chapter, three types of five-layer MMDFs that have LP, HP, and BP filter characteristics are designed through multi-objective ABC (MO-ABC) algorithm. Pareto optimality that allows to refine optimal solution in accordance with the multiple objective values is incorporated to ABC so as to adapt multi-objective ability. The materials are selected from a frequency-dependent artificial material database including complex permittivity and permeability. The MMDF is designed to operate at the frequency range of 2–18 GHz. The LP filter passes the band of

2–10 GHz and stops the band of 10–18 GHz, and the HP filter stops the band of 2–10 GHz and passes the band of 10–18 GHz. The BP filter stops both bands of 2–8 GHz and 12–18 GHz and passes the band of 8–12 GHz. Two objective functions are constituted to be minimized the total reflections (TRs) at the pass and stop bands. Global optimal MMDFs are selected within the Pareto optimal solutions by ensuring the trade-off between the two objective functions. Eventually, the frequency characteristics of the designed MMDFs are elaborately examined to demonstrate the performance of MO-ABC.

16.2 MO-ABC Algorithm

ABC modeled by observing the collective foraging behavior of the natural honey bees was initially originated for single objective optimization problems. Pareto optimality is a simple and easy approach to adapt multi-objective strategy to ABC. Pareto optimality obtains the optimal solutions which are the fittest ones according to either objective with respect to all objectives among the possible solutions.

16.2.1 Pareto Optimality Algorithm

Engineering design problems are often burdensome multi-objective optimizations. Since objective functions are dependent on decision vectors (design variables), they are hence related to each other, as well. In other words, while one objective is improved by changing the decision vector, the others may deteriorate [26]. Therefore, the results of all objective functions must be taken into consideration in finding out optimal solutions with respect to all objectives. Given that every decision vector has an outcome in terms of each objective function. A decision vector may dominate the other vectors in accordance with one objective function value, whereas it may not dominate according with the other objective function values. Hence, a set of decision vectors that dominate all other vectors for entire objective space should be considered. Pareto optimality whose pseudocode is given in Algorithm 1 is the most effective way that is incorporated to the metaheuristic algorithms, allowing to independently determine a diverse and uniform optimal solution set (non-dominated solutions) so-called Pareto front. Assuming that there is objective space $\mathrm{OF}(x) = [\mathrm{of}_1, \mathrm{of}_2, \ldots, \mathrm{of}_N]$ with N ($k = 1, 2, \ldots, N$: $N \geq 2$) objective functions $\mathrm{of}_k(x)$. The decision vector is $x_j = [x_1, x_2, \ldots, x_d]$ where $j = 1, 2, \ldots, d$ called decision space and search space $x_i = [x_1, x_2, \ldots, x_{NP}]$ with $i = 1, 2, \ldots, NP$ where NP is number of population.

Algorithm 1 Pseudocode of Pareto optimality

$\boldsymbol{x}^*$ and $\boldsymbol{x}^\#$ are decision vectors
If $\boldsymbol{x}^*$ dominates $\boldsymbol{x}$
    $\boldsymbol{x}^\# = \boldsymbol{x}^*$
    $i = 0$
else if $\boldsymbol{x}^\#$ dominates $\boldsymbol{x}^*$
    $i = i + 1$
else if $\boldsymbol{x}^*$ and $\boldsymbol{x}^\#$ are non–dominated (Pareto optimal) solutions
    if rand < 0.5
    $\boldsymbol{x}^\# = \boldsymbol{x}^*$
    $i = 0$
    end if
end if

16.2.2 ABC Algorithm

ABC algorithm was developed by simulating the self-organizing and collective foraging intelligence of the honey bee colonies [9, 27]. The bees forage as three colonies: employed bee, onlooker bee, and scout bee. The phases of ABC are also referred to the same name of colonies. In colonies, artificial bees search for quality nectar sources. ABC whose pseudocode given in Algorithm 2 proceeds iteratively through these phases. The total number of bees in the colony whose number is the *NP* is considered as two equal colonies: the employed and onlooker bees. Each bee works for a nectar source of which location stands for a candidate solution. The qualities of nectar sources correspond to the fitness of the candidate solutions. Number of *NP*/2 candidate solutions are evaluated at both employed and onlooker bee phases, separately. All employed bees are initially appointed as scout bees to discover new nectar sources. Then the employed bees search around the nectar sources and transfer information regarding the qualities of nectar sources to the onlooker bees. According to this information, the onlooker bees search again in vicinities of the nectar sources determined by the employed bees. If a more quality nectar source could not be found around the nectar source after a specified number of trying "*limit,*" it then abandoned and a totally new nectar source is defined instead of the old one.

Algorithm 2 Pseudocode of MO-ABC

Begin
Initialization
Set control parameters: limit, *NP* and *NI*
Generate randomly initial populations (decision vectors) with *NP* (16.1)
Compute initial objective vectors and fitness vectors (16.2)
Employed bee phase

7. Change randomly selected decision vectors (candidate solutions) (16.3)
8. Compute the new objective vectors and fitness vectors (16.2)
9. Compute the probability vectors (16.4)
10. *Refine Pareto optimal solutions (Algorithm 1)*
11. **Onlooker bee phase**
12. Change randomly selected candidate solutions (16.3)
13. Compute the new objective vectors and fitness vectors using (16.2)
14. *Refine Pareto optimal solutions (Algorithm 1)*
15. If a solution cannot be improved after trying as number of "*limit*"
16. **Scout bee phase**
17. Replace the old solution with a new generated solution (16.1)
18. Compute the new objective vectors and fitness vectors (16.2)
19. Record the best solution obtained so far
20. End

In ABC algorithm, at the initialization, candidate solution vectors (decision vectors) with *NP* x_{ij} are randomly defined within the decision space constrained between $x_j^{\max}$ and $x_j^{\min}$ by using the following mathematical operator.

$$x_{ij} = x_j^{\min} + \text{rand}(0, 1)(x_j^{\max} - x_j^{\min}) \tag{16.1}$$

The objective vectors $of_i(x)$ and accordingly the fitness vectors fit_i assessing the quality of candidate solutions are determined for all objective functions to evaluate the quality of candidate solutions as follows

$$\text{fit}_i = \begin{cases} \frac{1}{1+\text{of}_i(x)} & \text{if } \text{of}_i(x) \geq 0 \\ 1 + \text{abs}[\text{of}_i(x)] & \text{if } \text{of}_i(x) < 0 \end{cases} \tag{16.2}$$

In employed bee phase, new solution vector v_{ij} is produced near the before founded candidate solutions x_{ij} and x_{kj}, which are randomly selected from the population using the following operator.

$$v_{ij} = x_{ij} + \phi_{ij}(x_{ij} - x_{kj}) \tag{16.3}$$

The objective vectors $\text{of}_i(v)$ and fitness vectors fit_i are repeatedly evaluated for the new decision vector. The Pareto optimal solutions are obtained using Pareto optimality (Algorithm 1). A probability vectors prob_i through the fitness vectors are evaluated for all objectives via the operator (16.4). The Pareto optimal solutions are then obtained depending on Pareto optimality (Algorithm 1).

$$\text{prob}_i = \frac{\text{fit}_i}{\sum_{n=1}^{NP/2} \text{fit}_n} \tag{16.4}$$

In onlooker bee phase, new solutions are regenerated in the vicinity of the before obtained solutions x_{ij} and x_{kj} using the operator in the employed bee phase (16.3). The objective vectors $\mathrm{of}_i(v)$ and fitness vectors fit_i are then calculated via the operator (16.2) with similar manner in the employed bee phase. New Pareto optimal solutions are hence refined with regard to the new objective vector v_{ij} by repeating the processes in Algorithm 1.

In scout bee phase, a totally new candidate solution is stochastically regenerated via the same operator (16.1) in place of a solution that could not be improved after the number of trying "*limit*". The objective vectors $\mathrm{of}_i(v)$ and fitness vectors fit_i are therefore calculated via the operator (16.2). Finally, the last Pareto optimal solutions achieved so far are saved. Those three phases are iteratively proceeded up to a predefined number of iterations (NI).

16.3 Multi-objective EM Model of the MMDF

The MMDF is composed of multiple dielectric layers that stacked on each other. A MMDF structure with n layers $i = 0, 1, 2, \ldots, n$ is illustrated in Fig. 16.1. Each layer has thickness of d_i and material of m_i having complex permittivity ε_i and permeability

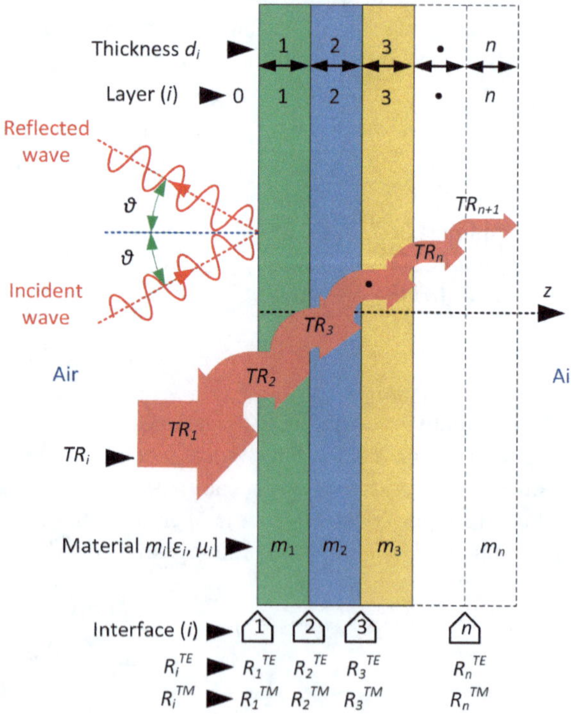

Fig. 16.1 Structure of the conceptual MMDF

μ_i. The propagating wave through the air strikes onto interface 1 with the incident angle of θ. The relationship between the incidence angle θ_{i-1} and transmitted incident waves angle θ_i at each interface can be determined by Snell's law as follows

$$\frac{\sin\theta_i}{\sin\theta_{i-1}} = \sqrt{\frac{\mu_{i-1}\varepsilon_{i-1}}{\mu_i\varepsilon_i}} \tag{16.5}$$

An EM model regarding the total reflection (TR) coefficient taking account of the oblique incident wave angle and comprising sub-reflections at the inner interfaces together with transverse electric (TE) and transverse (TM) at each layer is given below [15].

$$\mathrm{TR}_i^{\mathrm{TE}} = \frac{R_i^{\mathrm{TE}} + \left(\mathrm{TR}_{i+1}^{\mathrm{TE}}\right)\mathrm{e}^{-2jkz_id_i}}{1 + R_i^{\mathrm{TE}}\left(\mathrm{TR}_{i+1}^{TE}\right)\mathrm{e}^{-2jkz_id_i}} \tag{16.6}$$

$$\mathrm{TR}_i^{\mathrm{TM}} = \frac{R_i^{\mathrm{TM}} + \left(\mathrm{TR}_{i+1}^{\mathrm{TM}}\right)\mathrm{e}^{-2jkz_id_i}}{1 + R_i^{TM}\left(\mathrm{TR}_{i+1}^{\mathrm{TM}}\right)\mathrm{e}^{-2jkz_id_i}} \tag{16.7}$$

where $kz_i = \cos\theta_i\omega\sqrt{\mu_i\varepsilon_i}$ is the complex wave propagation number along the z-direction and $\omega = 2\pi f$ is the angular frequency, f is the frequency of the incident wave. The sole interface reflection (R) coefficients at the inner interfaces are given as

$$R_i^{\mathrm{TE}} = \frac{\mu_i kz_{i-1} - \mu_{i-1}kz_i}{\mu_i kz_{i-1} + \mu_{i-1}kz_i} \tag{16.8}$$

$$R_i^{\mathrm{TE}} = \frac{\varepsilon_i kz_{i-1} - \varepsilon_{i-1}kz_i}{\varepsilon_i kz_{i-1} + \varepsilon_{i-1}kz_i} \tag{16.9}$$

16.3.1 The Dual-Objective Functions for the Design of MMDFs

The three types of MMDF having LP, HP, and BP characteristics are designed through MO-ABC algorithm. The MMDFs are aimed to operate between 2 and 18 GHz. The band characteristics of the considered MMDFs are depicted in Fig. 16.2 [28]. In this regard, the objective functions of LP filter with cut-off frequency 10 GHz are constituted so as to pass the band of 2–10 GHz and to stop the band of 10–18 GHz. Conversely, those of HP filter are formed so that it passes the band of 10–18 GHz and stops the band of 2–10 GHz. Eventually, the BP filter with cut-off frequencies 8 GHz and 12 GHz is constructed to pass the band of 8–12 GHz and to stop the bands

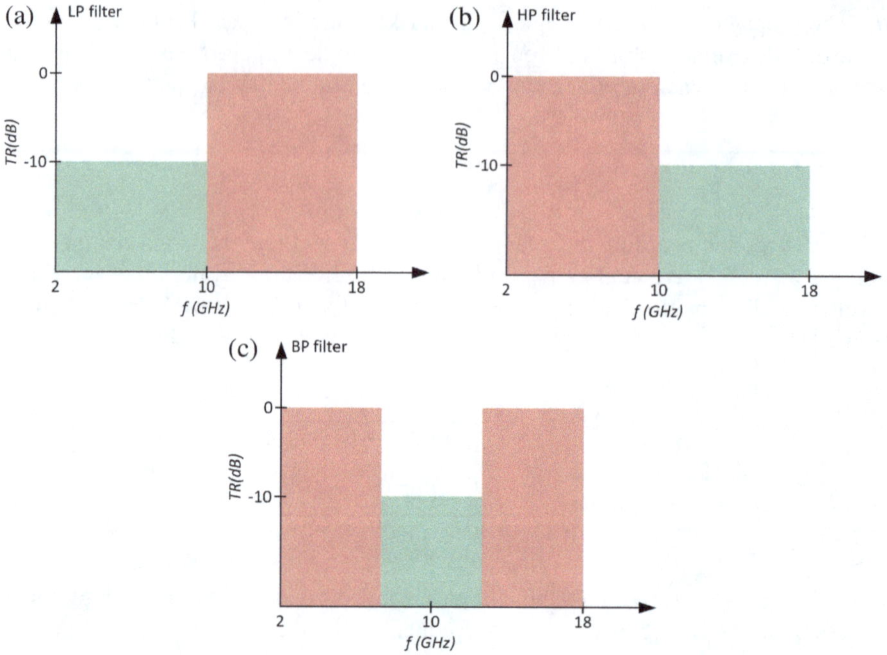

Fig. 16.2 Frequency band characteristics of the considered MMDFs: **a** LP filter, **b** HP filter, **c** BP filter

of 2–8 GHz and 12–18 GHz. The TR at 0.316 = −10 dB will be considered for a reference level of band pass region, meaning that 90% of the incident wave power pass through the MMDF.

The two objective functions regarding the mean of TR at the pass band and stop band are separately constituted to be minimizing. At pass band, it is aimed to minimize the first objective function of_1, however it is desired to maximize the second objective function of_2. Since the numerical value of TR varies between 0 and 1, the second objective function can be converted into be minimized by subtracting each $\mathrm{TR}^{\mathrm{TE}}$ and $\mathrm{TR}^{\mathrm{TM}}$ from 1. The final objective functions that should be minimized are constituted as follows

$$\mathrm{of}_1 = \frac{\sum_{f_l}^{f_u} \frac{\sum_{\theta_l}^{\theta_u}\left[\left|\mathrm{TR}_1^{\mathrm{TE}}(\theta, f_p)\right| + \left|\mathrm{TR}_1^{\mathrm{TM}}(\theta, f_p)\right|\right]}{2N_{\mathrm{ap}}}}{N_{\mathrm{fp}}} \tag{16.10}$$

$$\mathrm{of}_2 = \frac{\sum_{f_l}^{f_u} \frac{\sum_{\theta_l}^{\theta_u}\left[2 - \left|\mathrm{TR}_1^{\mathrm{TE}}(\theta, f_s)\right| - \left|\mathrm{TR}_1^{\mathrm{TM}}(\theta, f_s)\right|\right]}{2N_{\mathrm{ap}}}}{N_{\mathrm{fp}}} \tag{16.11}$$

where N_{ap} is the number of angle points for the range of $\theta_l \le \theta \le \theta_u$ and N_{fp} is the number of frequency points for the range of pass or stop bands. Therefore, the mean of the TR is obtained by divided them into N_{ap} and N_{fp}. Moreover, because there are two TRs related to TE and TM, first they divided into 2. f_p and f_s are referred to the frequency point falls in pass and stop regions, respectively. Hence, the range of f_p and f_s can be defined as follows

For the LP filter:

$$2\,\text{GHz} \le f_p \le 10\,\text{GHz} \tag{16.12}$$

$$10\,\text{GHz} \le f_s \le 18\,\text{GHz} \tag{16.13}$$

For the HP filter:

$$10\,\text{GHz} \le f_p \le 18\,\text{GHz} \tag{16.14}$$

$$2\,\text{GHz} \le f_s \le 10\,\text{GHz} \tag{16.15}$$

For the BP filter:

$$2\,\text{GHz} \le f_s \le 8\,\text{GHz} \tag{16.16}$$

$$8\,\text{GHz} \le f_p \le 12\,\text{GHz} \tag{16.17}$$

$$12\,\text{GHz} \le f_s \le 18\,\text{GHz} \tag{16.18}$$

16.4 The Designed MMDFs Through MO-ABC

Three types of five-layer MMDFs which are LP, HP, and BP filters are designed in this study to operate between 2 and 18 GHz. It is aimed that the LP, HP, and BP filters, respectively, pass the bands of 2–10 GHz, 10–18 GHz and 8–12 GHz, and stop the bands of 10–18 GHz, 2–10 GHz, and 2–8/12–18 GHz. Through MO-ABC, the Pareto optimal solutions each of which stands for design parameters of the MMDFs are found out within the all possible solutions. The material types to be used in the dielectric layer are picked up from a predefined material database. Global optimal MMDFs are then selected among the Pareto optimal solutions.

16.4.1 The Set Parameters and Material Database

The range of the incident angle is between 0 and 45°. The steps range for the angle and frequency are 15° and 0.2 GHz, respectively. Therefore, the N_{ap} would be 4. N_{fp} would be 41 for all bands of both designs of LP and HP filters and it would be 31 for both stop bands and 21 for the pass band of the BP filter. The thickness for each layer is searched between $d_l = 0$ mm and $d_u = 3$ mm. A frequency dependent material database which was initial defined in [29] are exploited in the multi-objective design of MMDFs. The database composed of 16 complex permittivity and permeability is given in Table 16.1. On the other hand, the control parameters of ABC such as NP, NI, and limit are set as 100, 1000, and 100, respectively.

16.4.2 The Performance Results of the Designed MMDFs

The objective functions involving TR^{TE} and TR^{TM} are computed at the incident angles of 0, 15, 30, and 45°. The material types $m[\varepsilon, \mu]$ which would be in the thickness range of 0–3 mm are searched within the material database given in Table 16.1 In the design of five-layer MMDFs, the obtained Pareto optimal solutions for the LP, HP, and BP filters are distributed over two-dimensional (2D) objective spaces in Fig. 16.3. The blue solid circle symbol indicates the Pareto optimal solutions, and the red star symbol stands for the selected global optimal solution (MMDF design).

The global optimal MMDF would be the trade-off solution with regard to both objective function values which are selected to be diagonally the nearest to the origin of the 2D objective space. The selected global optimal MMDFs with thickness, material types are given in Table 16.2. The TT of the global optimal LP, HP, and BP filters is 9.0799 mm, 5.5237 mm, and 6.1891 mm, respectively. The TRs versus frequency characteristic for MMDFs at 0 (normal incidence) 15, 30, and 45° are plotted in Fig. 16.4. It is seen that almost all TRs obey the desired MMDF characteristics. Yet the characteristics of HP filter are not coherent with an ideal filter as much as the other LP and HP filters. That is why the material database may not suitable for the HP filter. Only the TR^{TE} at high degrees such as 30 and 45 are not under −10 dB because the TR^{TE} is always higher the TR^{TM}.

In order to elaborately examine the TR results of the MMDFs, the maximum, average, and minimum values of the TRs at the pass and stop band regions are obtained for the TE and TM polarizations at 0, 15, 30, and 45° in Tables 16.3, 16.4 and 16.5. The performance of the multi-objective optimization scheme based on ABC is even clearly seen from the results in the tables. It is also verified that the TRs for TE are always higher than the TM. The average TRs are generally lower than −10 dB at the pass bands and higher than −3 dB at stop bands at which level 50% of the incident wave power reflects from the MMDF.

It can be inferred from results given in Fig. 16.4, Tables 16.3, 16.4 and 16.5 that ABC is even successful in multi-objective EM problems. Since ABC is one of the

Table 16.1 Predefined frequency-dependent artificial material database

Lossless dielectric materials ($\mu' = 1, \mu'' = 0$)				
Mat. #	ε'			
1	10			
2	50			
Lossy magnetic materials ($\varepsilon' = 15, \varepsilon'' = 0$)				
$\mu'(f) = \frac{\mu'(1\,\text{GHz})}{f^{\alpha}} \ldots \mu''(f) = \frac{\mu''(1\,\text{GHz})}{f^{\beta}}$				
Mat. #	$\mu'(1\text{ GHz})$	α	$\mu''(1\,\text{GHz})$	β
3	5	0.974	10	0.961
4	3	1.000	15	0.957
5	7	1.000	12	1.000
Lossy dielectric materials ($\mu' = 1, \mu'' = 0$)				
$\varepsilon'(f) = \frac{\varepsilon'(1\,\text{GHz})}{f^{\alpha}} \ldots \varepsilon''(f) = \frac{\varepsilon''(1\,\text{GHz})}{f^{\beta}}$				
Mat. #	$\varepsilon'\ (1\,GHz)$	α	$\varepsilon''\ (1\,GHz)$	β
6	5	0.861	8	0.569
7	8	0.778	10	0.682
8	10	0.778	6	0.861
Relaxation-type magnetic materials ($\varepsilon' = 15, \varepsilon'' = 0$)				
$\mu'(f) = \frac{\mu_m f_m^2}{f^2+f_m^2} \ldots \mu''(f) = \frac{\mu_m f_m f}{f^2+f_m^2}$ f and f_m in GHz				
Mat. #	μ_m		f_m	
9	35		0.8	
10	35		0.5	
11	30		1.0	
12	18		0.5	
13	20		1.5	
14	30		2.5	
15	30		2.0	
16	25		3.5	

latest metaheuristic algorithms, its implementations to multi-objective engineering design problems are not as much as the relatively former algorithms such as GA, PSO, and DE. It is observed that multi-objective variants of ABC have been attempted to developed as testing for benchmark functions [30, 31], and the results are better than those of the former algorithms. Therefore, the multi-objective versions of ABC should be implemented to more engineering problems in order to demonstrate its superiority over the former algorithms.

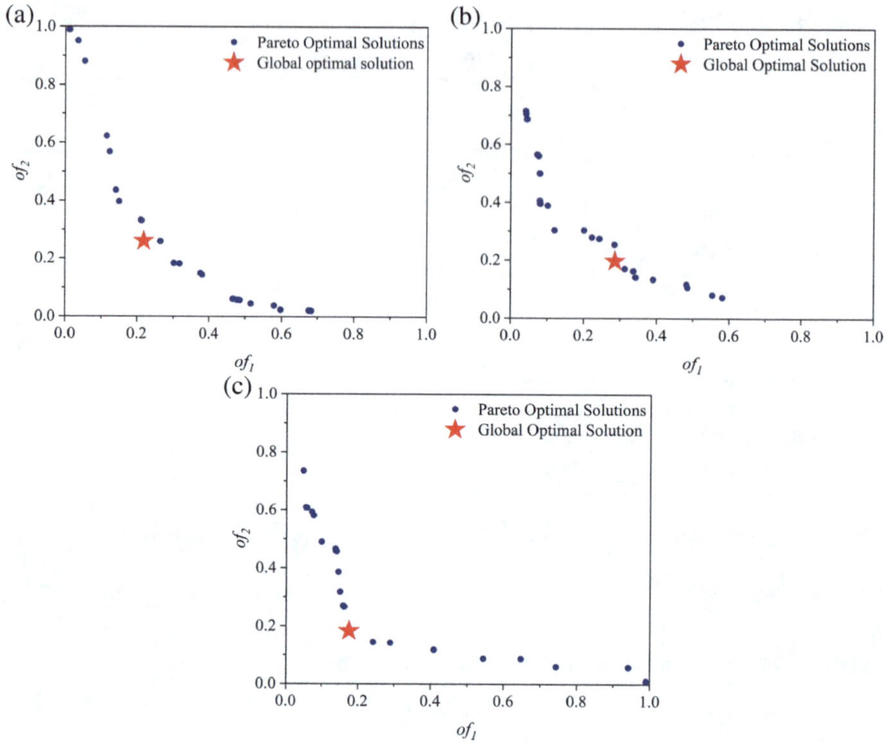

Fig. 16.3 2D Pareto objective space for the design of MMDFs: **a** LP filter, **b** HP filter, **c** BP filter

Table 16.2 Selected global optimal MMDFs among the pareto optimal solutions

Layer sequence	LP filter		HP filter		BP filter	
	Global optimal solution		Global optimal solution		Global optimal solution	
	of_1	of_2	of_1	of_2	of_1	of_2
	0.2185	0.2593	0.2855	0.1972	0.1751	0.1838
	Mat. #	Thickness (mm)	Mat. #	Thickness (mm)	Mat. #	Thickness (mm)
1	9	0.7118	1	0.2818	1	1.5615
2	8	3.0000	7	1.7758	6	0.3311
3	2	0.9224	8	0.2753	1	0.7746
4	8	3.0000	1	2.0800	2	0.9427
5	1	1.4457	13	1.1107	1	2.5793
TT (mm)	9.0799		5.5237		6.1891	

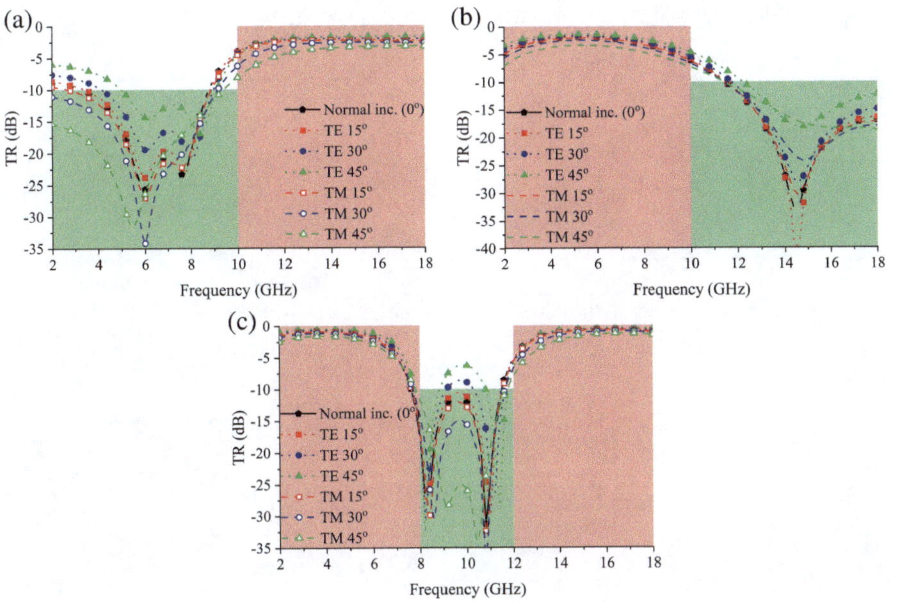

Fig. 16.4 TR plots of the global optimal solution for the design of MMDFs: **a** LP filter, **b** HP filter, **c** BP filter

Table 16.3 Maximum, average, and minimum TR of the LP filter at the pass and stop bands

Band	Polarization incidence		TR		
			Maximum	Average	Minimum
Pass	0°		−4.47	−14.86	−25.64
	TE	15°	−4.60	−14.35	−23.77
		30°	−4.97	−12.83	−20.87
		45°	−5.39	−10.38	−16.92
	TM	15°	−5.13	−15.38	−26.99
		30°	−6.93	−17.14	−34.17
		45°	−8.64	−18.76	−31.40
Stop	0°		−2.18	−2.36	−3.52
	TE	15°	−2.14	−2.32	−3.60
		30°	−1.99	−2.21	−3.80
		45°	−1.70	−1.98	−3.97
	TM	15°	−2.29	−2.51	−4.05
		30°	−2.62	−3.04	−5.62
		45°	−3.21	−4.00	−7.46

Table 16.4 Maximum, average and minimum TR of the HP filter at the pass and stop bands

Band	Polarization-incidence		TR		
			Maximum	Average	Minimum
Stop	0°		−1.87	−2.94	−5.66
	TE	15°	−1.80	−2.84	−5.51
		30°	−1.61	−2.56	−5.05
		45°	−1.30	−2.10	−4.25
	TM	15°	−1.99	−3.08	−5.80
		30°	−2.41	−3.56	−6.22
		45°	−3.25	−4.48	−6.96
Pass	0°		−6.49	−17.89	−33.35
	TE	15°	−6.33	−18.19	−40.28
		30°	−5.81	−16.20	−28.09
		45°	−4.91	−12.66	−18.09
	TM	15°	−6.62	−17.66	−30.31
		30°	−6.99	−16.82	−24.21
		45°	−7.58	−14.80	−18.66

Table 16.5 Maximum, average, and minimum TR of the BP filter at the pass and stop bands

Band	Polarization-incidence		TR		
			Maximum	Average	Minimum
Stop	0°		−0.65	−1.87	−13.04
	TE	15°	−0.62	−1.81	−12.82
		30°	−0.54	−1.62	−12.10
		45°	−0.41	−1.34	−10.71
	TM	15°	−0.70	−1.95	−12.57
		30°	−0.87	−2.21	−11.41
		45°	−1.23	−2.70	−9.95
Pass	0°		−6.63	−16.04	−31.27
	TE	15°	−6.89	−15.55	−29.46
		30°	−7.79	−14.27	−30.30
		45°	−6.24	−12.43	−32.56
	TM	15°	−7.12	−16.86	−32.37
		30°	−8.33	−19.14	−36.90
		45°	−9.37	−21.94	−34.00

16.5 Conclusions

Single objective ABC is a simple and robust metaheuristic optimization algorithm based on the collective foraging behavior of the natural honey bees. On the other hand, the multi-objective performance of ABC on engineering design problems especially on computational electromagnetic needs more implementations and demonstrations. Efficacy on the design of MMDF including expensive computations is anticipated through a multi-objective scheme. In this chapter, three types of MMDF which are LP, HP, and BP filters are designed through the MO-ABC scheme. Two objective functions involving mean TR at the pass band and stop band are constituted across an exact EM model based on the reflection mechanism taking into account even the incident wave angle with polarizations TE and TM. The Pareto optimal solutions are refined within the possible solutions for synchronously minimizing the objective functions. The global optimal MMDF designs are selected from the Pareto optimal solutions by providing the trade-off between the objective functions. The performance results of the designed MMDFs are investigated with regard to the frequency characteristics. Therefore, the designed MMDFs show near ideal filter characteristics thank to MO-ABC.

Acknowledgements The author is thankful to Dr. Deniz Ustun for valuable support and contribution to the computations carried out in this study.

References

1. Yang X-S (2020) Nature-Inspired optimization algorithms. Academic Press, Second
2. Goldberg DE (1989) Genetic algorithms in search, optimization, and machine learning. Addison-Wesley Longman Publishing Co., Boston
3. Storn R, Price K (1997) Differential evolution–a simple and efficient heuristic for global optimization over continuous spaces. J Glob Optim 341–359. https://doi.org/10.1023/A:1008202821328
4. Kennedy J, Eberhart R (1995) Particle swarm optimization. In: Proceedings of ICNN'95—international conference on neural networks. IEEE, pp 1942–1948
5. Dorigo M, Stützle T (2004) Ant colony optimization. MIT Press
6. Geem ZW, Kim JH, Loganathan GV (2001) A new heuristic optimization algorithm: harmony search. Simulation 76:60–68. https://doi.org/10.1177/003754970107600201
7. Yang XS, Deb S (2009) Cuckoo search via Lévy flights. In: NABIC 2009—Proceedings 2009 world congress on nature and biologically inspired computing, pp 210–214
8. Yang X-S (2010) A new metaheuristic bat-inspired algorithm
9. Karaboga D, Basturk B (2007) A powerful and efficient algorithm for numerical function optimization: artificial bee colony (ABC) algorithm. J Glob Optim 39:459–471. https://doi.org/10.1007/s10898-007-9149-x
10. Karaboga D, Gorkemli B, Ozturk C, Karaboga N (2014) A comprehensive survey: artificial bee colony (ABC) algorithm and applications. Artif Intell Rev 42:21–57. https://doi.org/10.1007/s10462-012-9328-0
11. Toktas A, Ustun D, Erdogan N (2020) Pioneer Pareto artificial bee colony algorithm for three-dimensional objective space optimization of composite-based layered radar absorber. Appl Soft Comput 96:1–12. https://doi.org/10.1016/j.asoc.2020.106696

12. Toktas A, Ustun D, Tekbas M (2020) Global optimisation scheme based on triple-objective ABC algorithm for designing fully optimised multi-layer radar absorbing material. IET Microwaves Antennas Propag 14:800–811. https://doi.org/10.1049/iet-map.2019.0868
13. Akdagli A, Toktas A (2010) A novel expression in calculating resonant frequency of H-shaped compact microstrip antennas obtained by using artificial bee colony algorithm. J Electromagn Waves Appl 24:2049–2061. https://doi.org/10.1163/156939310793675989
14. Ustun D, Ozdemir C, Akdagli A et al (2014) A powerful method based on artificial bee colony algorithm for translational motion compensation of ISAR image. Microw Opt Technol Lett 56:2691–2698. https://doi.org/10.1002/mop.28677
15. Toktas A, Ustun D (2020) Triple-objective optimization scheme using butterfly-integrated ABC algorithm for design of multilayer RAM. IEEE Trans Antennas Propag 68:5603–5612. https://doi.org/10.1109/TAP.2020.2981728
16. Akay B (2013) Synchronous and asynchronous pareto-based multi-objective artificial bee colony algorithms. J Glob Optim 57:415–445. https://doi.org/10.1007/s10898-012-9993-1
17. Harfouchi F, Habbi H, Ozturk C, Karaboga D (2018) Modified multiple search cooperative foraging strategy for improved artificial bee colony optimization with robustness analysis. Soft Comput 22:6371–6394. https://doi.org/10.1007/s00500-017-2689-1
18. Coello Coello CA (2018) Multi-objective optimization. In: Handbook of heuristics
19. Toktas A, Ustun D, Tekbas M (2019) Multi-objective design of multi-layer radar absorber using surrogate-based optimization. IEEE Trans Microw Theory Tech 67:3318–3329. https://doi.org/10.1109/TMTT.2019.2922600
20. Ustun D, Carbas S, Toktas A (2021) Multi-objective optimization of engineering design problems through pareto-based bat algorithm. In: Rajinikanth ND (ed) Applications of bat algorithm and its variants. Springer, Singapore, pp 19–43
21. Yang F, Rahmat-Sami Y (2019) Surface electromagnetics applications antenna microwave and optical engineering. Cambridge University Press, Cambridge
22. Bhartia P, Pramanick P (2016) Modern RF and microwave filter design. Artech House, Boston
23. Chakravarty S, Mittra R (2000) Design of microwave filters using a binary-coded genetic algorithm. Microw Opt Technol Lett 26:162–166. https://doi.org/10.1002/1098-2760(20000805)26:3%3c162::AID-MOP8%3e3.0.CO;2-B
24. Hoorfar A, Zhu J, Nelatury S (2003) Electromagnetic optimization using a mixed-parameter self-adaptive evolutionary algorithm. Microw Opt Technol Lett 39:267–271. https://doi.org/10.1002/mop.11187
25. Venkatarayalu NV, Ray T, Gan YB (2005) Multilayer dielectric filter design using a multiobjective evolutionary algorithm. IEEE Trans Antennas Propag 53:3625–3632. https://doi.org/10.1109/TAP.2005.858565
26. Steuer RE (1986) Multiple criteria optimization : theory, computation, and application. Wiley
27. Karaboga D (2010) Artificial bee colony algorithm. Scholarpedia 5:6915. https://doi.org/10.4249/scholarpedia.6915
28. Valkenburg V (1982) Analog filter design. Oxford University Press, Texas
29. Michielssen E, Sajer J-M, Ranjithan S, Mittra R (1993) Design of lightweight, broad-band microwave absorbers using genetic algorithms. IEEE Trans Microw Theory Tech 41:1024–1031. https://doi.org/10.1109/22.238519
30. Luo J, Liu Q, Yang Y et al (2017) An artificial bee colony algorithm for multi-objective optimisation. Appl Soft Comput J 50:235–251. https://doi.org/10.1016/j.asoc.2016.11.014
31. Akbari R, Hedayatzadeh R, Ziarati K, Hassanizadeh B (2012) A multi-objective artificial bee colony algorithm. Swarm Evol Comput 2:39–52. https://doi.org/10.1016/j.swevo.2011.08.001

Chapter 17
Multi-objective Sparse Signal Reconstruction in Compressed Sensing

Murat Emre Erkoc and Nurhan Karaboga

17.1 Introduction

According to the traditional sampling method, the frequency of sampling must be at least twice the maximum frequency of the signal to obtain the signal properly [1]. For practical systems, the sampling frequency can be much more than the signal frequency. This increases the cost and complexity of the hardware system [2]. According to the compressed sensing (CS) theory, certain signals can be recovered by using far fewer measurements than conventional methods use [3–6]. CS directly acquires compressed signals. In other means, it does the job of compression and sensing simultaneously. The basic principle of CS is the reconstruction of a sparse signal from very few linear and non-adaptive measurements by using sparse reconstruction methods. CS is a breakthrough topic that attracts the researchers in different branches of science and engineering. Therefore, lots of research papers exist in this field. Several of the fields where CS can be applied are imaging [7], radar target [8], communication [9], sensor network [10], and face recognition systems [11]. Particularly, CS caused significant contributions in medical imaging. Some companies have introduced CS method for faster magnetic resonance imaging (MRI) scans [12]. MRI can be a troublesome operation for some patients because patients need to be still during the scanning with MRI for a certain time. CS provides significant enhancements in timing by decreasing the number of samples [13].

The signal needs to be sparse for CS to be able to apply. Most of the entries of a sparse signal are zero. Some signals can be sparse in their original domain, while

M. E. Erkoc (✉) · N. Karaboga
Department of Electrical and Electronics Engineering, Erciyes University, Kayseri 38039, Melikgazi, Turkey
e-mail: merkoc@erciyes.edu.tr

N. Karaboga
e-mail: nurhan_k@erciyes.edu.tr

S. Carbas et al. (eds.), *Nature-Inspired Metaheuristic Algorithms for Engineering Optimization Applications*, Springer Tracts in Nature-Inspired Computing,
https://doi.org/10.1007/978-981-33-6773-9_17

some signals are sparse in transform domains. Another main process of CS is the measurement. The measurement is the process of taking very few samples from the original signals. Note that the measurement number is smaller than the length of the original signal. The measurement matrix needs to preserve the information of the signal. Thus, designing an efficient measurement matrix is still a promising research topic in CS. Moreover, the measurement matrix needs to satisfy the incoherence and restricted isometry property (RIP) conditions. The random Gaussian measurement matrix satisfies RIP and incoherence with high probability. Several types of efficient measurement matrices such as deterministic and structured have been studied in the literature [14]. Sparse signal reconstruction is another fundamental topic in CS. Sparse reconstruction algorithms rebuild the original signal from very few numbers of measurements. In literature, there have been many algorithms proposed to solve sparse reconstruction problems. In general, these algorithms can be classified as convex, non-convex, and greedy.

Lately, heuristic algorithms have been employed for optimizing various problems in literature [15–19]. Furthermore, multi-objective optimization algorithms (MOOA) have been presented to solve sparse recovery problem. Multi-objective optimization tries to solve the problem that has two or more objective functions. MOOAs are used in a wide range of fields such as antenna array design [20], data mining [21], bioinformatics [22], pattern recognition [23], wireless sensor network [24], image processing [25], and control systems and robotics [26].

In recent years, MOOAs have been used in CS to optimize the sparse reconstruction problem. They can optimize the sparsity and measurement error at the same time. They can eliminate some drawbacks of the other reconstruction algorithms. The sensitivity of regularization parameter and the necessity of prior knowledge of signal sparsity are some of the drawbacks of sparse reconstruction methods that MOOAs can eliminate. Furthermore, MOOAs can collaborate well with the classical methods for the efficient performance of the sparse reconstruction algorithm. Several of proposed methods in the literature are as following. A multi-objective sparse reconstruction (MOSR) method, called soft thresholding evolutionary multi-objective algorithm (StEMO), was proposed by Li et al. [27]. The method uses non-dominated sorting genetic algorithm II (NSGA II) as MOOA, and it optimizes the sparse reconstruction problem. Additionally, it uses iterative soft thresholding (IST) [28] method as a local search method. Another method, called MOEA/D-PGRASP, uses MOEA/DE algorithm to optimize the MOSR problem [29]. The method uses a probability-based greedy randomized search procedure as the local search technique. Another method based on hybrid evolutionary approach is called linear Bregman evolutionary algorithm (LBEA), proposed by Yan et al. [30]. Differential evolution algorithm is used to optimize the multi-objective problem, and linearized Bregman method is employed to improve local search capability of the proposed method. In the proposed approach, kink method [31] is used to obtain the knee point from Pareto front (PF). A preference-based multi-objective evolutionary algorithm for sparse reconstruction called sparse preference-based local search (SPLS) was presented in [32]. SPLS employs multi-objective evolutionary algorithm based on decomposition (MOEA/D) algorithm to optimize the multi-objective problem. Each of IST, iterative hard thresholding [33], and iterative half thresholding (IHT) [34] methods can be used as the local search method in SPLS. A multi-objective approach based on decomposition and Gaussian mixture clustering, called MOEA/D-GMC, for sparse optimization was proposed in another study [35]. In the method, a two-phase iterative search method is employed as the local search procedure. Additionally, MOEA/D-GMC finds the final solution by clustering via Gaussian mixture models. Yue et al. proposed a MOSR method, called multi-objective particle swarm optimization algorithm iterative half thresholding (MOPSO-iHalfT) [36]. A robust knee point selection method was designed to get the final solution for MOPSO-iHalft. MOPSO-iHalfT uses IHT for improving local search capability of the presented method.

The second section of this study explains the general principles of multi-objective optimization methods. In the third section, the fundamental elements of CS and MOSR in CS are explained. Then, the application of NSGA II algorithm in sparse signal reconstruction is presented in section four. In addition, the experiments of electrocardiogram (ECG) signal compression based on CS have been conducted to evaluate the reconstruction quality of the NSGA II-based method. Finally, the fifth section is the conclusion.

17.2 Multi-objective Optimization

An optimization problem can be defined as finding unknown parameter values satisfying certain limitations. The first step in the optimization process is defining a set of parameters called decision parameters. Then, depending on these parameters, a cost function to be minimized or a profit function to be maximized and the constraint functions related to the problem must be defined.

The optimization problems can be classified as single-objective and multi-objective optimization problems (MOOP). Single-objective optimization is the task of finding the optimal solution of problem which involves only one objective function. Some problems can have more than one cost function. These problems are called MOOP. MOOAs are used to optimize the problems with more than one objective. MOOAs try to optimize all the objectives simultaneously and can provide a variety of trade-off solutions among the objectives. For MOOP, there is not one single solution; instead there is a set of solutions. In general, these objectives are conflict with each other. Multi-objective optimization aims to solve the MOOP defined in Eq. 17.1 [37].

$$\min F(\boldsymbol{x}) = (f_1(\boldsymbol{x}), f_2(\boldsymbol{x}), \dots f_n(\boldsymbol{x}))^T s.t.\boldsymbol{x} \in \Omega \tag{17.1}$$

In Eq. 17.1, Ω is a decision space and $\boldsymbol{x} \in \Omega$ is a decision vector. *F(x)* consists of n objective functions f_i: $\Omega \rightarrow R$, $i = 1, \dots n$. In multi-objective optimization, the superiority of a solution over the other solutions is determined by using dominance. A vector $\boldsymbol{q} = (q_1, \dots, q_n)^T$ dominates another vector $\boldsymbol{v} = (\boldsymbol{v}_1, \dots, \boldsymbol{v}_n)^T$, denoted as $\boldsymbol{q} \preceq \boldsymbol{v}$, if and only if $\forall i \in \{1, 2, \dots, n\}$, $\boldsymbol{q}_i \leq \boldsymbol{v}_i$ and $\boldsymbol{q} \neq \boldsymbol{v}$ [37]. The example dominance test is shown in Fig. 17.1.

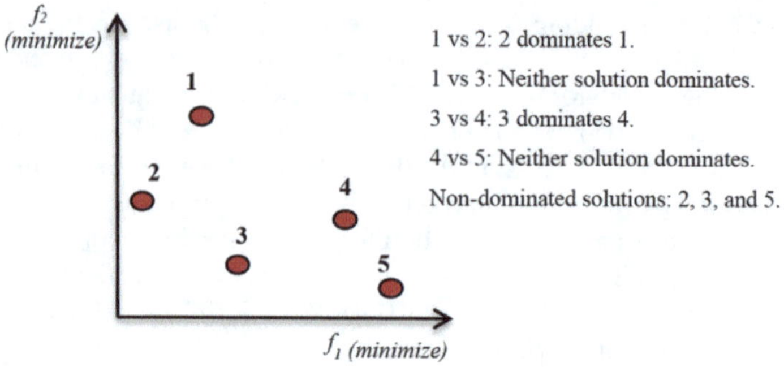

Fig. 17.1 Examples of the dominance test

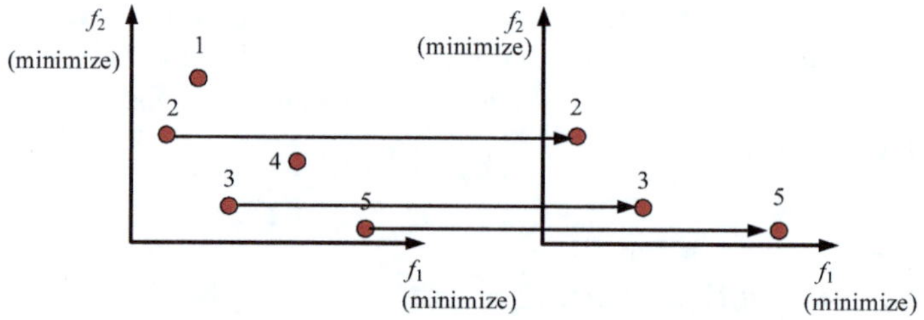

Fig. 17.2 Non-dominated solutions

The solutions not dominated by any other member of the solution are called non-dominated solutions. A solution $\boldsymbol{x}$ can be described as Pareto optimal solution if and only if there is no $\boldsymbol{x}' \in \Omega$ such that $\boldsymbol{x}' < x$. In Fig. 17.2, the solutions 2, 3, and 5 are non-dominated set of solutions. The 1 and 4 are dominated by at least one solution.

The boundary defined by the set of all point mapped from the Pareto optimal solutions is defined as the PF. The goal of multi-objective optimization is to obtain solutions as close as possible to the PF.

There have been several algorithms proposed to solve MOOPs. Some of them are following: vector evaluated genetic algorithm (VEGA) [38], multi-objective genetic algorithm (MOGA) [39], NSGA [40], Pareto archived evolution strategy (PAES) [41], Pareto envelope-based selection algorithm (PESA) [42], NSGA II [43], multi-objective particle swarm optimization (MOPSO) [44], and MOEA/D [45].

On the contrary to single-objective optimization, MOOAs try to find a set of solutions. In order to find a final solution from the set of solutions, a decision-maker needs to decide to obtain the final solution. A simple PF with a knee is illustrated in Fig. 17.3.

In Fig. 17.3, solution *S* is on the knee region. A decision-maker needs to decide according to the importance of the objective functions. Knee point in PF is an important region where small improvement in one of the objectives of the solution on the region causes a larger deterioration of the other objective relative to the other Pareto optimal solutions located away from the knee region. If there is no special preference

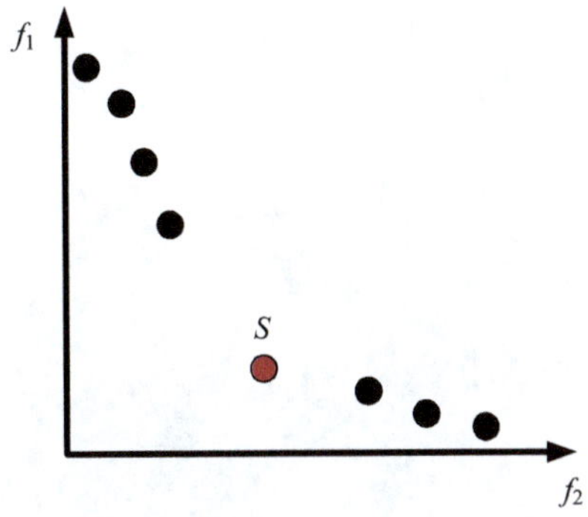

Fig. 17.3 Simple PF with a knee point

of a decision-maker, knee point is a first choice solution in the PF. Several knee point selection methods have been proposed in the literature [46]. In a study done by Das, a normal boundary intersection method was proposed to obtain the knee point [47]. In another study, an angle-based technique was presented by Branke [48]. Another method, designed for knee point selection, computes the weighed sum of the two conflict objectives and selects the point with the maximum value as a knee point [49]. Zhang proposed a knee point finding method based on the distance of each solution on the PF to the extreme line [50].

17.3 Compressed Sensing

CS method can be analyzed in three fundamental processes. These are making the signal sparse, measurement/sampling, and recovery of the sparse signal. A signal needs to be sparse or compressible to be able to apply the CS. A signal is called sparse if a large number of its entries are zero and called compressible if a large number of the entries are very small compared to the other entries. A signal $\boldsymbol{x} \in R^N$ can be represented as the linear combination of basis vectors as Eq. 17.2.

$$\boldsymbol{x} = \sum_{i=1}^{N} s_i \boldsymbol{\psi}_i \quad \text{or} \quad \boldsymbol{x} = \boldsymbol{\psi} \boldsymbol{s} \tag{17.2}$$

The vector $\boldsymbol{s}$ is another representation of the signal $\boldsymbol{x}$ in transform domain. The signal $\boldsymbol{x}$ is called *k-sparse* if only k number of s_i coefficients in Eq. 17.2 are nonzero and the other coefficients are zero. Most signals are sparse or compressible in its original domain or representation domain. Fourier, wavelet and discrete cosine transform can be used to obtain the sparse representation of signals. For example, ECG and image signals are sparse or compressible with respect to the basis of wavelets. Figure 17.4 illustrates *cameraman* image and wavelet coefficients of the image.

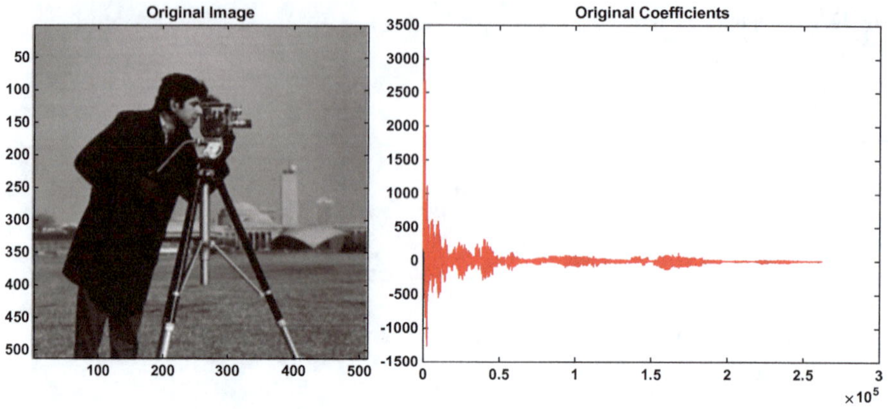

Fig. 17.4 Original cameraman image and its wavelet coefficients

As illustrated in Fig. 17.4, the number of large coefficients is less than the number of small coefficients. In addition, the large coefficients have a significant contribution to the construction of the image. When some small coefficients are discarded, there is not too much loss in the quality of the image. Figure 17.5 shows the discarded coefficients, and *cameraman* image which is reconstructed from these discarded coefficients.

In Fig. 17.5, the 90.8% of the wavelet coefficients of the image are discarded, and the image is reconstructed by using the remaining 9.2% of coefficients. As seen in Figs. 17.4 and 17.5, the reconstructed image obtained from the largest 9.2% coefficients is nearly indistinguishable from the original image. Additionally, SNR of the reconstructed image is 37.1804 and the structural similarity index measure (SSIM) [51] is 0.9705.

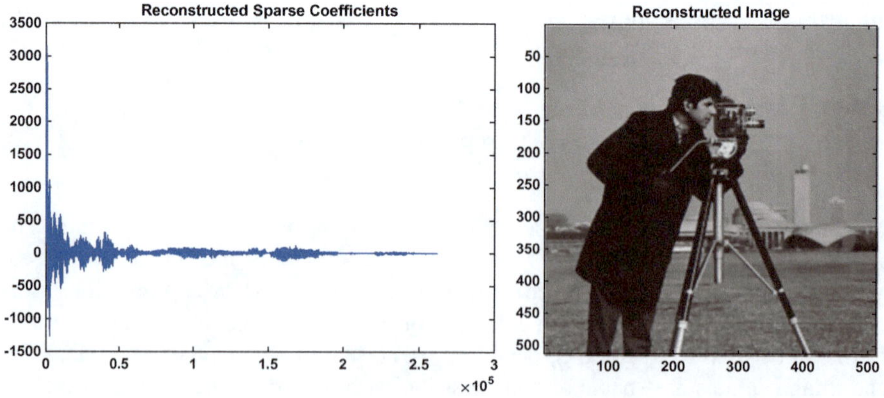

Fig. 17.5 Reconstructed wavelet coefficients and reconstructed image

Another main issue in CS is the measurement. The measurement in CS can be described as taking M measurement from N length test signals by using the measurement matrix. Note that M is smaller than N. This can be expressed mathematically in Eq. 17.3.

$$\boldsymbol{y} = A\boldsymbol{x} \tag{17.3}$$

where A is an MxN measurement/sensing matrix and $\boldsymbol{y} \in R^M$ is the observation vector. When noise is considered in the measurement process, Eq. 17.3 becomes as in Eq. 17.4.

$$\boldsymbol{y} = A\boldsymbol{x} + \boldsymbol{n}, \quad \boldsymbol{n} \in R^M \tag{17.4}$$

In the measurement process, the signal can be sampled with very few numbers of measurements. The measurement matrix needs to preserve the information in $\boldsymbol{x}$ signal for perfect reconstruction. Therefore, the measurement matrix is required to satisfy certain conditions such as RIP and incoherence. A matrix A satisfies the RIP if there exists a constant $\delta_k(0, 1)$ such that

$$(1 - \delta_k)||\boldsymbol{x}||_2^2 \leq ||A\boldsymbol{x}||_2^2 \leq (1 + \delta_k)||\boldsymbol{x}||_2^2 \tag{17.5}$$

for any k-sparse $\boldsymbol{x}$. When this property holds, any k-sparse vectors cannot be in the null space of the measurement matrix, and it is possible to recover the signal. In other words, all subsets of k-columns obtained from the measurement matrix are nearly orthogonal [6]. Another important issue related to the measurement matrix and the representation matrix is the incoherence. Coherence is the measure of the correlation between any two elements of measurement matrix and representation matrix. A is a measurement matrix with row vectors $A_1, A_2, A_3 \ldots .A_m$, and ψ is a representation matrix with column vectors $\psi_1, \psi_2 \ldots \psi_N$ The coherence of these two matrices is calculated as in Eq. 17.6.

$$\mu(A, \psi) = \sqrt{n}. \max_{1 \leq k,j \leq n} A_k \psi_j \tag{17.6}$$

In Eq. 17.6, $\mu(A, \psi)$ is between 1 and $\sqrt{n}$. When the coherence is 1, the correlation between the two matrices is minimum. In addition, when this value is $\sqrt{n}$, the correlation is maximum. In CS theory, the coherence between the measurement matrix and representation matrix needs to be minimum for the reconstruction of the sparse signal. The example of low coherence measurement matrix and basis matrix pairs is spike and Fourier basis. These two matrices are incoherence for all dimensions [6].

Random Gaussian matrices satisfy these conditions with high probability [52]. However, random matrices need to be transmitted with the signal because the same random matrix cannot be produced at the receiver. This is not efficient for practical applications. For that reason, efficient measurement matrix designs such as deterministic and structured have been proposed in the literature [14].

After the measurement process, the signal is reconstructed from the underdetermined system. Sparse signal reconstruction algorithms are employed to reconstruct the sparse signal. In general, they use as a prior knowledge of the sparsity level of the signal, measurement matrix, and observation vector. The signal can be recovered from the observation vector by solving the underdetermined system of Eq. 17.3. Equation 17.3 has the infinite number of possible solutions due to becoming an underdetermined system. The sparse reconstruction problem of the signal can be expressed as in Eq. 17.7.

$$\min_{x} ||\boldsymbol{x}||_0 \; s.t. \; \boldsymbol{y} = A\boldsymbol{x} \tag{17.7}$$

In general, the problem in Eq. 17.7 can be converted into following l_0 regularization problem as Eq. 17.8.

$$\min_{x \in R^N} \left\{ ||\boldsymbol{y} - A\boldsymbol{x}||_2^2 + \lambda ||\boldsymbol{x}||_0 \right\} \tag{17.8}$$

where $||\boldsymbol{x}||_0$ is called l_0-norm that counts the number of nonzero entries of the signal. However, the solution of this l_0-norm optimization is NP-hard and requires exhaustive enumeration. Several methods have been employed to deal with this problem. These methods are matching pursuit [53], orthogonal matching pursuit [54], compressed sampling matching pursuit [55], and iterative hard thresholding algorithm [33]. In order to deal with this problem more effectively, l_0-norm replaced with l_1-norm convex optimization problem in Eq. 17.9.

$$\min_{x \in R^N} \left\{ ||\boldsymbol{y} - A\boldsymbol{x}||_2^2 + \lambda ||\boldsymbol{x}||_1 \right\} \tag{17.9}$$

L_1-norm problem can be solved by using convex optimization techniques. L_1 minimization algorithms are among the most popular sparse signal reconstruction algorithms and widely used in CS. Additionally, l_1 minimization algorithms are robust to noise. Several l_1 minimization algorithms are basis pursuit [56], least absolute shrinkage and selection operator [57], IST [28], primal dual interior point method, and homotopy [58]. Additionally, l_1-norm regularization can be converted into l_q-norm as in Eq. 17.10.

$$\min_{x \in R^N} \left\{ ||\boldsymbol{y} - A\boldsymbol{x}||_2^2 + \lambda ||\boldsymbol{x}||_q^q \right\} \text{where } 0 < q < 1 \tag{17.10}$$

This is called the non-convex approach. This approach can optimize the sparse reconstruction problem with a fewer number of measurements than the convex approach can. Moreover, a weaker version of RIP condition is sufficient for perfect reconstruction in non-convex approach [14]. The implementation of the non-convex approach, likewise convex approach, is difficult for large-sized problems. Note that finding an optimal regularization parameter (λ) in Eq. 17.10 is difficult and critical because the performance of the reconstruction algorithms is very sensitive to the regularization parameter and poor choice of the regularization parameter strongly affects the performance of the algorithms. The sparse reconstruction problem can be converted into a multi-objective optimization problem to overcome the drawbacks of choosing the regularization parameter [27]. The sparse reconstruction problem can be designed as a multi-objective optimization problem as in Eq. 17.11.

$$\min_x F(x) = \min_x \left(||x||_0, ||Ax - y||_2^2 \right) \tag{17.11}$$

In Eq. 17.11, the first term is the sparsity inducing term, and the second term is the measurement error term of the reconstruction. In general, the proposed studies in the literature mentioned in the 'Introduction' part of this study use Eq. 17.11 as the objective function and try to optimize the magnitude and sparsity of the signal simultaneously. For that reason, there is no need to know the prior knowledge of signal sparsity in MOSR methods. Besides this, these methods use the advantage of local search methods such as iterative hard thresholding, IST, and IHT. The local search method can enhance the acceleration of the convergence of MOSR algorithms.

17.4 Multi-objective Sparse Reconstruction

Lately, MOOAs have been started to be employed in the reconstruction of sparse signals. Using MOOAs in CS has several advantages such as eliminating dependencies on some sensitive algorithm parameters and dependencies on prior knowledge of the signal sparsity. In general, knowing the information of the signal sparsity in advance is not possible in real-time applications. Most of the sparse signal reconstruction algorithms need to have sparsity information in advance, before optimizing sparse reconstruction problem. MOOA in the sparse reconstruction optimizes the measurement and sparsity terms, and it is not necessary to know the information of the signal sparsity in advance. Additionally, MOOAs use the benefit of their global search capability when optimizing the sparse reconstruction problem. There exist several multi-objective methods presented to solve the sparse signal optimization problem. Most of these algorithms use a local search method to accelerate the convergence and to improve the performance of the reconstruction. Widely used local search methods in the literature are IHT, IST, and iterative hard thresholding methods. Apart from these methods, linearized Bregman method has been used in the literature [30]. In addition, a final solution identification method is needed to obtain a solution from PF. The widely used final solution selection methods in the literature are weighted sum method, angle-based method, and kink method. Besides, clustering via Gaussian mixture models and preference-based method on the true k-sparse solution methods have been used as the final solution identification technique in the literature [32, 35]. A framework to the MOSR method based on NSGA II algorithm is as follows:

1. Initialize the population.
2. Sort the initialized population based on non-dominated domination principle.
3. Assign the crowding distance value.
4. Select the individual in population by using tournament selection with crowded comparison operator.
5. Perform genetic operators, crossover, and mutation.
6. Apply recombination and selection processes.
7. Apply local search method based on IHT.
8. Use final solution identification method.

In this study, NSGA II algorithm is employed for optimization of sparse reconstruction problem, and IHT algorithm is used as the local search method. The steps of the IHT method used in this study are as following [34, 59].

Inputs:

- A: Measurement matrix.
- y: Observation vector.
- $\boldsymbol{x}_{\boldsymbol{initial}}$: Initial solution.
- k: Sparsity level of the signal.
- nls: Iteration number of the local search.

Output:

- $\boldsymbol{x}$: Sparse signal.

Steps

(1) Set $t = 0$ and the current solution $\boldsymbol{x}^{(t)} = \boldsymbol{x}_{\boldsymbol{initial}}$
(2) Generate a trial solution $\tilde{\boldsymbol{x}}$ by gradient descent as follows:

$$\tilde{\boldsymbol{x}} = \boldsymbol{x}^{(t)} + \mu \times A^T \left(A\boldsymbol{x}^{(t)} - \boldsymbol{b} \right) \tag{17.12}$$

(3) Sort the components of the trial solution as in Eq. 17.13.

$$|\tilde{\boldsymbol{x}}_{i_1}| \geq |\tilde{\boldsymbol{x}}_{i_2}| \geq \ldots \geq |\tilde{\boldsymbol{x}}_{i_n}| \tag{17.13}$$

(4) The regularization parameter is set to Eq. 17.14.

$$\lambda_t = \frac{\sqrt{96}}{9} |\tilde{\boldsymbol{x}}_{i_{\bar{k}+1}}|^{3/2} \tag{17.14}$$

(5) Some smallest components of the trial signal are set to zeros to obtain the sparsity.

$$x_i^{(t+1)} = \begin{cases} \psi_2(\tilde{x}_i) & \text{if } |\tilde{x}_i| > \frac{\sqrt[3]{54}\lambda_t^{3/2}}{4} \\ 0 & \text{otherwise} \end{cases} \tag{17.15}$$

$$\psi_2(\tilde{x}_i) = \left(\frac{2}{3}\right)\tilde{x}_i\left(1 + \cos\left(\left(\frac{2\pi}{3}\right) - \left(\frac{2\phi(\tilde{x}_i)}{3}\right)\right)\right) \tag{17.16}$$

$$\phi(\tilde{x}_i) = \arccos\left(\left(\frac{\lambda_t}{8}\right)\left(\frac{|\tilde{x}_i|}{3}\right)^{-3/2}\right) \tag{17.17}$$

(6) If $t > nls$, then return $\boldsymbol{x}$. Otherwise, set $t = t + 1$ and go to **Step** 2.

Moreover, to verify the effectiveness of the designed method, it is implemented in ECG signal compression in this study.

17.4.1 ECG Signal Compression

The ECG data compression is necessary for extracting information from heart signals stored long term. In addition, there exist some wearable ECG recorder that needs to be energy efficient and doing simple tasks. In some applications, the ECG data is transmitted on communication channel. ECG data compression is a vital task for efficient use of the bandwidth and storage capacity. There exists numerous ECG data compression techniques proposed in the literature [60–62]. In recent years, CS methods have been started to be used in ECG data compression. The advantages of CS method in ECG data compression are following [63]: (1) CS provides less computational burden for the encoder. (2) The location of the wavelet coefficients of the ECG signals does not need to be encoded. In this study, NSGA II-based CS method is used in ECG data compression. NSGA II is employed in sparse reconstruction of the compressed ECG signals.

In this section of the study, the ECG samples recorded from MIT–BIH arrhythmia database are used in this ECG signal compression example used to evaluate the performance of the NSGA II algorithm [64]. The recorded samples are 100, 101, 103, 117, and 207. The signal length of all ECG samples used in this study is 3328. These ECG signals are not sparse in the original domain. Therefore, the wavelet transform is applied to obtain the sparse representation of the signals. The coefficients that have little contribution to the signal are discarded to obtain the sparse representation of the signal. Then, CS-based compression is applied to the coefficients of the ECG signal. In this study, the sparse reconstruction algorithm based on NSGA II algorithm is applied to reconstruct the sparse signal in CS. SNR and percentage root means square difference (PRD) are used as performance metrics of the reconstruction. SNR is calculated by using Eq. 17.18, and PRD is calculated by using Eq. 17.19.

$$\mathrm{SNR} = -20\log_{10}(\mathrm{PRD}/100) \tag{17.18}$$

$$\mathrm{PRD} = \sqrt{\frac{\sum_{i=0}^{N-1}(\boldsymbol{x}(n) - \tilde{\boldsymbol{x}}(n))^2}{\sum_{i=0}^{N-1}\boldsymbol{x}^2(n)}} \tag{17.19}$$

According to the following studies, the obtained performance metrics can be interpreted as shown in Table 17.1 [65, 66].

The original ECG signal of the record number 100 is given in Fig. 17.6. As shown in Fig. 17.6, the signal is not sparse. Therefore, wavelet transformation is applied by using a two-level Sym4 wavelet family.

The obtained coefficients of the ECG sample 100 are presented in Fig. 17.7. The number of nonzero coefficients of the signal is 3328. The wavelet coefficient that has little contribution to the signal is discarded from the coefficients to get the sparse representation. After discarding small coefficients below a threshold value, the number of the obtained nonzero coefficients is 1208.

Table 17.1 Evaluation of the performance results [66]

Quality	'Very good'	'Good'	'Undetermined'
PRD	$0 < \mathrm{PRD} < 2\%$	$2\% < \mathrm{PRD} < 9\%$	$\mathrm{PRD} \geq 9\%$
SNR	$\mathrm{SNR} > 33$ dB	20 dB $< \mathrm{SNR} < 33$ dB	$\mathrm{SNR} \leq 20$ dB

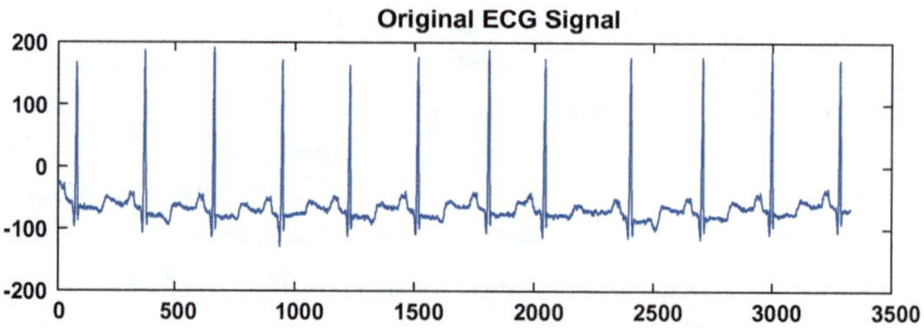

Fig. 17.6 Original signal recorded from MIT–BIH arrhythmia sample number 100

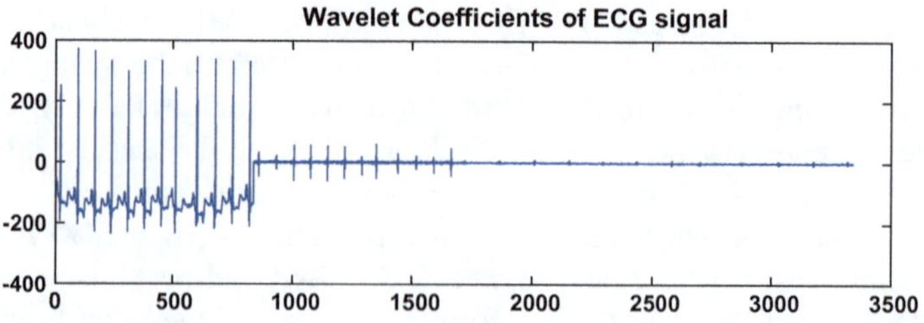

Fig. 17.7 Wavelet coefficients of the sample 100

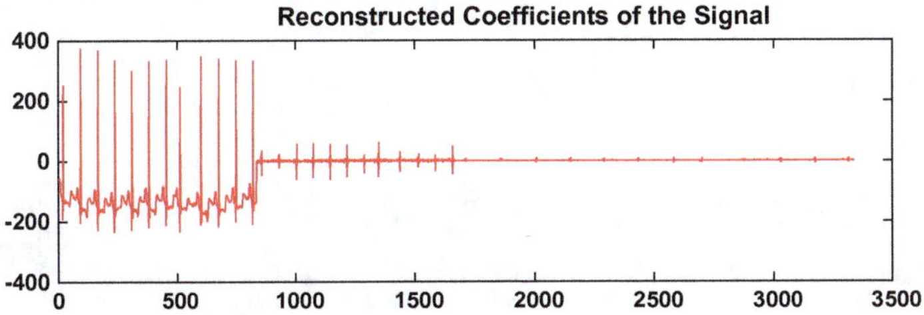

Fig. 17.8 Reconstructed wavelet coefficients of the sample 100

In this case, the sparse reconstruction method based on NSGA II reconstructs a signal with sparsity value 1208, a length of 3328, and measurement number 2327. After NSGA II-based sparse reconstruction method is applied to this signal, the reconstructed coefficients are shown in Fig. 17.8.

The reconstruction error (RE) of this operation is 0.0142. As seen in Figs. 17.7 and 17.8, the difference between original coefficients and reconstructed coefficients is very small and indistinguishable, although many coefficients are discarded. After reconstruction of coefficients, the ECG signal is reconstructed from the reconstructed coefficients by using inverse wavelet transform. The reconstructed signal and original signal after compression are illustrated in Fig. 17.9.

The PRD and SNR of the reconstructed ECG signal are 1.4197 and 36.9558, respectively. According to Table 17.1, the reconstruction performance of the sample 100 ECG signal after compression can be expressed as 'very good'.

The original ECG signal of record number 101 is demonstrated in Fig. 17.10. The ECG signal is not sparse as shown in Fig. 17.10. Therefore, wavelet transform is applied by using two-level Sym4 wavelet family.

The obtained coefficients of the ECG sample 101 are given in Fig. 17.11. The number of nonzero coefficients of the signal is 3328. After discarding coefficients that have small contribution to the signal, the number of the obtained nonzero coefficients is 1273.

In this case, the sparse reconstruction method based on NSGA II reconstructs a signal with sparsity value 1273, a length of 3328, and measurement number 2327.

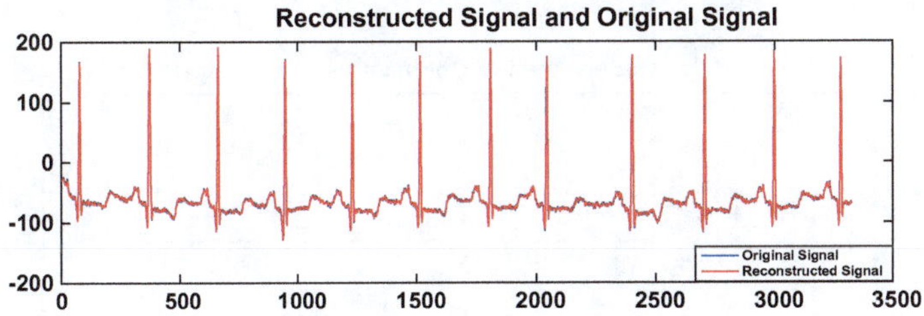

Fig. 17.9 Reconstructed and original signal of sample 100

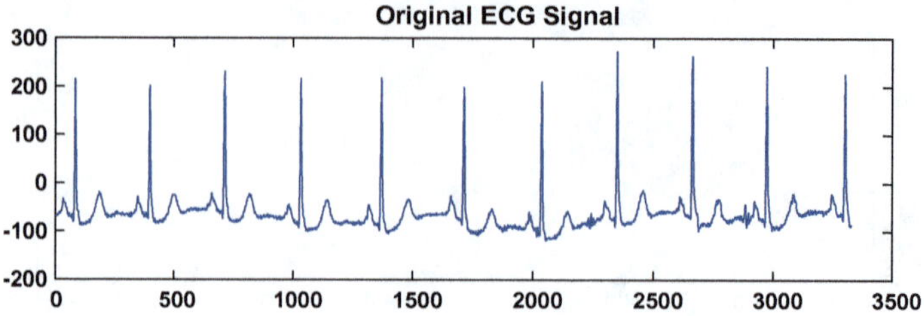

Fig. 17.10 Original signal recorded from MIT–BIH arrhythmia sample number 101

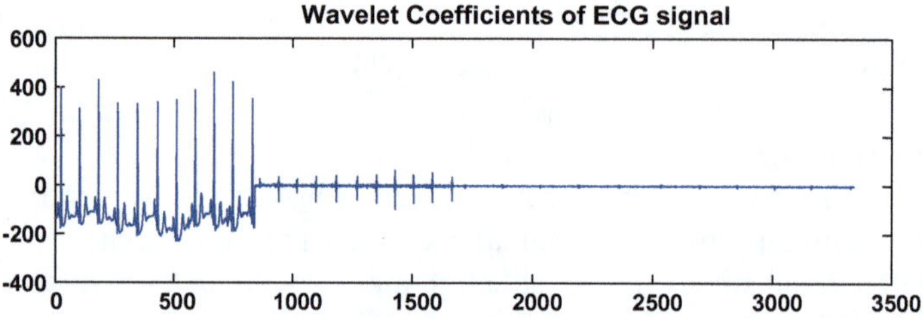

Fig. 17.11 Wavelet coefficients of the sample 101

After NSGA II-based sparse reconstruction method is applied to this signal, the reconstructed coefficients are seen in Fig. 17.12.

The RE of this operation is 0.0139. Although many coefficients are discarded, the difference between original coefficients and reconstructed coefficients is very small and indistinguishable as seen in Figs. 17.11 and 17.12. After reconstruction of coefficients, the signal is obtained from the reconstructed coefficients by using the inverse wavelet transform. The reconstructed signal and the original signal after compression are illustrated in Fig. 17.13.

The PRD and SNR of the reconstructed ECG signal are 1.3888 and 37.1473, respectively. According to Table 17.1, the reconstruction performance of the sample 101 ECG signal after compression can be expressed as ‘very good’.

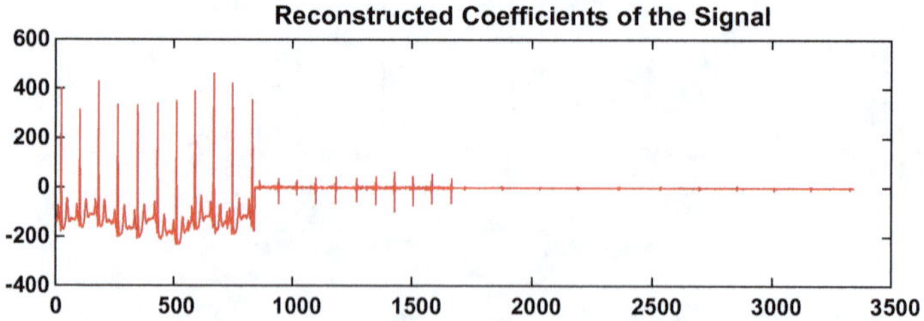

Fig. 17.12 Reconstructed wavelet coefficients of the sample 101

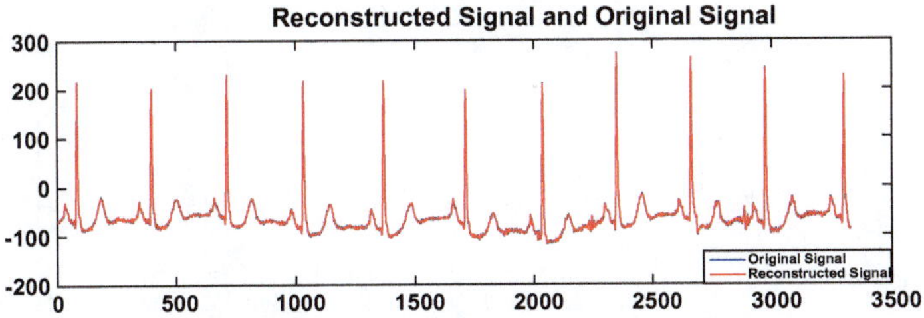

Fig. 17.13 Reconstructed and original signal

The original ECG signal of record number 103 is shown in Fig. 17.14. The signal is not sparse in the original domain. Therefore, the wavelet transformation is applied by using two-level Sym4 wavelet family.

The obtained coefficients of the ECG sample 103 are shown in Fig. 17.15. The number of nonzero coefficients of the signal is 3328. The wavelet coefficient that has small contribution to the signal is discarded from the coefficients for the aim of obtaining the sparse representation. After discarding small coefficients below a threshold value, the number of the obtained nonzero coefficients is 1174.

In this case, the sparse reconstruction method based on NSGA II reconstructs a signal with sparsity value 1174, a length of 3328, and measurement number 2327. After NSGA II-based sparse reconstruction method is applied to this signal, the reconstructed coefficients are given in Fig. 17.16.

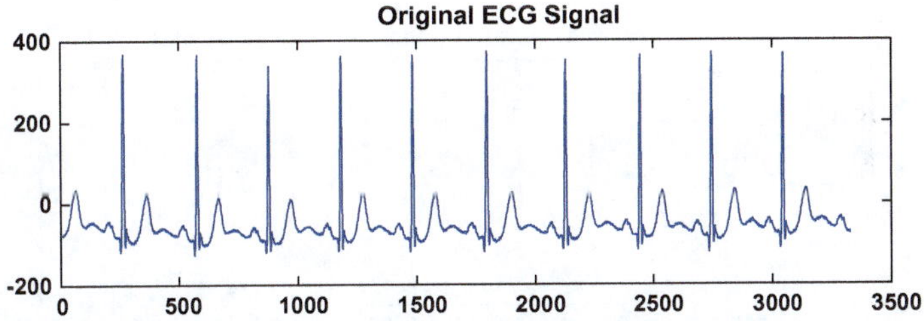

Fig. 17.14 Original signal recorded from MIT–BIH arrhythmia sample number 103

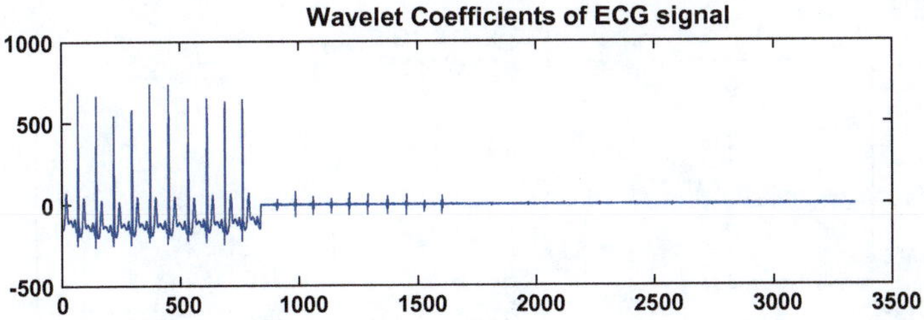

Fig. 17.15 Wavelet coefficients of the sample 103

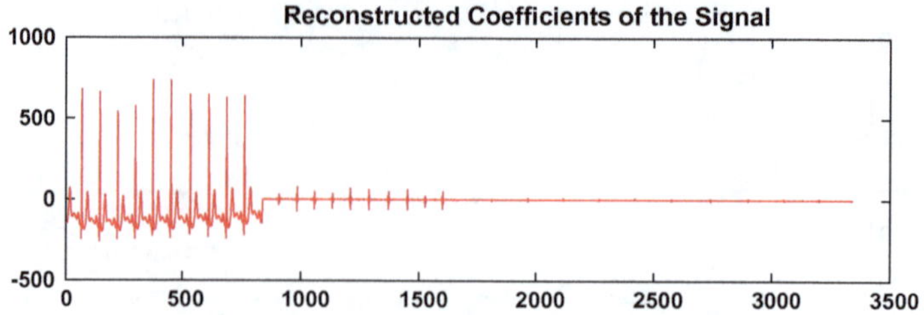

Fig. 17.16 Reconstructed wavelet coefficients of the sample 103

The RE of this operation is 0.0140. As seen in Figs. 17.15 and 17.16, the difference between original coefficients and reconstructed coefficients is very small and indistinguishable. After reconstruction of coefficients, the ECG signal is rebuilt from the reconstructed coefficients. The reconstructed signal and original signal after compression are shown in Fig. 17.17.

The PRD and SNR of the reconstructed ECG signal are 1.3981 and 37.0893, respectively. According to Table 17.1, the reconstruction performance of this ECG signal after compression can be expressed as 'very good'.

The original ECG signal of record number 117 is shown in Fig. 17.18. As shown in Fig. 17.18, the signal is not sparse. Therefore, the wavelet transformation is applied by using two-level Sym4 wavelet family.

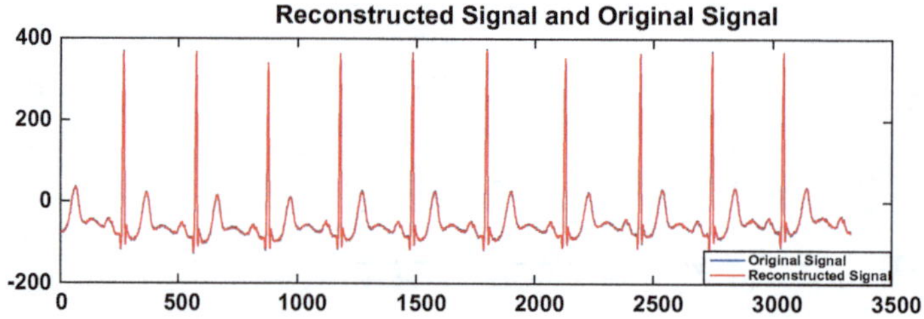

Fig. 17.17 Reconstructed and original signal 103

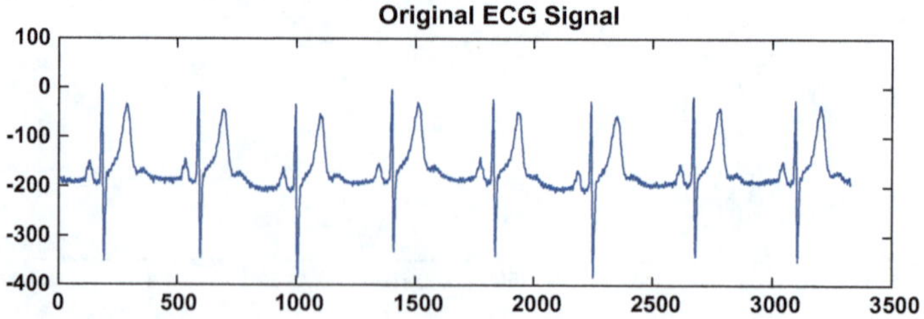

Fig. 17.18 Original signal recorded from MIT–BIH arrhythmia sample number 117

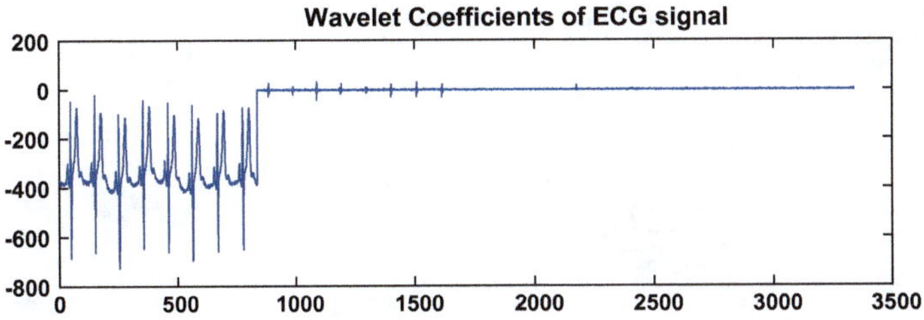

Fig. 17.19 Wavelet coefficients of the sample 117

The obtained coefficients of the ECG sample 117 are shown in Fig. 17.19. The number of nonzero coefficients of the signal is 3328. The wavelet coefficient that has little contribution to the signal is discarded from the coefficients to make the coefficients sparse. After discarding small coefficients below a threshold value, the number o f he obtained nonzero coefficients is 1062.

In this case, the sparse reconstruction method based on NSGA II reconstructs a signal with sparsity value 1062, a length of 3328, and measurement number 2327. After NSGA II-based sparse reconstruction method is applied to this signal, the reconstructed coefficients are shown in Fig. 17.20.

The RE of this operation is 0.0062. As seen in Figs. 17.19 and 17.20, the difference between original coefficients and reconstructed coefficients is very small and indistinguishable from the figures. After reconstruction of coefficients, the ECG signal is reconstructed from the reconstructed coefficients by using the inverse wavelet transform. The reconstructed signal after the compression and original signal is illustrated in Fig. 17.21.

The PRD and SNR of the reconstructed ECG signal are 0.6261 and 44.0668, respectively. According to Table 17.1, the reconstruction performance of this ECG signal after compression can be said as 'very good'. Compared to the other ECG signal tested in this study, the reconstruction results of this signal are a bit better than others. This is caused from the lower level of the sparsity.

The original ECG signal of sample 207 is shown in Fig. 17.22. The signal is not sparse as seen in Fig. 17.22. For that reason, the wavelet transform is applied by using two-level Sym4 wavelet family.

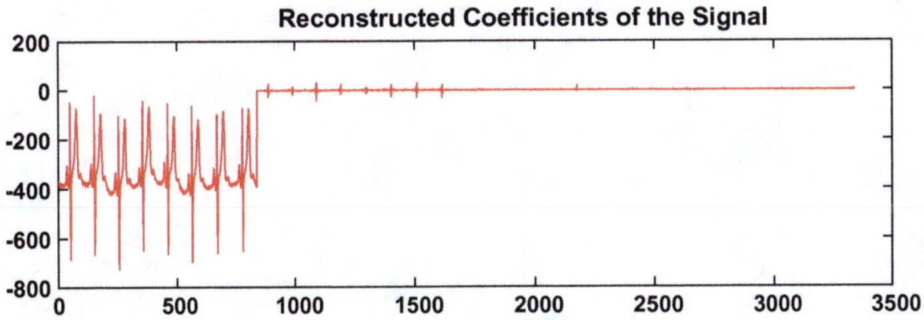

Fig. 17.20 Reconstructed wavelet coefficients of the sample 117

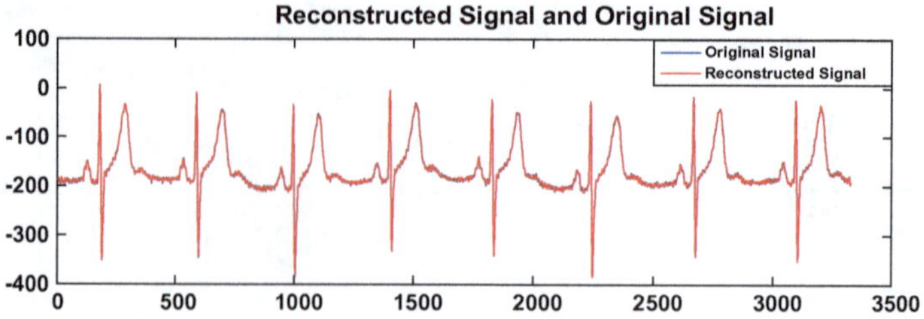

Fig. 17.21 Reconstructed and original signal 117

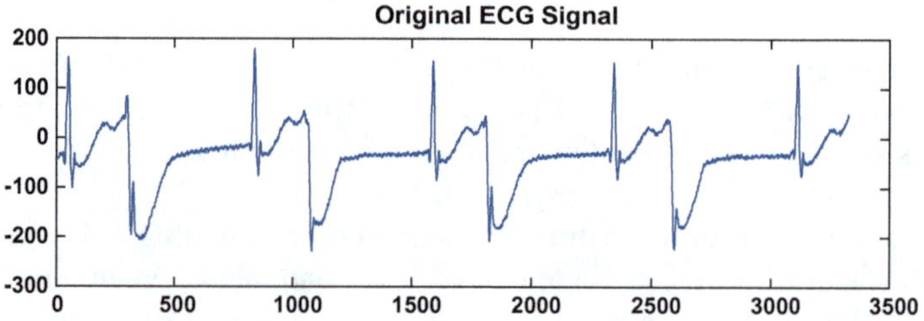

Fig. 17.22 Original signal recorded from MIT–BIH arrhythmia sample number 207

The obtained coefficients of ECG sample 207 are given in Fig. 17.23. The number of nonzero coefficients of the signal is 3328. The wavelet coefficient that has the small contribution to the signal is discarded from the coefficients to obtain the sparse representation of the signal. After discarding small coefficients below a threshold value, the number of the obtained nonzero coefficients is 1297.

In this case, the sparse reconstruction method based on NSGA II reconstructs a signal with sparsity value 1297, a length of 3328, and measurement number 2327. After NSGA II-based sparse reconstruction method is applied to this signal, the reconstructed coefficients are demonstrated in Fig. 17.24.

The RE of this operation is 0.0161. As seen in Figs. 17.23 and 17.24, the difference between original coefficients and reconstructed coefficients is very small and indistinguishable, although many coefficients are discarded. After reconstruction of

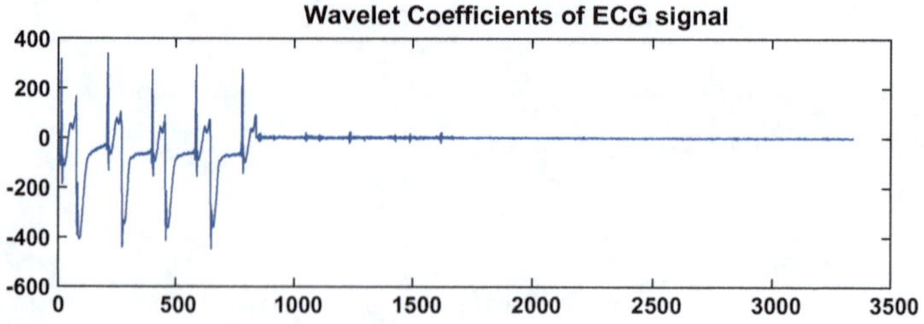

Fig. 17.23 Wavelet coefficients of the sample 207

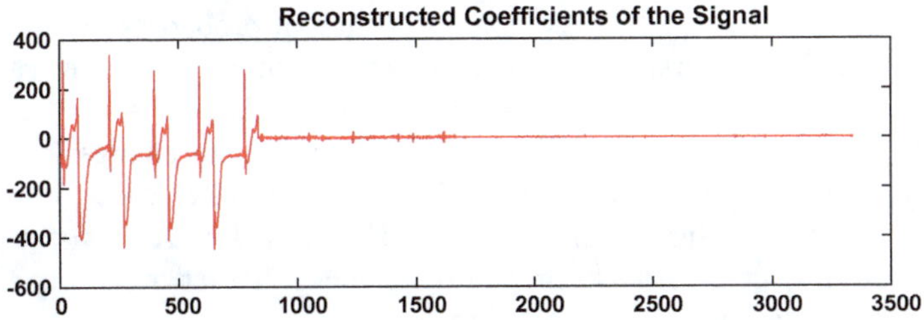

Fig. 17.24 Reconstructed wavelet coefficients of the sample 207

coefficients, the ECG signal is reconstructed from the reconstructed coefficients by using the inverse wavelet transform. The reconstructed signal after compression and original signal are shown in Fig. 17.25.

The PRD and SNR of the reconstructed ECG signal are 1.6113 and 35.8565, respectively. According to Table 17.1, the reconstruction performance of sample 207 ECG signal after compression can be defined as 'very good'. Compared to the other ECG signal tested in this study, the reconstruction results of this signal are a bit worse than others. This is caused from the high level of sparsity.

The PRD, SNR, and RE values belong to the ECG signals are summarized in Table 17.2.

According to Tables 17.1 and 17.2, PRD, SNR, and RE results of compressions by using NSGA II-based CS are 'very good' for all samples of the ECG signal in this study. As it is seen from the examples that the performance results of ECG samples are close to each other. The reason for the small difference between reconstructions

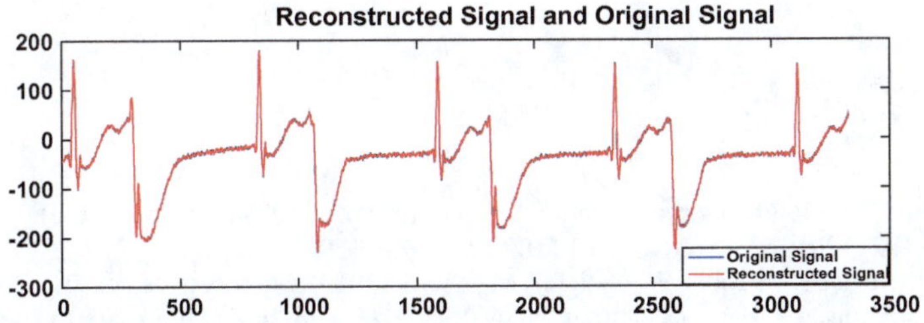

Fig. 17.25 Reconstructed and original signal 207

Table 17.2 PRD, SNR, and RE values belong to the ECG signals

Number of record sample	PRD	SNR(dB)	RE
100	1.4197	36.9558	0.0142
101	1.3888	37.1473	0.0139
103	1.3981	37.0893	0.0140
117	0.6261	44.0668	0.0062
207	1.6113	35.8565	0.0161

of the samples is that every ECG sample has different properties. Therefore, each sample has a different sparsity level. The property of the signal might vary by age and gender. As seen in the examples, all ECG samples have sparse representation by using the wavelet transform. The representation domain changes with respect to signal type. For example, the wavelet, Gabor, and discrete cosine transform can be used for electroencephalogram signals. Additionally, for the image signals, the wavelet, or discrete cosine transform can be applied. Fourier transform is suitable for sparse representation of sinusoidal signals.

17.5 Conclusion

The multi-objective optimization methods are used to optimize the problems having more than one objective function simultaneously. They can be employed to optimize the problems in many fields such as science, engineering, and industry. Recently, they have been used in solving the sparse reconstruction problem in CS. In this study, firstly, multi-objective optimization in general is explained. Then, the basic principles of CS and MOSR are explained, and the MOSR methods in the literature are summarized. In addition, the sparse reconstruction method based on NSGA II is designed. The efficiency of the designed sparse reconstruction method is examined in CS-based ECG signal compression. The CS-based compression technique is applied to five different ECG signals recorded from Physionet database. PRD, SNR, and RE are used as the performance metrics of NSGA II-based reconstruction method. According to the obtained results from this study, NSGA II-based sparse reconstruction technique is efficient for CS-based ECG compression.

References

1. Shannon CE (1984) Communication in the presence of noise (reprinted). Proc IEEE 72(9):1192–1201. https://doi.org/10.1109/Proc.1984.12998
2. Qaisar S, Bilal RM, Iqbal W, Naureen M, Lee S (2013) Compressive sensing: from theory to applications, a survey. J Commun Netw 15(5):443–456. https://doi.org/10.1109/Jcn.2013.000083
3. Donoho DL (2006) Compressed sensing. IEEE Trans Inf Theory 52(4):1289–1306. https://doi.org/10.1109/Tit.2006.871582
4. Candes EJ, Romberg J, Tao T (2006) Robust uncertainty principles: exact signal reconstruction from highly incomplete frequency information. IEEE Trans Inf Theory 52(2):489–509. https://doi.org/10.1109/Tit.2005.862083
5. Candes EJ (2008) The restricted isometry property and its implications for compressed sensing. CR Math 346(9–10):589–592. https://doi.org/10.1016/j.crma.2008.03.014
6. Candes EJ, Wakin MB (2008) An introduction to compressive sampling. IEEE Signal Process Mag 25(2):21–30. https://doi.org/10.1109/Msp.2007.914731
7. Otazo R, Candes E, Sodickson DK (2015) Low-Rank plus sparse matrix decomposition for accelerated dynamic MRI with separation of background and dynamic components. Magn Reson Med 73(3):1125–1136. https://doi.org/10.1002/mrm.25240

8. Patel VM, Easley GR, Healy DM, Chellappa R (2010) Compressed Synthetic Aperture Radar. IEEE Journal of Selected Topics in Signal Processing 4(2):244–254. https://doi.org/10.1109/Jstsp.2009.2039181
9. Sharma K, Lagunas E, Chatzinotas S, Ottersten B (2016) Application of Compressive Sensing in Cognitive Radio Communications: A Survey. IEEE Communications Surveys and Tutorials 18(3):1838–1860. https://doi.org/10.1109/Comst.2016.2524443
10. Liu XY, Zhu YM, Kong LH, Liu C, Gu Y, Vasilakos AV, Wu MY (2015) CDC: Compressive Data Collection for Wireless Sensor Networks. IEEE Trans Parallel Distrib Syst 26(8):2188–2197. https://doi.org/10.1109/Tpds.2014.2345257
11. Nagesh P, Li BX (2009) A Compressive Sensing Approach for Expression-Invariant Face Recognition. Paper presented at the Cvpr: 2009 IEEE Conference on Computer Vision and Pattern Recognition, Vols 1–4
12. https://www.siemens-healthineers.com/magnetic-resonance-imaging/clinical-specialities/compressed-sensing
13. Lustig M, Donoho DL, Santos JM, Pauly JM (2008) Compressed sensing MRI. IEEE Signal Process Mag 25(2):72–82. https://doi.org/10.1109/Msp.2007.914728
14. Rani M, Dhok SB, Deshmukh RB (2018) A Systematic Review of Compressive Sensing: Concepts, Implementations and Applications. IEEE Access 6:4875–4894. https://doi.org/10.1109/Access.2018.2793851
15. Erkoc ME, Karaboga N (2019) Evolutionary algorithms for sparse signal reconstruction. SIViP 13(7):1293–1301. https://doi.org/10.1007/s11760-019-01473-w
16. Karaboga N (2009) A new design method based on artificial bee colony algorithm for digital IIR filters. Journal of the Franklin Institute-Engineering and Applied Mathematics 346(4):328–348. https://doi.org/10.1016/j.jfranklin.2008.11.003
17. Ozdemir G, Karaboga N (2019) A review on the cosine modulated filter bank studies using meta-heuristic optimization algorithms. Artif Intell Rev 52(3):1629–1653. https://doi.org/10.1007/s10462-017-9595-x
18. Kockanat S, Karaboga N (2015) The design approaches of two-dimensional digital filters based on metaheuristic optimization algorithms: a review of the literature. Artif Intell Rev 44(2):265–287. https://doi.org/10.1007/s10462-014-9427-1
19. Kockanat S, Karaboga N (2015) A novel 2D-ABC adaptive filter algorithm: A comparative study. Digit Signal Proc 40:140–153. https://doi.org/10.1016/j.dsp.2015.02.010
20. Panduro MA, Brizuela CA, Covarrubias D, Lopez C (2006) A trade-off curve computation for linear antenna arrays using an evolutionary multi-objective approach. Soft Comput 10(2):125–131. https://doi.org/10.1007/s00500-004-0434-z
21. Alatas B, Akin E, Karci A (2008) Modenar: Multi-objective differential evolution algorithm for mining numeric association rules. Appl Soft Comput 8(1):646–656. https://doi.org/10.1016/j.asoc.2007.05.003
22. Janson S, Merkle D, Middendorf M (2008) Molecular docking with multi-objective particle swarm optimization. Appl Soft Comput 8(1):666–675. https://doi.org/10.1016/j.asoc.2007.05.005
23. Reynolds AP, de la Iglesia B (2009) A multi-objective GRASP for partial classification. Soft Comput 13(3):227–243. https://doi.org/10.1007/s00500-008-0320-1
24. Konstantinidis A, Charalambous C, Zhou AM, Zhang QF (2010) Multi-objective Mobile Agent-based Sensor Network Routing using MOEA/D. Paper presented at the 2010 IEEE Congress on Evolutionary Computation (Cec)
25. Lazzerini B, Marcelloni F, Vecchio M (2010) A multi-objective evolutionary approach to image quality/compression trade-off in JPEG baseline algorithm. Appl Soft Comput 10(2):548–561. https://doi.org/10.1016/j.asoc.2009.08.024
26. Wozniak P (2011) Preferences in multi-objective evolutionary optimisation of electric motor speed control with hardware in the loop. Appl Soft Comput 11(1):49–55. https://doi.org/10.1016/j.asoc.2009.10.015

27. Li L, Yao X, Stolkin R, Gong MG, He S (2014) An Evolutionary Multiobjective Approach to Sparse Reconstruction. IEEE Trans Evol Comput 18(6):827–845. https://doi.org/10.1109/Tevc.2013.2287153
28. Daubechies I, Defrise M, De Mol C (2004) An iterative thresholding algorithm for linear inverse problems with a sparsity constraint. Commun Pure Appl Math 57(11):1413–1457. https://doi.org/10.1002/Cpa.20042
29. Zhou Y, Kwong S, Guo HN, Zhang X, Zhang QF (2017) A Two-Phase Evolutionary Approach for Compressive Sensing Reconstruction. IEEE Transactions on Cybernetics 47(9):2651–2663. https://doi.org/10.1109/Tcyb.2017.2679705
30. Yan B, Zhao Q, Wang ZH, Zhao XY (2017) A hybrid evolutionary algorithm for multiobjective sparse reconstruction. SIViP 11(6):993–1000. https://doi.org/10.1007/s11760-016-1049-4
31. Mierswa I, Wurst M (2006) Information preserving multi-objective feature selection for unsupervised learning. Paper presented at the Gecco 2006: Genetic and Evolutionary Computation Conference, Vol 1 and 2
32. Li H, Zhang QF, Deng JD, Xu ZB (2018) A Preference-Based Multiobjective Evolutionary Approach for Sparse Optimization. IEEE Transactions on Neural Networks and Learning Systems 29(5):1716–1731. https://doi.org/10.1109/Tnnls.2017.2677973
33. Blumensath T, Davies ME (2009) Iterative hard thresholding for compressed sensing. Applied and Computational Harmonic Analysis 27(3):265–274. https://doi.org/10.1016/j.acha.2009.04.002
34. Xu ZB, Chang XY, Xu FM, Zhang H (2012) L-1/2 Regularization: A Thresholding Representation Theory and a Fast Solver. IEEE Transactions on Neural Networks and Learning Systems 23(7):1013–1027. https://doi.org/10.1109/Tnnls.2012.2197412
35. Li H, Sun JY, Meng DY, Zhang QF (2019) A Multiobjective Approach Based on Gaussian Mixture Clustering for Sparse Reconstruction. IEEE Access 7:22684–22697. https://doi.org/10.1109/Access.2019.2898987
36. Yue CT, Liang J, Qu BY, Han YH, Zhu YS, Crisalle OD (2020) A novel multiobjective optimization algorithm for sparse signal reconstruction. Signal Processing 167. doi:Artn 107292
37. Zhou AM, Qu BY, Li H, Zhao SZ, Suganthan PN, Zhang QF (2011) Multiobjective evolutionary algorithms: A survey of the state of the art. Swarm and Evolutionary Computation 1(1):32–49. https://doi.org/10.1016/j.swevo.2011.03.001
38. Schaffer JD Multiple objective optimization with vector evaluated genetic algorithms. In: 1st International Conference on Genetic Algorithms, 1985. pp 93–100
39. Fonseca CM, Fleming PJ (1993) Genetic Algorithms for Multiobjective Optimization - Formulation, Discussion and Generalization. Proceedings of the Fifth International Conference on Genetic Algorithms:416–423
40. Siinivas N, Deb K (1994) Multiobjective Optimization Using Nondominated Sorting in Genetic Algorithms. Evol Comput 2(3):221–248
41. Knowles JD, Corne DW (2000) Approximating the Nondominated Front Using the Pareto Archived Evolution Strategy. Evol Comput 8(2):149–172. https://doi.org/10.1162/106365600568167
42. Corne DW, Knowles JD, Oates MJ (2000) The Pareto Envelope-Based Selection Algorithm for Multiobjective Optimization. Berlin, Heidelberg. Parallel Problem Solving from Nature PPSN VI. Springer, Berlin Heidelberg, pp 839–848
43. Deb K, Pratap A, Agarwal S, Meyarivan T (2002) A fast and elitist multiobjective genetic algorithm: NSGA-II. IEEE Transactions on Evolutionary Computation 6 (2):182–197. doi:Pii S 1089-778x(02)04101-2, https://doi.org/10.1109/4235.996017
44. Coello CAC, Lechuga MS (2002) MOPSO: A proposal for multiple objective particle swarm optimization. Paper presented at the Cec'02: Proceedings of the 2002 Congress on Evolutionary Computation, Vols 1 and 2
45. Zhang QF, Li H (2007) MOEA/D: A multiobjective evolutionary algorithm based on decomposition. IEEE Trans Evol Comput 11(6):712–731. https://doi.org/10.1109/Tevc.2007.892759

46. Liang JJ, Zhu XP, Yue CT, Li ZH, Qu BY (2018) Performance Analysis on Knee Point Selection Methods for Multi-objective Sparse Optimization Problems. Paper presented at the 2018 IEEE Congress on Evolutionary Computation (Cec)
47. Das I (1999) On characterizing the "knee" of the Pareto curve based on Normal-Boundary Intersection. Structural Optimization 18(2–3):107–115. https://doi.org/10.1007/s001580050111
48. Branke E, Deb K, Dierolf H, Osswald M Finding knees in multi-objective optimization. In: Parallel Problem Solving from Nature - Ppsn Viii, 2004. pp 722–731. doi:https://doi.org/10.1007/978-3-540-30217-9_73
49. Li H, Fan YY, Zhang QF, Xu ZB, Deng JD A Multi-phase Multiobjective Approach Based on Decomposition for Sparse Reconstruction. In: 2016 IEEE Congress on Evolutionary Computation (Cec), 2016. pp 601–608
50. Zhang XY, Tian Y, Jin YC (2015) A Knee Point-Driven Evolutionary Algorithm for Many-Objective Optimization. IEEE Trans Evol Comput 19(6):761–776. https://doi.org/10.1109/Tevc.2014.2378512
51. Wang Z, Bovik AC, Sheikh HR, Simoncelli EP (2004) Image quality assessment: From error visibility to structural similarity. IEEE Trans Image Process 13(4):600–612. https://doi.org/10.1109/Tip.2003.819861
52. Baraniuk RG (2007) Compressive sensing. IEEE Signal Processing Magazine 24 (4):118- + . https://doi.org/10.1109/msp.2007.4286571
53. Mallat SG, Zhang ZF (1993) Matching Pursuits with Time-Frequency Dictionaries. IEEE Trans Signal Process 41(12):3397–3415. https://doi.org/10.1109/78.258082
54. Tropp JA, Gilbert AC (2007) Signal recovery from random measurements via orthogonal matching pursuit. IEEE Trans Inf Theory 53(12):4655–4666. https://doi.org/10.1109/Tit.2007.909108
55. Needell D, Tropp JA (2009) CoSaMP: Iterative signal recovery from incomplete and inaccurate samples. Applied and Computational Harmonic Analysis 26(3):301–321. https://doi.org/10.1016/j.acha.2008.07.002
56. Chen SSB, Donoho DL, Saunders MA (1998) Atomic decomposition by basis pursuit. SIAM J Sci Comput 20(1):33–61. https://doi.org/10.1137/S1064827596304010
57. Tibshirani R (1996) Regression shrinkage and selection via the Lasso. Journal of the Royal Statistical Society Series B-Methodological 58(1):267–288. https://doi.org/10.1111/j.2517-6161.1996.tb02080.x
58. Donoho DL, Tsaig Y (2008) Fast Solution of l(1)-Norm Minimization Problems When the Solution May Be Sparse. IEEE Trans Inf Theory 54(11):4789–4812. https://doi.org/10.1109/Tit.2008.929958
59. Li H, Fan YY, Zhang QF, Xu ZB, Deng JD (2016) A Multi-phase Multiobjective Approach Based on Decomposition for Sparse Reconstruction. Paper presented at the 2016 IEEE Congress on Evolutionary Computation (Cec)
60. Mamaghanian H, Khaled N, Atienza D, Vandergheynst P (2011) Compressed Sensing for Real-Time Energy-Efficient ECG Compression on Wireless Body Sensor Nodes. IEEE Trans Biomed Eng 58(9):2456–2466. https://doi.org/10.1109/Tbme.2011.2156795
61. Singh A, Dandapat S (2017) Block sparsity-based joint compressed sensing recovery of multi-channel ECG signals. Healthcare Technology Letters 4(2):50–56. https://doi.org/10.1049/htl.2016.0049
62. Lu ZT, Kim DY, Pearlman WA (2000) Wavelet compression of ECG signals by the set partitioning in hierarchical trees algorithm. IEEE Trans Biomed Eng 47(7):849–856. https://doi.org/10.1109/10.846678
63. Polania LF, Carrillo RE, Blanco-Velasco M, Barner KE (2011) Compressed Sensing Based Method for Ecg Compression. 2011 IEEE International Conference on Acoustics, Speech, and Signal Processing:761–764. https://doi.org/10.1109/icassp.2011.5946515
64. Goldberger AL, Amaral LAN, Glass L, Hausdorff JM, Ivanov PC, Mark RG, Mietus JE, Moody GB, Peng CK, Stanley HE (2000) PhysioBank, PhysioToolkit, and PhysioNet - Components of a new research resource for complex physiologic signals. Circulation 101(23):E215–E220. https://doi.org/10.1161/01.Cir.101.23.E215

65. Zigel Y, Cohen A, Katz A (2000) The weighted diagnostic distortion (WDD) measure for ECG signal compression. IEEE Trans Biomed Eng 47(11):1422–1430. https://doi.org/10.1109/Tbme.2000.880093
66. Izadi V, Shahri PK, Ahani H (2020) A compressed-sensing-based compressor for ECG. Biomedical Engineering Letters 10(2):299–307. https://doi.org/10.1007/s13534-020-00148-7

Chapter 18
Optimal Allocation of Flexible Alternative Current Transmission Systems: An Application of Particle Swarm Optimization

Akiko Takahashi, Hirotaka Takano, and Shigeyuki Funabiki

18.1 Introduction

Electric power distribution systems consist of distribution substations, distribution feeders, and several regulators and have extremely important roles in delivering electrical power to the consumers. The distribution feeders, e.g., 6.6 kV feeders in Japan, have been expanded radially from the distribution substations in association with the spread of power-consuming area. With the aim of enhancing the power supply reliability, the distribution feeders are sectionalized into several load sections, and each of the load sections is connected to different feeders, which is the multi-sectionalized and multi-connected distribution systems [1, 2]. The distribution systems, as are well known, become one of the largest and most complicated infrastructures for activities in our societies. In recent years, the growth in penetration of renewable energy-based generation systems, especially photovoltaic generation (PV) systems, accompanies new challenges in operations of the distribution systems and is revolutionizing them [3–5].

In traditional electric power grids, centralized power generation systems supply electrical power generated from non-renewable energy sources through extensive

A. Takahashi (✉) · S. Funabiki
Department of Information and Communication System, Graduate School of Natural Science and Technology, Okayama University, 3-1-1, Tsushima-naka, Kita-ku, Okayama-shi, Okayama 700-8530, Japan
e-mail: takahashi-a@okayama-u.ac.jp

S. Funabiki
e-mail: sfunabiki@ec.okayama-u.ac.jp

H. Takano
Department of Electrical, Electronic and Computer Engineering, Gifu University, 1-1, Yanagido, Gifu-shi, Gifu 501-1193, Japan
e-mail: takano@gifu-u.ac.jp

S. Carbas et al. (eds.), *Nature-Inspired Metaheuristic Algorithms for Engineering Optimization Applications*, Springer Tracts in Nature-Inspired Computing,
https://doi.org/10.1007/978-981-33-6773-9_18

transmission lines to the consumers. This situation causes high dependence on fossil fuels and thus may lead to high fossil fuel costs. With a view to improve the situation, installation of the PV systems has been attracting attention around the world. The PV systems assist to provide the electrical power in the sites close to the consumers, while their outputs bring voltage rises in the distribution systems and thus emphasize voltage variations among the distribution feeders. Furthermore, the PV installation intensifies uncertainties in voltage regulation of the distribution systems' operations because the PV outputs strongly depend on the weather conditions and cannot be monitored or controlled from the operator side.

According to the advancement of communication technologies, automation of operations in the distribution systems has been progressing. Remote and automatic voltage regulation is a key item in the automation. When the undesirable voltage changes occur in the distribution feeder, load ratio control transformers (LRTs) adjust them remotely and automatically into acceptable range of the voltages. In addition, step voltage regulators (SVRs), which are normally installed on long-distance distribution feeders, support the voltage regulating operation in corporation with the LRTs. By the functions of these voltage regulators, the distribution systems' operators maintain the quality and the reliability in power supply until now. However, the PV penetration, as already mentioned, currently brings difficulties in the voltage regulation, and their countermeasures are crucially required.

Installation of flexible alternative current transmission systems (FACTS) is expected to relax the above difficulties [6–8]. The FACTS devices are static power electronic devices installed in AC transmission systems to encourage power transfer capability, stability, and controllability through series and/or shunt compensation. Static VAR compensator (SVC) is one of the typical FACTS devices. The SVCs continuously and quickly adjust the leading/lagging reactive power, and therefore, available for the voltage regulating operations in the distribution systems [9, 10]. Although various optimization algorithms have been proposed to determine appropriate sets of the sizes (capacities) and the locations (connection points) of the SVCs [11–18], there is no established solution method. This is because sizing and allocation problems of the SVCs become complicated optimization problems.

This chapter, first, presents a problem formulation to obtain the optimal combination of sizes and locations of the SVCs with a view to maintaining the distribution voltage appropriately. In the problem formulation, the generation opportunity of PV systems is emphasized as the objective function. If the voltages in the distribution system cannot be maintained in their acceptable ranges, the PV systems might be suppressed by their outputs or disconnected from the distribution system. The objective function militates against these undesired situations in consideration of implementation and running costs of the SVCs. Next, particle swarm optimization (PSO) is applied to the formulated problem. In application of the PSO, to enhance its search ability, a reflectance adjusting mechanism at the boundary in the solution space and ideas of mutation and elitism is employed. The authors named PSO with all of the above strategies as RAPSO-ME [19–21]. Finally, validity of the problem framework and usefulness of the RAPSO-ME are verified through numerical simulations and discussions on their results.

18.2 Distribution Voltage Regulation and Its Issue

In Japan, the standard values of distribution voltages are set to 100 V or 200 V on the consumer side, and acceptable ranges of the voltage variation are set from 95 to 107 V or from 182 to 222 V, respectively. LRTs and SVRs control the distribution voltages appropriately based on the line drop compensation (LDC) method. In the LDC method, the feeder voltages are estimated with reference to passing feeder currents and set point values at the voltage adjusting points. Figure 18.1 shows a conceptual illustration of the voltage regulation in Japanese distribution systems.

In Fig. 18.1, the LRTs facilitate remotely boosting the sending voltage higher during peak periods of power consumption than that of off-peak periods. In addition, the SVRs, especially in the long-distance feeders, adjust the voltages in the middle of distribution feeders to satisfy the voltage acceptable ranges, 101 ± 6 V or 202 ± 20 V, in the low-voltage distribution lines.

If the PV systems' output becomes more dominant in the power sources, the reverse power flow from the PV systems causes the voltage rises. As a result, the distribution voltages can exceed values of their upper limit. Moreover, the growth in PV penetration complicates the voltage distributions and may lead to the resulting voltage violation or cause malfunction of the LDC method. This is because the distribution system operators cannot know the voltage distributions precisely along each feeder. In these situations, the PV systems might be suppressed by their outputs or disconnected from the distribution system. However, such countermeasures bring negative impact on benefits of the PV owners, and therefore, must be avoided as much as possible. Figure 18.2 summarizes the issues in the distribution voltage regulation.

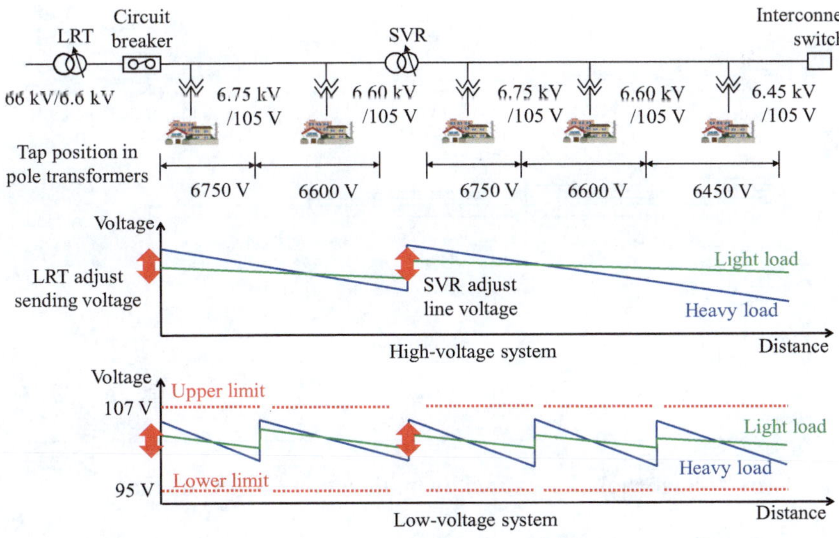

Fig. 18.1 Voltage regulation in Japanese distribution systems

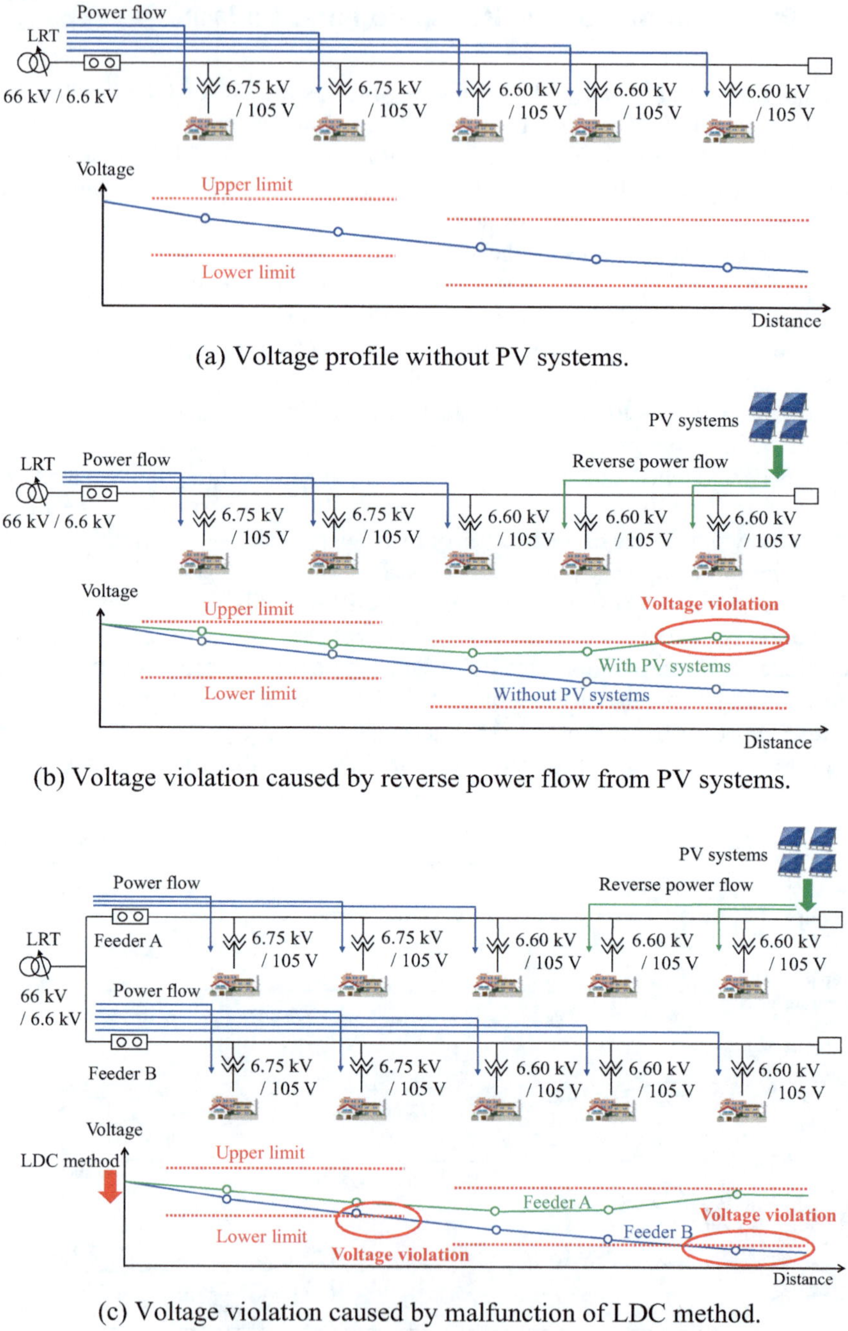

Fig. 18.2 Traditional voltage regulation and its issues

On the other hand, SVCs can support the voltage regulation in the distribution systems by changing its leading/lagging reactive power output quickly and consecutively. The SVCs monitor the voltage at their own connection point and keep its value in a certain value even though the voltage fluctuation occurs in the distribution systems. That is, appropriate installation of the SVC becomes an effective alternative to the traditional voltage regulation in the distribution systems if the PV installation keeps growing continuously.

18.3 Target Optimization Problem

In the real world, we cannot select size of the SVCs as the continuous variables, and therefore, the optimization variables are defined as

$$Q_i \in \mathbb{Z}, \text{ for } \forall i, \tag{18.1}$$

where Q_i is the capacity of aggregated SVC (or the total capacity of SVCs) installed in the section i $(i = 1, 2, \ldots, N)$, and an element of the vector Q. N is the number of aggregated SVCs.

Now, the main objective of this study is to determine the set of sizes (capacities) and locations (connection points) of SVCs for maximizing the reduction in generation opportunity loss of the PV systems (income). On the other hand, we need to consider the total cost for installing the SVCs (expense) in the optimization. Under the circumstances, the objective function is formulated as

$$\max_{Q} f, \tag{18.2}$$

$$f = LR(Q) - (EC(Q) + IC(Q)). \tag{18.3}$$

The first term of Eq. (18.3) represents the benefit brought by the loss reduction of generation opportunity of PV systems. By the following equation, the generation opportunity loss reduction is converted into the price.

$$LR(Q) = \sum_{t=1}^{T} \alpha_t \left(p_t(q_t) - p_t^* \right), \tag{18.4}$$

where T is the end of time period; α_t is the unit price in selling electricity; $q_{i,t}$ is the output of aggregated SVC installed in the section i, and an element of the vector q_t; $p_t(q_t)$ is outputs of PV systems after installing SVCs; p_t^* is outputs of PV systems before installing SVCs.

The second term of Eq. (18.3) expresses the initial equipment cost invested for operation of the installed SVCs, and is calculated as

$$EC(Q) = (1 + \varepsilon + \delta) \cdot \sum_{t=1}^{T} \sum_{i=1}^{N} \rho_i \gamma_i Q_i, \tag{18.5}$$

where ε is the ratio of auxiliary equipment cost; δ is the ratio of maintenance cost; ρ_i is the unit rate of interest; γ_i is the unit price of the aggregated SVC installed in the section i.

In Eq. (18.5), the auxiliary equipment and the maintenance costs depend on the equipment cost of the SVCs. The unit rate of interest in Japan can be calculated as

$$\rho_i = \tau_i \frac{\eta}{1 - (1 + \eta)^{-\tau_i}}, \quad \text{for } \forall i, \tag{18.6}$$

where η is the yearly rate of interest; τ_i is the payback time of the SVC installed in the section i.

In [18], the unit price of the SVC ($/kVAR) is experimentally calculated as Eq. (18.7), and it is used in the numerical simulations of this chapter.

$$\gamma_i = 0.03 Q_i^2 - 0.3051 Q_i + 127.38. \tag{18.7}$$

Lastly, the third term of Eq. (18.3) calculates the implementation cost of SVC installation as

$$IC(Q) = \sum_{t=1}^{T} \sum_{i=1}^{N} \beta_{i,t} Q_i, \tag{18.8}$$

where $\beta_{i,t}$ is the unit price of implementation cost of the aggregated SVC installed in the section i.

Operational constraints in the distribution systems are formulated as

$$0 \le q_{i,t} \le q_i^{\max}, \quad \text{for } \forall i, \quad \forall t \tag{18.9}$$

$$\underline{V_i} \le V_i(q_{i,t}) \le \overline{V_i}, \quad \text{for } \forall i, \quad \forall t, \tag{18.10}$$

where $q_i^{\max}$ is the maximum output of the aggregated SVC installed in the section i; $V_i(q_{i,t})$ is the voltage in the section i; $\overline{V_i}$ is the upper limit of voltage fluctuation in the section i; $\underline{V_i}$ is the lower limit of voltage fluctuation in the section i.

18.4 Particle Swarm Optimization-Based Solution Method

The target problem formulated in Sect. 18.3 is a complicated optimization problem to determine the set of sizes and locations of the SVCs. Since it is extremely difficult to solve the target problem exactly, intelligent optimization-based solution methods have been attracting attention as a realistic alternative. In fact, several metaheuristics and evolutionary computation techniques have been applied to the similar problem frameworks [11–18]. The authors select a PSO as the basis of solution method. In addition, improvement strategies are introduced to enhance the search ability of the PSO.

18.4.1 Particle Swarm Optimization

PSO is a population-based stochastic computational method [22] that optimizes (or approximately optimizes) a target problem by iteratively trying to improve a solution candidate with regard to a give measure of solution quality, which is the fitness function. Algorithm of the original PSO was developed through studies on the simulated social behavior of birds or fishes. In the framework of PSO, individuals of the society are called 'particle'. These studies showed that some animals are able to share information among their group, and this ability confers a survival advantages in the group. The PSO was derived by such concept of swarm intelligence and is well known as a metaheuristic algorithm that is appropriate to optimize nonlinear continuous functions.

The PSO algorithm has many similarities with genetic algorithm (GA); however, unlike the GA, the standard PSO has no evolution operators for crossover and mutation. An initial set of randomly created solutions (particles), which is the initial swarm, propagates in the design of the search space toward the global optimal solution over a number of iterations. All members of the swarm fit and share information for their search places. Each particle k has a position x_k^n and a velocity v_k^n in iteration n $(n = 1, 2, \ldots, W)$ and flies through on the solution space for finding their best positions based on Eqs. (18.11) and (18.12).

$$v_k^n = \omega v_k^{n-1} + c_1\theta_1\left(x_k^* - x_k^{n-1}\right) + c_2\theta_2\left\{\left(\min_{k \epsilon S} x_k^*\right) - x_k^{n-1}\right\}, \tag{18.11}$$

$$x_k^n = x_k^{n-1} + v_k^n, \tag{18.12}$$

where ω. is the inertia weight factor which controls the iteration size of the PSO; c_1 and c_2 are the cognitive factors that represent the trust for each particle and the swarm; θ_1 and θ_2 are the random numbers in the range, [0, 1]; x_k^* is the personal best (pbest) for particle k; $\min_{k \epsilon S} x_k^*$ is the swarm best (gbest); S. is the set of all particles.

Equation (18.11) denotes that there are three different contributions to a particle's movement in an iteration, so there are three terms in this equation. The first term of Eq. (18.11) is a product between parameter ω and previous velocity of the particle. When higher values are set as ω, the particles search the solution space globally (exploration), while locally when lower values are set (exploitation). The second term is the individual cognition term, which is calculated by means of the difference

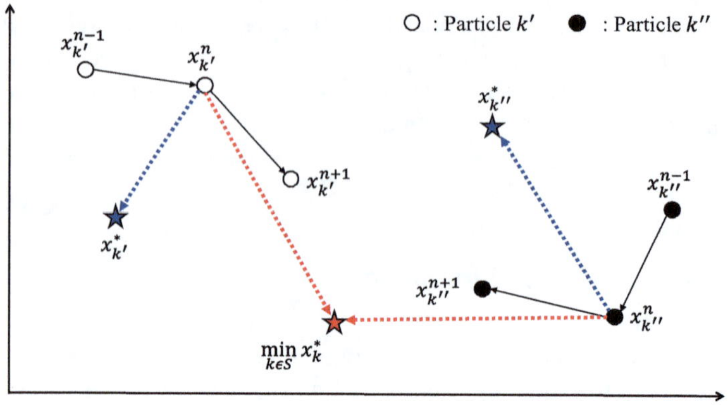

Fig. 18.3 Update of position and velocity of particles in two-dimensional search space

between the particle's best position and its current position. Increase of this term means that the particle is attracted by its own best. The parameter c_1 weights the importance of particle's own previous experiences. T other parameter θ_1 plays an important role, as it avoids premature convergences, increasing the most likely global optima. Finally, the third term represents the social learning mechanism. By this term, all particlesn the swarm are able to share the information of the best position achieved regardless of which particle had found it. Its format is the same as the second term. Thus, the difference acts as an attraction for the particles to the best position found until iteration n. Similarly, c_2 weights the importance of the global learning of the swarm, and θ_2 plays exactly the same role as θ_1.

Figure 18.3 illustrates a conceptual illustration of the update process in a particle's position and in its velocity.

In this chapter, the position of particles is expressed as

$$x_k^n = Q. \tag{18.13}$$

As already defined, we can select size of the SVCs only as the discrete variable. Although the PSO has succeeded in many continuous optimization problems, it has still some difficulties to treat the discrete optimization problems [23]. As a countermeasure for this issue, the ceiling function is introduced in the PSO algorithm.

$$Q_i = \lceil Q_i^\dagger \rceil, \quad \text{for} \quad \forall i, \tag{18.14}$$

where $Q_i^\dagger$ is the SVC capacity actually found by the PSO.

During the iterative process, the fitness function of particles is valued to measure their superiority. With the aim of handling the constraints, the penalty method, which replaces a constrained optimization problem by a series of unconstrained problems, is applied. The fitness function of PSO is formulated as

$$\text{Fitness} = f + \varepsilon \text{VIO}, \tag{18.15}$$

$$\text{VIO} = \sum_{t=1}^{T} \sum_{i=1}^{N} \begin{bmatrix} \max\left(\frac{V_i(q_{i,t}) - \overline{V_i}}{\overline{V_i}}, \frac{\underline{V_i} - V_i(q_{i,t})}{\underline{V_i}}, 0\right) \\ + \max\left(q_{i,t} - q_i^{\max}, 0\right) \end{bmatrix}, \tag{18.16}$$

where ε (< 0) is the weighting factor ($-\varepsilon \text{VIO} \gg f$).

18.4.2 *Improved Particle Swarm Optimization (RAPSO-ME)*

In this study, first, the ideas of mutation and elitism are adopted to the PSO algorithm. These ideas are often used in the other metaheuristic algorithms. Mutation is a genetic operator used to maintain genetic diversity from one generation of a population to the next. This idea may change the particle's position entirely from the previous one. On the other hand, elitism strategy provides a means for reducing genetic drift by ensuring that superior particles, which is the elites, among candidates for selection are permitted to copy their characteristics to the next generation. By adopting these ideas, we can expect that the PSO has diversity into behavior of the swarm.

Next, the authors employ a reflectance adjusting mechanism. In practical applications of the PSO, the search spaces are often high-dimensional and bounded with a view to keeping the physical meaning of the parameters. In [24], it is mathematically proved that boundary violation increases as dimensionality increases. References [25–27] show possibility that improper bound-handling schemes can prevent the search process of PSO algorithms. Typical bound-handling schemes are the random, reflection, and absorption. The random scheme is adopted as the default setting in many PSO algorithms. If a particle flies out the boundary of a parameter, a random value drawn from a uniform distribution between the lower and the upper boundaries of the parameter is assigned. The reflection scheme acts like a mirror and reflects the projection of the particle's displacement when the particle flies out the boundary. On the other hand, under the absorption scheme, a particle flying outside of the boundary is relocated at the boundary in that dimension such that the particle is absorbed by the boundary.

The proposed reflectance adjusting mechanism combines the reflection and the absorption schemes. By updating reflectivity with increase of the iteration, the bound-handling scheme is controlled in dynamic between these schemes. Now, the positions and the velocities of particles are redefined as

$$x_k^n = \begin{cases} \overline{x_k} - r^n \cdot \left|x_k^n - \overline{x_k}\right|, (\overline{x_k} < x_k^n) \\ \underline{x_k} - r^n \cdot \left|x_k^n - \underline{x_k}\right|, (x_k^n < \underline{x_k}), \end{cases}, \tag{18.17}$$

$$v_k^{n-1} = r^n\left(-v_k^n\right), \tag{18.18}$$

where $\overline{x_k}$ is the upper boundary; $\underline{x_k}$ is the lower boundary.

Table 18.1 Specifications of benchmark problems

j	$g_j(z)$	$f_j(z^*)$	Boundary of z
1	$\sum_{d=1}^{D} z_d^2$	0	(–5.12, 5.12)
2	$\sum_{d=1}^{D-1} \left(100(z_{d-1} - z_d^2) + (1 - z_d^2)^2\right)$	0	(–2.048, 2.048)
3	$\sum_{d=1}^{D} (z_d^2 + 10\cos(2\pi z_j) + 10)$	0	(–5.12, 5.12)
4	$418.9829D - \sum_{d=1}^{D} z_d \sin\sqrt{\lvert z_d \rvert}$	0	(–512, 512)

The reflectivity, r^n, in Eqs. (18.17) and (18.18) is defined as Eq. (18.19) and thus decreases in contrast to increase of the iteration.

$$r^n = r_{\max} - \left(1 - \frac{\sigma\left(J_k^n\right)}{\sigma\left(J_k^1\right)}\right) \cdot (r_{\max} - r_{\min}), \tag{18.19}$$

where $r_{\max}$ and $r_{\min}$ are the maximum and the minimum assumable values of reflectivity; J_k^n is the personal best of the particle k in iteration n; $\sigma()$ is the standard deviation.

18.4.3 Validation of Improved Particle Swarm Optimization

Preliminary numerical simulations were carried out on the benchmark problems to validate performance of the improved PSO that is RAPSO-ME. Table 18.1 describes specifications of the target benchmark problems. Here, z_d is the optimization variable ($d = 1, 2, \ldots, D$), and an element of the vector z; z^* is the optimal solution; $g_j(z)$ is the objective function ($j = 1, 2, 3, 4$).

In Table 18.1, $g_1(z)$ and $g_2(z)$ are unimodal functions, and $g_3(z)$ and $g_4(z)$ are multimodal functions. Also, these functions are very popular as the sphere function, the Rosenbrock function, the Rastrigin function, and Schwefel function, respectively. Parameters of the RAPSO-ME are shown in Table 18.2. These parameters were set by trial and errors.

Table 18.3 shows the results of benchmark problems for 30 trials. As for reference, transitions of the average values of standard deviation and the average reflectivities in the Rosenbrock and the Schwefel functions are displayed in Figs. 18.4 and 18.5.

As shown in Table 18.3, the RAPSO-ME found sufficiently small values in each benchmark problem. In the Rosenbrock function $g_2(z)$, the standard deviation of

Table 18.2 Parameters of PSO

Total number of dimensions, D	10
Total number of particles, S	100
Maximum iteration, N	1000
Total number of elite particles	5
Mutation probability	0.01
Maximum reflectance, $r_{\max}$	1.00
Minimum reflectance, $r_{\min}$	0.01
Inertia weight factor, ω	0.819
Cognitive factor, c_1	1.414
Cognitive factor, c_2	1.414

Table 18.3 Results of preliminary numerical simulations

	$g_1(z)$	$g_2(z)$	$g_3(z)$	$g_4(z)$
Best	1.7589×10^{-15}	0.0231	4.4629×10^{-10}	1.2728×10^{-4}
Mean	2.4775×10^{-8}	3.4291	1.5969	4.9214×10^{-4}
Worst	4.6439×10^{-7}	13.2916	3.9798	0.0091
Standard deviation	8.7446×10^{-8}	3.5416	1.0976	0.0016

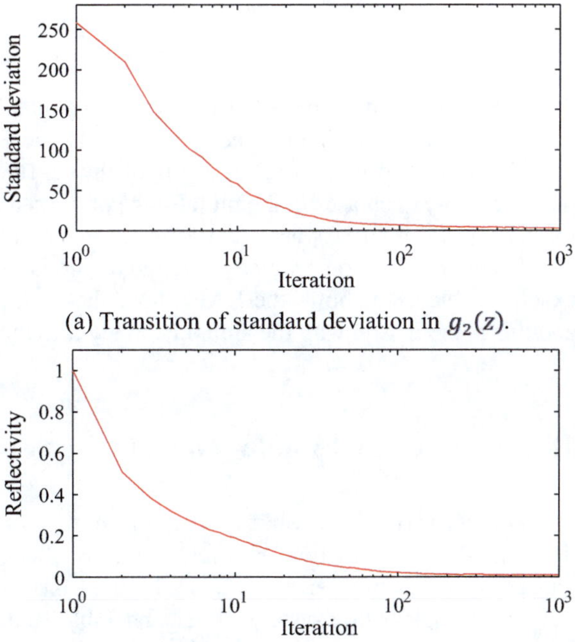

(a) Transition of standard deviation in $g_2(z)$.

(b) Transition of reflectivity in $g_2(z)$.

Fig. 18.4 Search process of RAPSO-ME in Rosenbrock function $g_2(z)$

Fig. 18.5 Search process of RAPSO-ME in Schwefel function $g_4(z)$

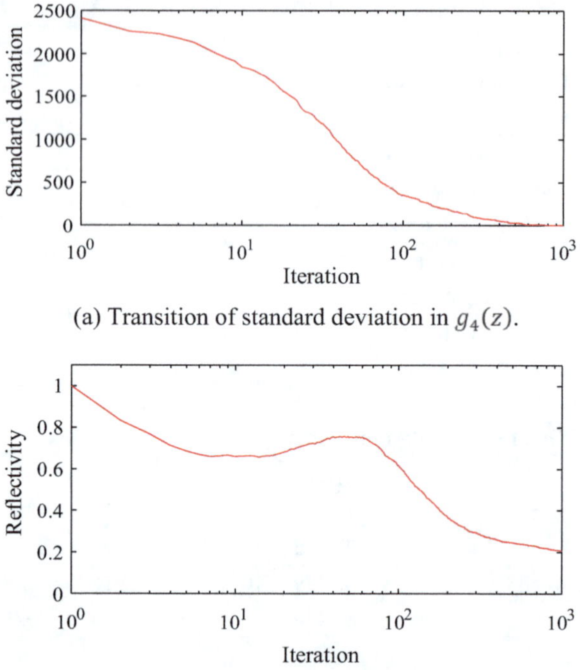

(a) Transition of standard deviation in $g_4(z)$.

(b) Transition of reflectivity in $g_4(z)$.

the fitness function of personal best (pbest) gradually decreases, and the reflectivity decreases accordingly (see Fig. 18.4). In the Schwefel function $g_4(z)$, each particle holds information of the value of fitness function of pbest, and therefore, the reflectivity decreases in the initial stage of the search (see Fig. 18.5). In contrast, the reflectivity increases around 100 iterations. Normally, it is difficult to categorize actual optimization problems into the unimodal function or the multimodal one exactly. Table 18.3 shows the RAPSO-ME has compatibility with both functions, and this is the reason why the authors applied it to the formulated problem.

18.5 Numerical Simulation and Discussion on Its Result

The authors verified the validity of the formulated optimization problem through daily numerical simulation on a real scale distribution system model. This model was constructed by referring to [21]. Figure 18.6 and Table 18.4 show specifications of the distribution system model, and Table 18.5 describes specifications of the newly installed SVCs. In Fig. 18.6, the load sections were aggregated into each node. Although the distribution system model has one LRT and one SVR, voltage regulation of the SVR is emphasized in the numerical simulations. Power conditioning

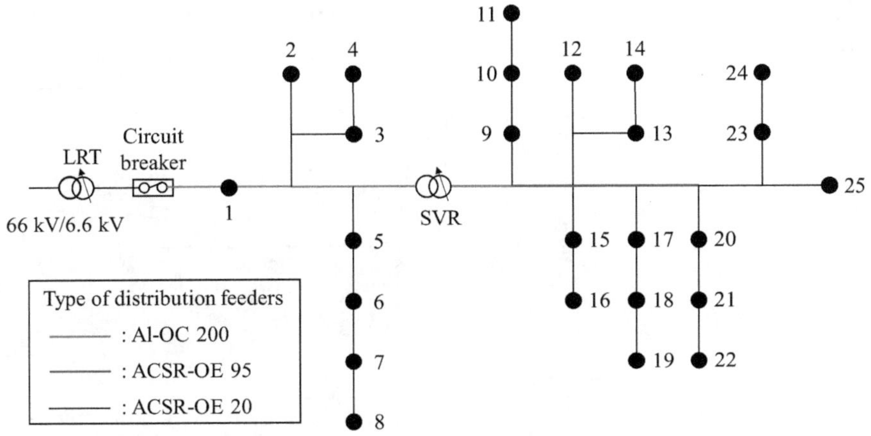

Fig. 18.6 Numerical simulation model

Table 18.4 Specifications of Fig. 18.6

Rated capacity of distribution substation	20 MVA
Total number of load sections	25
Reference value of voltage	6600 V (1.0 p.u.)
Dead band in SVR	±0.015 p.u.
Operation cycle of SVR	60 s
Target voltage of each SVC	1.000 p.u.
Operation cycle of each SVR	2 s
Total number of residences in each load section	65
Total number of residences having PV system in each load section	20
Total number of all-electric residence in each load section	15
Rated power of each PV output	5.0 kW
Rated capacity of each PCS equipped in PV systems	5.0 kVA
Dead band in each PCS	0.025 p.u.
Operation cycle of each PCS	60 s
Acceptable voltage range in low-voltage distribution lines	±0.030 p.u.

Table 18.5 Specifications of SVC

Unit price in selling electricity, α	14 JPY/kWh
Ratio of auxiliary equipment cost, ε	0.1
Ratio of maintenance cost, δ	0.01
Yearly rate of interest, η	0.04
Payback time of SVC, τ	10 year
Unit price of implementation cost of aggregated SVC, β	5.0×10^6 JPY

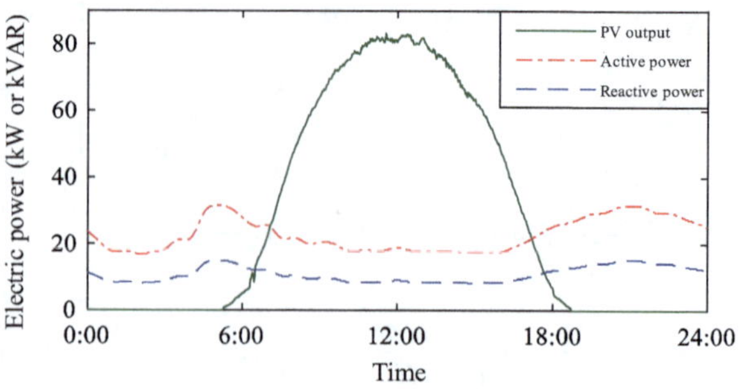

Fig. 18.7 Profiles of active and reactive power and PV output in each load section

Table 18.6 Numerical results (simply extended to ten years)

	Node number i	Node number i
	7	25
Capacity of installed SVC, Q_i^*	60 kVAR	60 kVAR
Income brought by SVC installation, f	215.6×10^6 JPY	
Reduction of generation opportunity loss, $LR(Q^*)$	228.0×10^6 JPY	
Cost for SVC installation, $EC(Q^*) + IC(Q^*)$	12.4×10^6 JPY	

subsystems (PCSs) in each PV system (500 PCSs) corporate the voltage regulation. Here, the authors assume that newly installed SVCs support to maintain the voltages at their connection points in a certain value. Figure 18.7 displays profiles of the active power and the reactive power consumptions and the PV output in each load section (each node in Fig. 18.6). Parameters of the RAPSO-ME were set as the same of preliminary numerical simulations.

Application result of the RAPSO-ME is summarized in Table 18.6. Figure 18.8 shows the voltage profiles in the distribution system model, Fig. 18.9 displays the output profiles of the PV systems, and Fig. 18.10 is the profiles of reactive power output of the installed SVCs.

As shown in Table 18.6, the optimal solution was to install two aggregated SVCs (60 kVAR of SVCs at Sects. 7 and 25, respectively) in the distribution system model. It can be understood that we can expect income by installing SVCs even though their installation required large costs. In Figs. 18.8, 18.9 and 18.10, the PV outputs were suppressed to satisfy the distribution voltages appropriately in the distribution system model before installing SVCs. On the other hand, after the optimal SVC installation, the generation opportunity loss of PV systems was relaxed dramatically. Therefore, the authors concluded that the problem formulation and the proposed method are useful.

Figure 18.11 displays transitions of the fitness function of the swarm best (gbest) and the reflectivity during the search process of RAPSO-ME. In Fig. 18.11a, value of the fitness function approximately converged at 20 iterations. From Fig. 18.11b, we can understand that the reflectivity gradually converged in association with increase

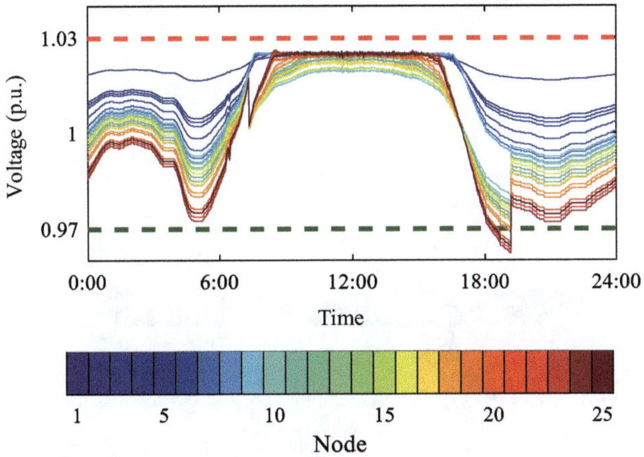

(a) Voltage profiles in each load section before SVC installation.

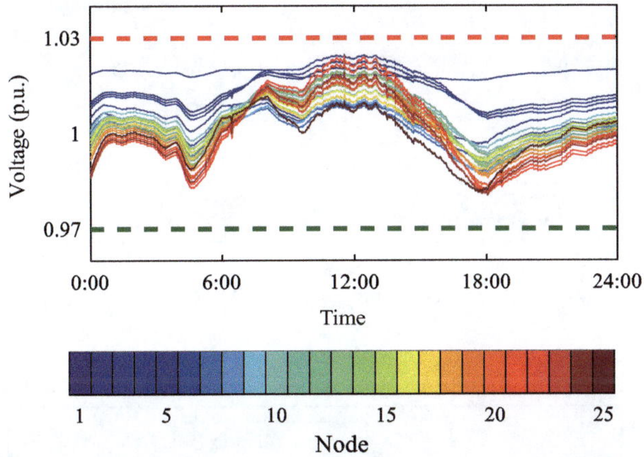

(b) Voltage profiles in each load section after SVC installation.

Fig. 18.8 Comparison of voltage profiles

of the iteration. These results show that the RAPSO-ME functioned appropriately in the target optimization problem.

18.6 Conclusions

This chapter presented a problem formulation and its solution method to obtain the optimal sizes and locations of newly installed SVCs. In the problem formulation, the generation opportunity of PV systems was encouraged in consideration of the balance with the costs required in the SVC installation. In the solution method, the authors

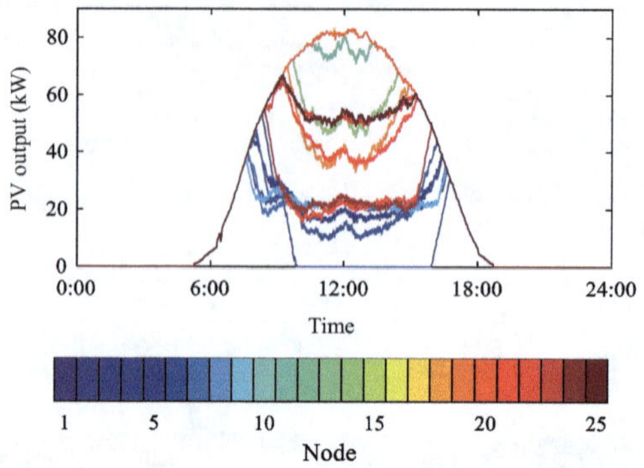

(a) Output profiles of PV systems in each load section before SVC installation.

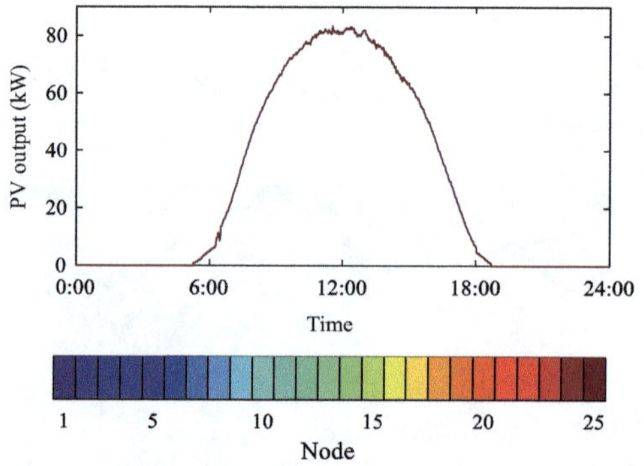

(b) Output profiles of PV systems in each load section after SVC installation.

Fig. 18.9 Comparison of output profiles of PV systems

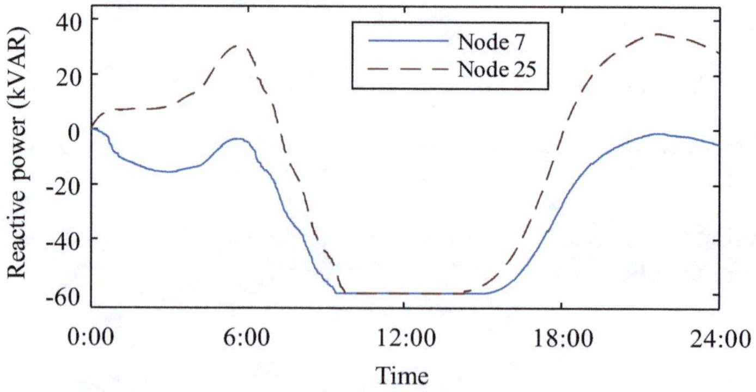

Fig. 18.10 Profiles of reactive power output of SVCs

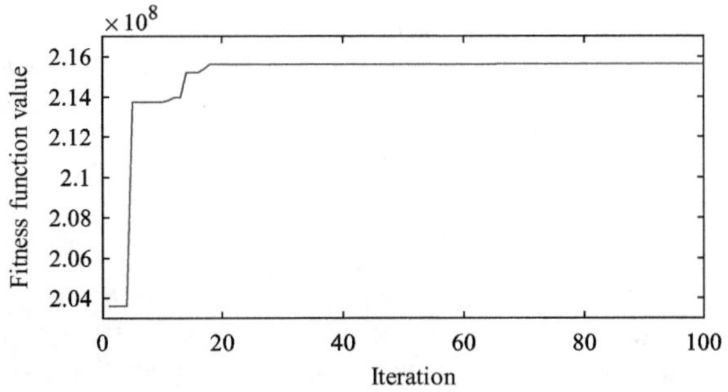

(a) Value of fitness function of swarm best (gbest) in each iteration.

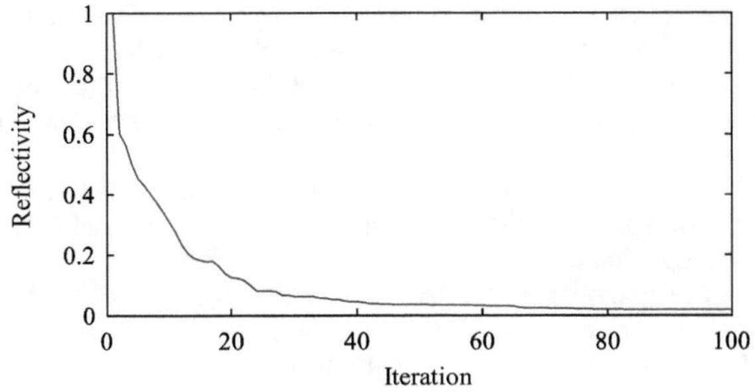

(b) Value of reflectivity in each iteration.

Fig. 18.11 Optimization process of RAPSO-ME

employed a reflectance adjusting mechanism at the boundary in the solution space and ideas of mutation and elitism to improve the search ability of PSO. Performance of the improved PSO (RAPSO-ME) was examined on the benchmark problems. In the preliminary numerical simulations, the RAPSO-ME showed good results, and its effectiveness was confirmed. Moreover, the RAPSO-ME was applied to the optimization problem of SVC sizing and allocation. Through the numerical simulation and discussion on its result, we could understand that the problem formulation and its solution method were useful.

Acknowledgements The authors would like to acknowledge the support of Mr. Junya Nishimura.

References

1. Sekine Y (1991) Haiden Gijutsu Sougou Manual (Integrated Manual for Distribution Technologies). Ohm-sha Publishing (in Japanese)
2. Takano H, Kawamura K, Iizaka T (2016) Industry practice and operational experience of key distribution applications of smart grid technologies. In: Liu CC, McAuthur S, Lee SJ (eds) Smart grid handbook, vol 3, pp 967–994

3. Tonkoski R, Lopes L (2011) Impact of active power curtailment on overvoltage prevention and energy production of PV inverters connected to low voltage residential feeders. Renew Energy 36(12):3566–3574
4. De Cagnano E, Tuglie ML, Mastromauro RA (2011) Online optimal reactive power control strategy of PV inverters. IEEE Trans Ind Electron 58(10):4549–4558
5. Takahashi S, Takada Y, Tsuji M, Kamiya E (2012) Method of optimal allocation of SVR in distribution feeders with renewable energy sources. J Int Council Electr Eng 2(2):159–165
6. Lahacani NA, Aouzellag D, Mendil B (2010) Contribution to the improvement of voltage profile in electrical network with wind generator using SVC devices. Renew Energy 35:243–248
7. Boynuegri AR, Vural B, Tascikaraoglu A, Uzunoglu M, Yumurtac R (2012) Voltage regulation capability of a prototype static VAR compensator for wind applications. Appl Energy 93:422–431
8. Bhaskar MA, Dash SS, Magdal R, Jerry J, Tamheed AM, Laksshman S, Aravindan S (2012) Application of integrated wind energy conversion system (WECS) and photovoltaic (PV) solar farm as STATCOM to regulate grid voltage during night time. Energy Procedia 14:1536–1541
9. Sugimoto J, Yokoyama R, Fujita G, Fukuyama Y (2006) Cooperative allocation of SVR and SVC for voltage fluctuation in case of connecting distributed generators. IEEJ Trans Power Energy 126(12):1191–1198 (in Japanese)
10. Hatta H (2015) Effect evaluation of reactive power control method according to PV power output on required capacity of SVC and distribution line loss. IEEJ Trans Power Energy 135(2):106–110 (in Japanese)
11. Hsiao YT, Liu CC, Chiang HD (1993) A new approach for optimal VAR sources planning in large scale electric power systems. IEEE Trans Power Syst 8:988–996
12. Sugimoto J, Tsukamoto Y, Tani Y, Yokoyama R (2005) Coordinated allocation and control of voltage regulators at the minimum cost based on reactive Tabu Search. In: Proceedings of the 5th WSEAS/IASME international conference on electric power systems, high voltages, electric machines, pp 625–631
13. Minguez R, Milano F, Zarate-Minano R, Conejo A (2007) Optimal network placement of SVC devices. IEEE Trans Power Syst 22(4):1851–1861
14. Singh JG, Singh SN, Srivastava SC (2007) An Approach for optimal placement of static VAR compensators based on reactive power spot price. IEEE Trans Power Syst 22(4):2021–2029
15. Mori H, Fujita H (2011) Development of hybrid-coded EPSO For optimal allocation of FACTS devices in uncertain smart grids. Procedia Comput Sci 6:429–434
16. Aghaei J, Gitizadeh M, Kaji M (2012) Placement and operation strategy of FACTS devices using optimal continuous power flow. Sci Iranica 19(6):1683–1690
17. Savic A, Durisic Z (2014) Optimal sizing and location of SVC devices for improvement of voltage profile in distribution network with dispersed photovoltaic and wind power plants. Appl Energy 134:114–124
18. Vanishree J, Ramesh V (2018) Optimization of size and cost of static VAR compensator using dragonfly algorithm for voltage profile improvement in power transmission systems. Int J Renew Energy Res 8(1):56–66
19. Ohba T, Takahashi A, Imai J, Funabiki S (2013) A comparative study of performance in particle swam optimization methods with reflection. In: Proceedings of IEEE power and energy society general meeting, 659
20. Aoki Y, Takahashi A, Funabiki S (2016) Individual representation for optimizing allocation of SVCs in high-voltage distribution systems. In: Proceedings of the 19th international conference on electrical machines and systems, DS4G-4–1
21. Nishimura J, Takahashi A, Imai J, Funabiki S (2019) Improvement method of reducing reflectance in RAPSO-ME and its application to optimal allocation of SVCs. The papers of technical meeting on "Power Engineering", IEE Japan-international workshop on power engineering in remote Islands, PE-19-64, pp 75–80
22. Clerc M (2006) Particle Swarm Optimization. ISTE Ltd
23. Lee S, Soak S, Oh S, Pedryczm W, Jeon M (2008) Modified binary particle swarm optimization. Prog Nat Sci 18:1161–1166

24. Helwig S and Wanka R (2007) Particle swarm optimization in high-dimensional bounded search spaces. In: Proceedings of IEEE swarm intelligence symposium, pp 198–205
25. Huang T, Mohan AS (2005) Hybrid boundary condition for robust particle swarm optimization. IEEE Antennas Wirel Propag Lett 4:112–117
26. Xu X, Rahmat-Samii Y (2007) Boundary conditions in particle swarm optimization revisited. IEEE Trans Antennas Propag 55(3):760–765
27. Chu W, Gao X, Sorooshian S (2011) Handling Boundary Constraints for particle swarm optimization in high-dimensional search space. Inf Sci 171:4569–4581